Studies in Computational Intelligence

Volume 1236

The series "Studies in Computational Intelligence" (SCI) publishes new developments and advances in the various areas of computational intelligence—quickly and with a high quality. The intent is to cover the theory, applications, and design methods of computational intelligence, as embedded in the fields of engineering, computer science, physics and life sciences, as well as the methodologies behind them. The series contains monographs, lecture notes and edited volumes in computational intelligence spanning the areas of neural networks, connectionist systems, genetic algorithms, evolutionary computation, artificial intelligence, cellular automata, self-organizing systems, soft computing, fuzzy systems, and hybrid intelligent systems. Of particular value to both the contributors and the readership are the short publication timeframe and the world-wide distribution, which enable both wide and rapid dissemination of research output.

Indexed by SCOPUS, DBLP, WTI AG (Switzerland), zbMATH, SCImago.

All books published in the series are submitted for consideration in Web of Science.

S. Rakesh Kumar · N. Gayathri ·
Vamsidhar Yendapalli · Alvaro Rocha
Editors

AI-Driven IoT Solutions in Agriculture: New Insights

Springer

Editors
S. Rakesh Kumar
Department of Computer Science
and Engineering
GITAM (Deemed to be University)
Bengaluru, Karnataka, India

N. Gayathri
Department of Computer Science
and Engineering
GITAM (Deemed to be University)
Bengaluru, Karnataka, India

Vamsidhar Yendapalli
Department of Computer Science
and Engineering
GITAM (Deemed to be University)
Bengaluru, Karnataka, India

Alvaro Rocha
ISEG
University of Lisbon
Lisbon, Portugal

ISSN 1860-949X ISSN 1860-9503 (electronic)
Studies in Computational Intelligence
ISBN 978-3-032-05602-3 ISBN 978-3-032-05603-0 (eBook)
https://doi.org/10.1007/978-3-032-05603-0

This Springer imprint is published by the registered company Springer Nature Switzerland AG
The registered company address is: Gewerbestrasse 11, 6330 Cham, Switzerland

Contents

The Role of AI and IoT in Transforming Modern Agriculture

Revuri Jaswanth, Gayathri Nagasubramanian, and Rakesh Kumar Sakthivel

Abstract An in-depth discussion of Artificial Intelligence (AI) and Internet of Things (IoT) in agriculture, and how AI and IoT are disrupting and transforming the agricultural industry and the practices of modern agriculture is explained in this chapter. It begins with an elaborate overview of AI and IoT as research fields, advancements in these sectors, and how they have begun converging in the agricultural industry. It also covers how AI and IoT have transformed precision agriculture, which has greatly enhanced real-time monitoring, data driven-decision-making, and automation features to assist with ongoing tasks such as irrigation, monitoring crop health, and livestock management. AI and IoT have improved efficiencies, increased productivity and sustainability, however there are issues with technical infrastructure, costs, human resources and skills training, and doing business in a trust based age of data. The chapter then gets into some future innovation and apps with focus on how Robotics, Predictive Analytics, and Blockchain add value to any agricultural supply chain. The content of the chapter accumulates with surrounding AI and IoT implications and impacts, and highlights the role of AI or IoT in shaping the future of agriculture, food systems, and design goals of the 2030 Agenda for Sustainable Development.

Keywords Artificial Intelligence (AI) · Internet of Things (IoT) · Data-driven decision making · Sustainable agriculture · Agricultural IoT sensors · AI-powered analytics · Farm automation · Supply chain transparency

R. Jaswanth
Department of AI &ML, GITAM University, Bangalore, India

G. Nagasubramanian · R. K. Sakthivel (✉)
Department of Computing Science and Engineering, GITAM University, Bangalore, India
e-mail: rsakthiv@gitam.edu

S. R. Kumar et al. (eds.), *AI-Driven IoT Solutions in Agriculture: New Insights*, Studies in Computational Intelligence 1236, https://doi.org/10.1007/978-3-032-05603-0_1

1 Introduction to AI and IoT in Agriculture

1.1 Overview of AI and IoT Technologies

Artificial Intelligence (AI) and Internet of Things (IoT) are transforming agriculture as we know it. AI is computer-based intelligence that provides reasoning and problem solving like humans and creates simple to complex behaviours. In agriculture, AI is being used for analysing complex data sets, predicting behaviours, and automating decisions. IoT consists of internet-enabled devices (e.g., sensors, cameras and actuators) that capture and share real-time data (referred to as smart devices). Together, AI and IoT create new capabilities by allowing the values of the environment to be captured and shared, but also making the process in the field smarter, faster, and more sustainable [1, 2].

1.2 Evolution of Digital Agriculture

The transition to digital agriculture began with the mechanization of farming in the twentieth century, focusing on the problem of workforce and improving productivity. The second leap forward was the adoption of information and communication technologies (ICT) in the form of GPS guided equipment, computerized farm management, and precision application equipment. Currently, we are in another leap forward the convergence of IoT and AI in what has been identified as precision agriculture, where efficiency will be improved through the use of real-time data, and intelligent automation, at every stage from initial preparation through harvest [3]. This transition is being driven to respond to the necessity to increase food production to meet the demands of a growing global population while utilizing few resources and reducing environmental impact.

1.3 Key Drivers for Adoption in Agriculture

A number of events are contributing to the adoption of AI and IoT in agriculture:

- Increased global demand for food: By 2050, the global population will reach close to 10 billion people, and farmers will need to produce more food with fewer resources on less land.
- Scarcity of resources: Water, arable land, and workers are all becoming less available and therefore, the need for technologies that get the most from inputs and maximize efficiency.

- Climate change: Unpredictable weather and the increased frequency of extreme weather events make it essential to manage food sources in an adaptive manner using data from sensors and other technology.
- Economic challenges: Market conditions are volatile and farmers' profits margins are tight. In this battle, innovations that save farmers cash are critical.

1.4 Core Technologies and Components

1.4.1 IoT Sensors and Real-Time Data Collection

IoT devices are the foundation of smart agriculture [11]. Wireless sensors on the farm are deployed in fields to monitor soil moisture, nutrient levels, pH, and temperature, giving farmers a more granular level of information about the crop and soil health. Wearables and biometric sensors track livestock health and behavior; and they can signal when livestock indicates early signs of illness or distress. These sensor systems can be used to carry out precise irrigation, fertilization, and pest control using exactly the right amount of resources, produce less waste, and ensure better yields.

1.4.2 Artificial Intelligence and Machine Learning Applications

AI systems analyze the massive amounts of information gathered by IoT devices, converting the raw data into intelligent practices. For example, machine learning algorithms can evaluate data patterns to predict disease outbreaks, or predict outcomes from the weather, or suggest optimal times to plant or harvest crops. Computer vision allows machines to identify crops from weeds; therefore, they should know on which weeds to apply herbicides, and they should recognize crops so machines could probably harvest crops without human intervention. These processes enable higher efficiency, but they also create lower impacts to the environment, including lower use of chemicals [4].

1.4.3 Drones, Robotics, and Automation

Drones with multispectral cameras cover ground quickly across large areas, identifying plant health problems that are undetectable by the naked eye. Autonomous ground robotic systems conduct the labor-intensive tasks of planting, weeding, and harvesting at a high degree of precision. As agriculture confronts labor shortages in its workforce, these autonomous systems will not only be able to conduct the work, but they will also operate around the clock for maximum efficiency.

1.4.4 Cloud Computing and Big Data Analytics

Cloud platforms store huge datasets created by IoT sensors and AI systems. Big data analytics identifies patterns and/or trends that help with future planning and can contribute to risk management. If data from multiple sources are integrated–weather stations, satellite imagery, market prices–producers can make informed decisions that maximize yield and profitability.

1.4.5 Connectivity: 5G and Communication Networks

When it comes to the real-time transfer of data and/or control of devices, dependable connectivity, and speed are essential. The arrival of 5G technology will enable faster, reliable communication between sensors and drones, and farm-management systems, allowing autonomous data-collection and automation that can be enabled in easily read remotely rural areas.

1.5 The Synergy of AI and IoT in Precision Agriculture

Digital agriculture really shines when artificial intelligence and the Internet of Things come together. IoT devices are a type of technology that collect real-time data from the field and AI systems can analyze that data to produce actionable recommendations and automate complex processes. As an example, suppose you have IoT soil moisture sensors that detect a dry condition in a small area of a field. The AI system that analyzed that data, reviewed weather forecasts, and then automatically activated the smart watering system within that area to dispense the exact amount of water it determined was required (no more, no less). This is what is referred to as precision agriculture, and it improves resource efficiency, crop yields and land sustainability.

1.6 Benefits of AI and IoT Integration in Agriculture

AI and IoT are being integrated into agriculture, creating a variety of benefits:

- Increased Efficiency and Productivity: Moreover, through automated monitoring and decision-making at every stage of crop and livestock production, we can produce more, with increased yields and in lower inputs.
- Data-Driven Decision Making: In addition, real-time collection and analysis of data will help us make informed interventions in a timely manner, with less guessing and more positive outcomes.

- Reduced Environmental Impact: Greenhouse gases, water, fertilizers and pesticides are applied only as specified (quantity), reducing pollution due to wastewater runoff, improving sustainable agriculture.
- Increased Supply Chain Transparency: Likewise, better access to data analysis (i.e., using blockchain technology) will allow agriculture to increase and formalize tracking ability, thus increasing food safety, reliability in food sourcing, and trust with consumers.
- Cost Savings: In conclusion, increased automation will allow for more efficiencies in resource use, and move costs further down their supply chain, which helps them remain competitive or level the playing field with bigger companies, in a volatile food commodity market.

1.7 The Future of Smart Agriculture

As we move forward, the fusion of AI and IoT is poised to advance many more innovations in agriculture. New technologies including edge computing, cutting-edge robotics, and next-generation sensors will continously enhance smart farming systems. As these technologies become cheaper and more accessible, even small-scale farmers will be able to access digital agriculture solutions and contribute meaningfully to still-expanding global food security and sustainable development.

2 Impact of AI and IoT on Agriculture

Artificial Intelligence (AI) and the Internet of Things (IoT) are revolutionizing the agriculture industry, touching all aspects of farming like, everything from planting, irrigation, harvesting, and everything in between in the supply chain. This transformation is more than the adoption of new technologies, it's a change in paradigm for how agricultural activities will be planned, executed, and optimized and in relation to meeting sustainable and economically viable demands for food globally.

AI and IoT are making a huge impact on agriculture, mainly through their ability to improve efficiencies and increase productivity. In agriculture, traditional farming practices have relied on manual labor and knowledge-based decision-making [1]. This style of operation is often riddled with uncertainties and inefficiencies. On the contrary, by deploying IoT sensors around their fields, farmers have real-time access to soil moisture, temperature, nutrient level, and crop health conditions. AI algorithms run real-time data through models that allow for precision farming which enables the efficient application of water, fertilizers, and pesticides in the amounts necessary; if required. When we think of antibiotic use on livestock, usage guided by precision farming reduces the use of important drugs like ketoprofen, and even impacts on antibiotic development in other livestock. Take irrigation for another example: smart irrigation systems can apply water only to the areas of a field that need it at the time, rather than having to apply water uniformly to the whole field—resulting them to conservatively use water and also spend less on irrigation costs. AI-powered weather prediction models are the future of farming as they analyze changing environmental conditions and historical data that will alert farmers so they can avoid potential disease outbreaks. These same capabilities will allow farmers to harness accurate weather models and to apply crop protection products. Using the above capabilities enables farmers to increase yield and reduce input costs and avoid significant crop loss and once again make farm management and decisions more efficient and sustainable.

AI and IoT are not only helping agriculture improve productivity, but they help the sector leverage data for decision-making changes. The data created by IoT is then used by AI for insights that can lead to actionable decision-making by farmers. For example, AI can predict best planting dates before a farmer knows when they are going to take place, which is sometimes a sequence of guesses informed by weather forecasts and soil condition. AI can also help farmers determine specific applications of fertilizer, or indicate how much nitrogen is already in the soil, as producers work toward a best decision. AI will also indicate the right time to harvest to provide the best quality crop. Moving farmers away from intuition-based decision-making, to evidence based decision-making, reduces the risk evaluation work farmers have to conduct, and gives farmers defined moments so there is no need to be reactive to circumstances. Further, AI-enabled systems (and other IoT systems) have a capability to be constantly learning from the real time data they are collecting, and the more they collect data over time, the better their predictive capabilities become. We will need AI to have the ability constantly adapt to become better, if we want farms to have the ability to respond to the variability of climate and markets. Another significant area with value for environmental sustainability is AI and IoT. Agriculture uses a large amount of our natural resources and contributes to greenhouse gas emissions, so it needs to make use of resources and save money as much as possible to lessen its environmental footprint. Farmers can benefit from precision agriculture techniques in conjunction with AI and IoT to lower the water use, fertilizers, and pesticides in agriculture which can limit soil degradation, water pollution, and loss of biodiversity, all while decreasing costs. Examples of precision agriculture that can save and reduce environmental damage include: smart sensors that can track soil health or moisture levels that can tell the farmer how much he needs to irrigate so he doesn't over-irrigate his crops, which can save water and minimize salinization; and AI pest control systems that can detect pest infestations at an earlier stage and suggest the correct amount of application to deal with that infestation instead of applying a chemical to the entire field. These tools can give farmers the sustainability options they want, but also are able to maintain ecosystems in the process of being productive.

Adopting AI and IoT also strengthens transparency and traceability along the agricultural supply chain. Today's consumers want more assurance about where their food comes from, how it was cultivated, its quality, and safety when consuming it. IoT devices allow products to be tracked from farm to fork, allowing IoT devices to monitor and record information on farming practices, handling, and transportation. AI can monitor the information provided by IoT devices for anomalies and compliance with safe agricultural practices. Blockchain technology, usually incorporated with IoT technology, also provides a trusted, immutable ledger that guarantees its authenticity and provenance. This additional transparency builds trust with consumers, opens up new market channels, and allows farmers to get higher prices for quality and sustainable production.

AI and IoT usage in agriculture is explicitly expanding, offering a multitude of economic opportunities. By lowering input costs and increasing outputs, these technologies increases farm profit and viability so that farmers can afford to sit on potential returns on investment on input costs prior to planting or growing of actual crops. AI and IoT also create new models to do business, for example, farmers paying for precision farming services or paying a subscription based on actual yields; the collection of data and analytics and then advisor models are new businesses using early stage AI. Moreover, automation is increasing and AI and robotics are partially reducing labour shortages, in farming these are becoming a growing concern particularly with an aging farmer demographic, and urban migration reducing an already small labour pool. Automated robots and drones are able to handle repeatable, time-consuming, labour intensive tasks with high precision and velocity. This also reduces farmers from having to deal with these low level tasks, to allow them more time for higher value strategic decision making. This continues to alter the labour force in agriculture, but with a different emphasis to earlier decades, in that it is shifting from low skilled human labour to more technically skilled technicians and data analysts.

Although AI and IoT technologies, as well as precision-agriculture technology, are transforming agricultural systems around the world, they are not without challenges. The initial costs associated with acquiring these types of technologies can be a barrier to access for smallholder farmers, particularly for people in less developed country contexts. For many farmers, there is also a significant skills gap, as people have not had opportunities to learn how to operate and maintain advanced word-level, digitally-enabled systems. Infrastructure limitations (especially in poorer, rural agricultural areas) may limit internet connectivity thereby limiting the extent to which farmers can adopt these technologies. Moreover, as farms become more connected and reliant on advanced digital monitoring systems and platforms, issues regarding data privacy and security will likely emerge. Governments, industry, and academia must collaborate to identify potential barriers such as affordability of the technology, training programs for human resources and developing more supportive policies [5].

As we look forward, the adoption and implemented impacts of AI and IoT in agriculture will deepen and broaden the capabilities and efficiencies for farmers, agricultural companies, and markets. As things stand the on-going reductions in costs of computing will likely accelerate with the arrival of edge computing technologies. Becoming wider available with lower prices and better adjusted for poor connectivity to the web will assist in supporting better decision-making in operational contexts. Edge computing removes limits imposed by latency, as data processed where it is created, improves the responsiveness of these systems.

AI and IoT are also likely to create synthetic biology applications and AI-based additive models materializing on-box applications for future crop improvements or pest resistance systems. As the world expands dense coverage build-outs of 5G networks towards more universal global connectivity, the roles and implications of these tools for farmers, will be to incorporate real-time remote, 3D monitoring in some of the most remote farming areas.

Understanding the impacts these technologies will have on access and affordability for farmers will unlock more of their potential, and better empower farmers around the world to produce more food on the same footprint, more sustainably with and through food systems.

3 Applications in Modern Agriculture

The combination of Artificial Intelligence (AI) and the Internet of Things (IoT) represents a new era in agricultural innovation, more commonly named smart farming or precision agriculture. This technology allows farmers to ensure that every step of the agricultural process is done with greater accuracy and efficiency than ever before. While traditional agriculture exemplifies the farming lifecycle from soil preparation to harvesting, AI and IoT applications will take agriculture from a manual, tedious, and inefficient process to a digital, automated, and sustainable process. In this section we will look into the major applications of AI and IoT in modern agriculture to demonstrate that they can increase productivity and sustainability in today's agricultural economy through maximum resource use [1].

Figure 1 outlines the relative distribution of IoT applications in smart farming. Precision irrigation and crop monitoring make up the largest share of the pie chart and are also the most well-known and used application areas, and they also directly relate to yield and resource efficiency. Livestock management, greenhouse automation, and supply chain tracking rounded out the most significant applications, and they highlight the extensiveness and versatility of IoT towards the agricultural value chain.

3.1 Precision Farming and Resource Optimization

Precision agriculture is the flagship application of artificial intelligence and the internet of things in agriculture. It focuses on using near real-time, high-resolution data to manage variability within a field and to optimize the use of inputs, such as water, fertilisers and pesticides. IoT sensors positioned in the soil continuously measure water content (moisture), nutrient levels, temperature, and pH. This level of fine detail is sent to AI-enabled platforms that analyse the information and provide zone-specific recommendations (i.e., recommendations tailored to distinct zones) on what and how much to apply.

For example, instead of the time-honoured method of irrigating and watering a whole farm, smart irrigation systems would only provide water where it is needed, based on what the soil moisture sensors are telling them [3]. This keeps water costs

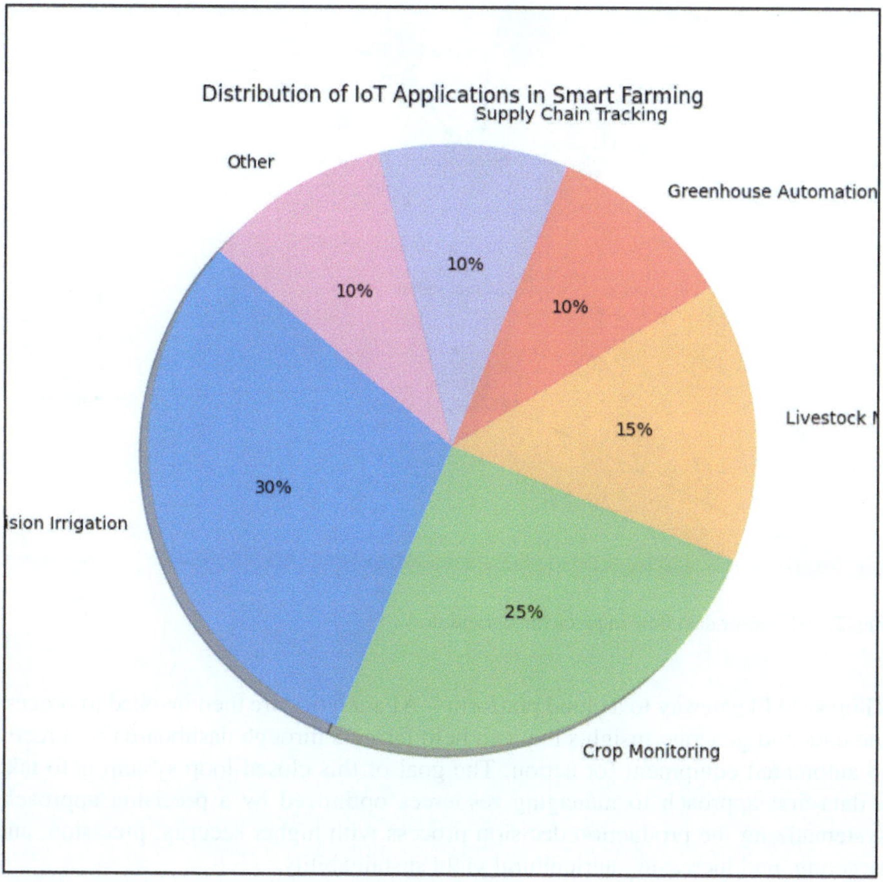

Fig. 1 Distribution of IoT applications in smart farming

down, while conserving water for the future. Likewise, AI-enabled microservice algorithms can analyse the soil nutrient data to determine exactly how much and what fertiliser to apply, thus preventing over-application of fertilisers which harm the environment. Along these lines, the application of precision farming can enhance crop yield as well as limit waste and environmental degradation, and make agricultural sector more sustainable.

In addition, AI-powered models take into account forecast data as well as historical data to make recommendations for optimal planting windows and estimates for optimal harvest dates. This level of precision puts farmers in a position to maximize the quality and quantity of their produce while also avoiding losses due to damaging weather events or pest outbreaks.

Figure 2 illustrates an example of data flow in a precision agriculture case enabled by IoT and AI. Sensors in the field (soil, weather, livestock) continuously collect data in real- time and relay the data from the field sensors through an Internet of

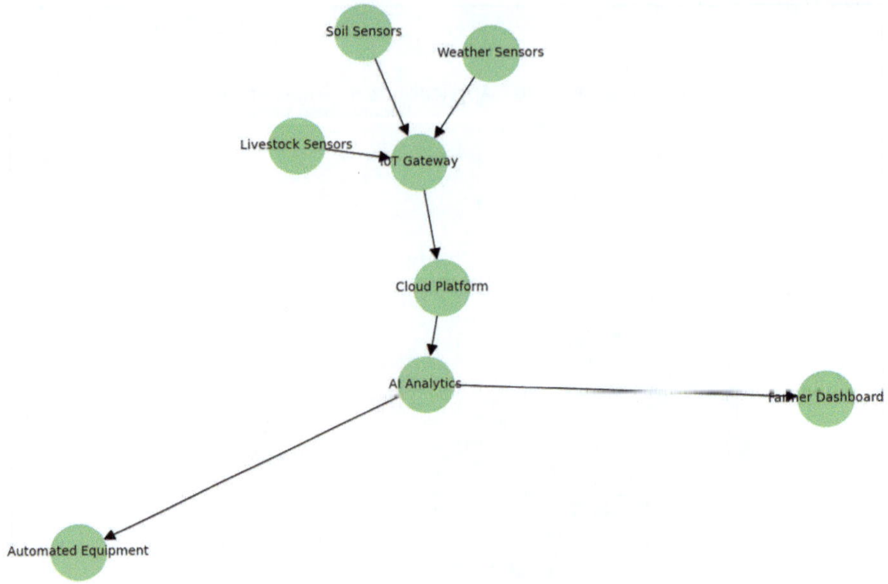

Fig. 2 IoT sensor data flow in precision agriculture

Things (IoT) gateway to a cloud platform—AI analytics are then invoked to process the data and generate insights that can help farmers through dashboards or directly to automated equipment for action. The goal of this closed-loop system is to take a data-first approach to managing resources optimized by a precision approach, systematizing the production decision process with higher security, precision, and targeting, and increasing agricultural yield sustainability.

3.2 *Crop Monitoring and Disease Prediction*

Detecting crop stress, disease, and pest infestations early in the farming lifecycle is essential to preventing yield loss. AI and IoT technologies are transforming crop monitoring by providing continuous, real-time monitoring. Drones with multispectral and thermal imaging cameras fly over the fields and gather high-resolution images that reveal small changes in plant color, temperature, and moisture content—indicators of stress or disease not readily observable to the human eye.

The AI-based computer vision system processes the images to identify specific diseases, insect infestations, or nutrient deficiencies with a higher degree of accuracy. AI model that is trained on large datasets of different crops in order to differentiate between a fungal infection and insect consumption can target the problem. The systems will provide early warnings of the relative positions of treatments, allowing farmers to apply treatments directly to where they are needed and avoid excessive use of chemicals, as well as preventing the further spread of disease.

To supplement aerial monitoring, ground-based IoT sensors can provide the environmental factors essential for managing conditions conducive to disease, stress, and infestations, such as humidity and temperature. AI algorithms analyze the ground-based sensor data to anticipate outbreaks before symptoms present themselves, which gives the farmer another necessary time advantage to protect the crop.

3.3 Livestock Management and Health Monitoring

Continuous health monitoring and behavioral analysis for livestock farming is another area where AI and IoT are having a great impact. Wearable IoT devices are outfitted on the animals to allow them to collect their vital signs (heart-rate, body temperature, and activity levels). This information is sent to AI systems that can find anomalies related to distress, illness, or reproductive cycles.

For example, sudden declines in an animal's activity or a sudden rise in temperature could mean that a disease is about to begin. This could prompt early veterinary intervention and could mean not only financial savings from treatment, but savings on animal loss as well. AI-powered behavioural analysis can indicate distress (like heat stress or discomfort) so that the farmer can take action (increasing water availability, changing ventilation, changing feeding cycle etc.) to relieve that distress.

These technologies, in terms of health management are also allowing for sustainable feeding management through consumption and growth rates recorded for each individual animal. Automatic feeding systems allow farmers to be more efficient in their feeding management purchase. Given the real-time data from the IoT devices, we can adjust the amount of feeding provided on daily basis (or each feeding) to add the ultimate nutrition and reduce waste.

When implemented, this technology means improved welfare of the animal, improved production of the animal, and better sustainability for livestock farming.

3.4 Smart Irrigation and Water Management

Water scarcity poses a significant challenge in agriculture, and the responsible management of water is crucial to sustainable agriculture. With the advances in AI and IoT technologies, smart irrigation systems are being developed capable of reducing water consumption through the use of real time data from soil moisture sensors, weather forecasts, and crop water requirements models.

The IoT sensors also allow for continuous then monitoring of soil moisture at various depths to create an overview of how much water may be available in the root zone. The AI algorithms can analyze weather conditions simultaneously with the sensor data to elucidate when and how much irrigation should apply. For example, if heavy rain is forecast, the system can delay an irrigation event so as not to oversaturate the soil, or if there is a dry spell, the irrigation event can be turned up and applied in the areas where it is required [6].

The smarts of the irrigation systems can be automated to carry out the actions of pumps and valves remotely. This decreases labor, increases critical timing and volume of water reach those plants, and can enable the significant savings of water, energy consumption and resource waste, both feeding directly into the appealing goal of ecological sustainability and economic efficacy.

3.5 *Automated Planting, Harvesting, and Sorting*

Tasks once done manually, such as planting, harvesting, sorting and grading produce, can now be done through automation utilizing AI and robotics. For example, autonomous tractors and autonomous seeders produce GPS and AI to plant seeds to within a centimeter, providing spatial control and depth to enhance germination and crop growth for optimal agricultural value. Increased precision will reduce wasted seed and generate equitable sized crops.

Robots equipped with AI and computer vision are also capable of detecting ripe fruit or vegetables and can take fruit or vegetables without damaging the crop. Further, harvesting robots grow and harvest continually without interruption. With robots doing the work of labor's, they will also speed up the harvest and provide a solution to industry labor shortages in agricultural regions and production systems. Automated harvesting can also reduce post-harvest losses through a reduction in delays and handling damage.

Post-harvest and after harvest, AI driven sorting and grading systems use more cameras or sensors to sort product by size, shape, color, and quality as it goes through systems. This also allows processors to get product consistently and at a faster rate providing a solution to the agricultural supply chain labor and produce challenges faced by farmers and distributors, in maintaining the quality and speed to meet consumer demands.

4 Impact of AI and IoT on Agriculture

The intersection of Artificial Intelligence (AI) and the Internet of Things (IoT) has had a transformative influence on agriculture, leading to a revolution in the way food can be produced, managed and delivered. This section will highlight the benefits of AI and IoT, through recognition of their holistic implications for agriculture in terms of improved efficiency and productivity, data-driven decision-making, sustainability, transparency in supply chains, and socio-economic implications, as well as acknowledging the challenges that need to be dealt with to facilitate their full impact.

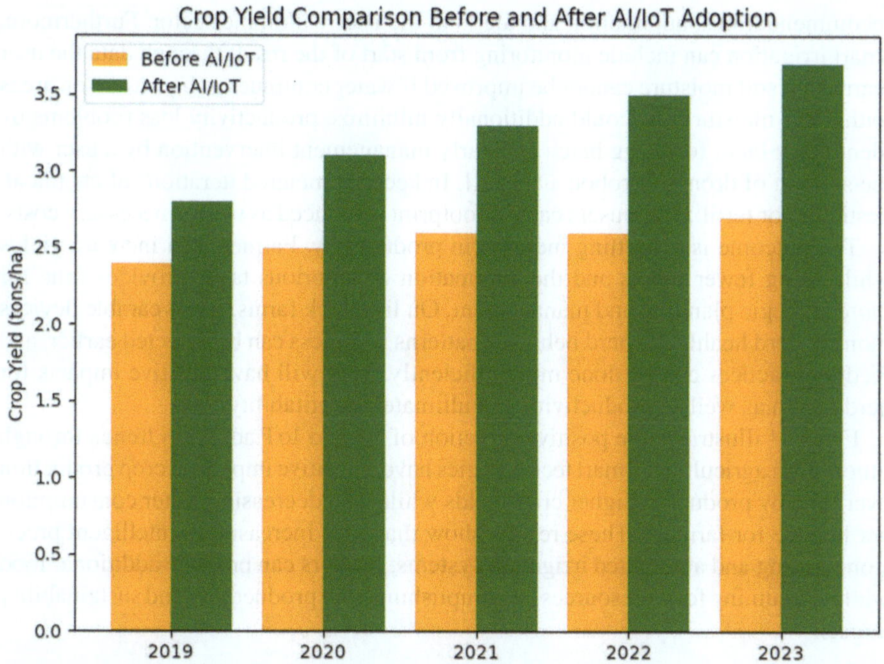

Fig. 3 Comparison of crop yields before and after AI/IoT adoption

Figure 3 graph illustrates average crop yields five years before and five years after the adoption of AI and IoT technologies. The indicated yields clearly indicate evidence of a substantial growth pattern in yields after employing precision agricultural tools. This explicates how tools are data-driven, automated decision-making in efficiency and productivity when it comes to farming.

4.1 Increased Efficiency and Productivity

The most immediate impact of AI and IoT in agriculture is likely increasing efficiency and productivity. "Traditional" farming involved a good deal of manual labor, with a lot of time spent working "in the field." Often, farm management applied one systemic solution for all fields instead of nearing precision measures for individual plants or animals (e.g., using an entire sprayer capacity to fertilize multiple field reports) Solely generational labor often translates into recklessness for chemicals, irrigation, or fertilizer. Today, however, with AI and IoT use, 'agriculture' has adapted to become 'precision agriculture' [4, 5].

IoT sensors are capable of monitoring a myriad of things (e.g. soil moisture, nutrients, temperature and weather) on a continuous basis. The system will typically transmit the information to the tab one using AI, which provides comprehensive recommendations, automation and alerts to the user within the sector. Furthermore, smart irrigation can include monitoring from start of the result through data the user learns that soil moisture cannot be improved if water continues to be wasted on areas with a 0% moisture. AI could additionally minimize productivity loss problems by identifying larva form egg hatch and early management intervention by a user with the support of drones or robots using AI. In keeping metered iterations of chemical, pesticides or fertilizers, a users carbon footprint is reduced as well as necessary costs.

The outcome is a startling increase in productivity. Farmers can increase yields while using fewer inputs and the automation of laborious tasks provides time for more strategic planning and management. On livestock farms, IoT wearable devices monitor herd health and herd behavior patterns so illness can be detected earlier, and feeding practices can be done more efficiently. This will have positive impacts for herd health as well as productivity and ultimately profitability.

Figure 4 illustrates the positive direction of AI and IoT adoption trends through time within agriculture. Smart technologies have a positive impact on crop production over time by producing higher crop yields while also decreasing water consumption per hectare for farmers. These results show that with increasingly intelligent precision farming and automated irrigation systems, farmers can produce additional food while consuming fewer resources, accomplishing both productivity and sustainability targets.

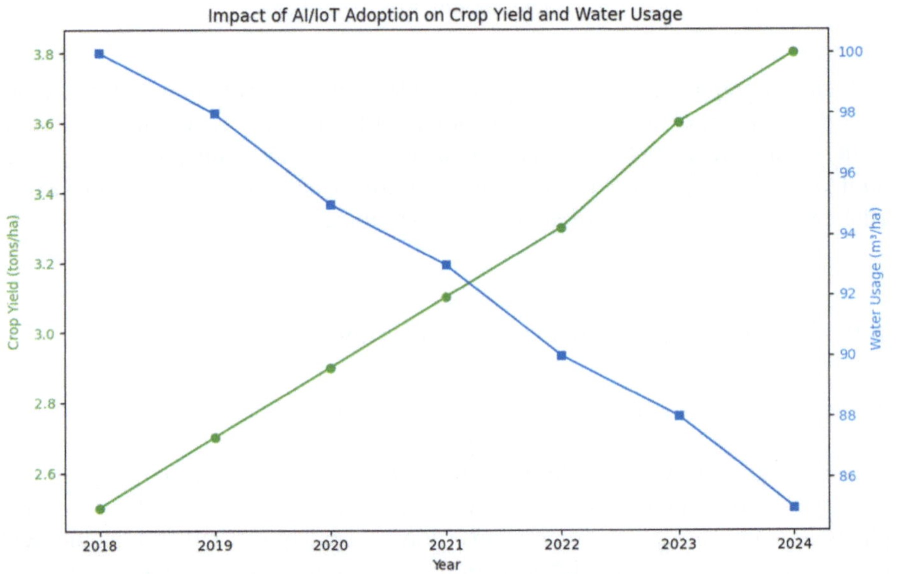

Fig. 4 Impact of AI/IoT adoption on crop yield and water usage

4.2 Data-Driven Decision Making

AI and IoT has triggered perhaps one of the most important transitions we will see, a shift from an intuition-based decision-making model to a data-based decision-making model. These have always been critical decisions for farmers, but in the past they were often informed only by experience and first-hand observations. Experience, of course, is valuable, but today's farms produce so much more data and more complex data than any one individual could possibly handle and make sense of.

AI systems can analyze very large datasets, applying sophisticated analytical techniques to identify patterns in data and predicting outcomes. By using data from IoT sensors, along with weather forecasts, optical and satellite imagery and market trends, AI platforms can ultimately help to advise a farmer with very precise recommendations based on their unique and specific conditions. An example: AI would be able to recommend optimal planting times based on soil conditions and weather forecasts; or recommend the best varieties to plant based on the particular region; or even forecasts of demand that can help farmers plan production and sales, if you will.

Data-based decision making is all about reducing uncertainty and risk for farmers. It is about enabling them to make better-informed decisions, to strive for maximum yield and minimum cost, and hence the goal of a financially sustainable profit margin. It also helps producers think about the longer-term future of the farm, resource use efficiency, sustainability, and other factors that contribute to healthy soils, safe food, and the least environmental impact possible.

4.3 Reduced Environmental Impact and Sustainability

Agriculture is a significant consumer of natural resources and a significant contributor to environmental consequences, such as greenhouse gas emissions, water use, and chemical runoff. AI and IoT technologies can offer ways to reduce agriculture's impact on the environment and potentially make more sustainable practices available.

The use of AI and IoT will support a more sustainable way to farm through the use of precision agriculture. AI and IoT-based solutions can reduce the amount of water, fertilizers, and pesticides used in agriculture by encouraging farmers to only apply inputs when needed and where needed to prevent waste, pollution, and disruption of natural ecosystems. Smart irrigation technology, for example, uses sensors to determine when to apply water to an area and apply it directly to the root zone of the plants; AI pest management would provide information related to the specific pest infestation enabling targeted intervention and much less reliance on broad-spectrum chemical pesticides.

Aside from optimizing inputs, AI and IoT technologies facilitate sustainable land management practices. Soil health sensors provide farmers with real-time data on soil nutrient levels and organic matter. This will help farmers make decisions around crop rotation and cover cropping to maintain the fertility of soils and avoid soil erosion. AI tools aimed at analyzing farming practices will support farmers in adapting to the situation and changing environmental conditions that are outside of their control, such as drought and extreme weather events, by evaluating and analyzing data over a longer time horizon [6].

Not only are these developments important for reducing the footprint of agriculture today, but they will contribute to the adaptation and resilience of farming systems in the future, despite environmental disruption.

4.4 *Improved Supply Chain Transparency*

The convergence of AI and IoT goes well beyond the farm gate—the agricultural supply chain will see transformative change as well. Consumers, retailers and regulators are demanding more transparency and traceability in food systems due to growing concerns about food safety, quality and sustainability.

IoT devices track and trace agricultural commodities from the farmgate to the consumer. The data captured with IoT for agricultural monitoring and traceability includes field operations, storage specifications, transportation dynamics, processing information and distribution protocols. AI examines that data for anomalies, ensures compliance with economic and safety standards, and better logistical operations. For example, temperature and humidity monitoring through IoT enabled devices can monitor the storage to transport of perishable goods, while signalling to the supply chain when those goods are spoiled or at risk of being compromised (and therefore surveilled for food quality, safety and transparency).

Blockchain technology, when paired with AI and IoT technology, further enhances the traceability and transparency of the agricultural supply chain by providing a permanent record of every transaction and movement throughout the supply chain. This builds trust among stakeholders that together hold accountability for the agricultural commodity transparency and traceability. It allows consumers to verify the origin of their food and the value associated with it; farmers and food producers can seek to access premium income opportunities by demonstrating compliance to sustainable and quality benchmark standards.

4.5 *Economic Benefits and Market Growth*

AI and IoT are having a large economic footprint in agriculture. Effectively, AI and IoT are increasing efficiency and productivity to help farmers reduce costs and ultimately grow their profits as well as yield. Automating labor intensive tasks ends reliance on seasonal or migrant workers as much of farming labor shortages occur in regions that are demographic locked into a small labor pool.

AI and IoT enable innovation and new business models as well, like subscription based produced-oriented data driven advisory platform, leasing/farming of precision equipment through sharing amongst several users, democratizing access and value, providing a lower barrier to for smallholder farmers, providing an experience for inclusive growth, rural development.

As for the broader agricultural sector, an increase in competitiveness, need for innovation, and market access will increase through AI and IoT. Once the technology matures and becomes cheaper, it will not be too long to expect hyper accelerated adoption and it will be an indirect driving force for economic growth associated with job creation in support services, ag—agri-techs, data analytics, equipment manufacturer and more improvements like supply chains.

4.6 Societal and Global Implications

The influence of artificial intelligence (AI) and the Internet of Things (IoT) in agriculture addresses larger societal and global challenges. Efficiently and sustainably producing food with technologies like AI and IoT positively contributes to food security and poverty alleviation, especially in countries where agriculture represents a key sources of livelihood. AI and IoT technologies are also supported by these innovations to achieve global sustainability goals, such as the United Nations Sustainable Development Goals (SDGs), in the sense that they promote responsible resource use, climate resilience, and economic development.

However, AI and IoT innovations also raise important considerations related to equity, access, and the future of work in agriculture. Promoting the benefits of AI and IoT innovations to smallholder and marginalized farmers will require the development of policies, investment in infrastructure, and capacity-building initiatives.

5 Challenges and Considerations

Incorporating Artificial Intelligence (AI) and the Internet of Things (IoT) into agricultural practices has opened up incredible potential for productivity, sustainability, and efficiency. With that potential, there are challenges as well. The pros of digitization in agriculture are multiple, but as the agriculture sector maximizes on AI and IoT, navigating various challenges becomes complex and farmers can be left trying to figure out how to use these innovations without addressing the challenges. This subsection will look at some main challenges and considerations related to the application of AI and IoT in agriculture, addressing important technical and infrastructural issues, economic issues, workforce changes, data security and privacy, and social and ethical concerns [7].

5.1 Technical and Infrastructural Barriers

One of the major limitations of AI and IoT for agriculture is ensuring the level of infrastructure that is necessary for robustness and reliability. Commercial AI and IoT solutions in agriculture require sufficient infrastructure relating to internet connectivity, electricity access, and related hardware: sensors, drones, or automated machines. For AI and IoT to work in farming applications, all these requirements must be adhered to, and in many cases we are improving in developed areas, but in many rural areas, even in developing countries, the infrastructure issue is a very large barrier.

Limited broadband access is an important limitation for IoT because it means that there is a much lower ability for real time transmission of data from IoT devices. This means the advanced technologies of precision agriculture that rely on real time data are not used to their full potential. Poor electricity access means even if farmers invested in automation systems, or data centers, are also limited by unreliable electricity. Also consider that the environments on farms also define reliability and robustness of any technology—these environments are often harsh with dust, moisture and extreme temperature which can degrade accuracy in equipment such as sensors or hardware, or even working all together.

These are some key limitations to enabling AI and IoT on farms, so we would need to invest in the rural infrastructure, which means advances in 5G, ruggedizing hardware, or enabling more decentralized data processing, for instance in edge computing.

5.2 High Initial Investment and Cost Factors

Investing in new IoT and AI technologies requires a substantial amount of cash upfront. The cost associated with purchasing and installing sensors, drones, automated equipment, and data management could deter even fairly well-to-do producers from purchasing and installing even a modest IoT environment, even smallholder and family farmers that are in profit can be on a thin, less-than-$2000 per year profit margin. There can also be ongoing costs (e.g. software subscription, maintenance, ongoing technical support).

While the long term benefits of enhanced efficiency and productivity through IoT can justify the upfront expenditures, they still serve as an upfront barrier to widespread adoption. However, many producers may not have access to alternative, affordable, or subsidized financing mechanisms to diffuse these costs, and as the pace of technological change accelerates, what is new in manufacturing or software development does not last long before it too is obsolete [7].

5.3 *Skills Gap and Workforce Transformation*

Digitalizing agricultural systems requires an entirely new skill set and expertise. Use, maintenance, and interpretation of data from AI-driven and IoT-enabled systems demands technical knowledge that many farmers and agricultural workers may not have or have never seen. Traditional farming skills, traditionally learned through a hands-on, experiential basis and 'feeling it in their gut' now need to be paired with skills in data analytics, programming and troubleshooting of digital systems.

This skills gap levies a significant challenge through the adoption of technological systems particularly in regions where there is limited availability of education and training. Farmers without adequate support may not have the capacity to realize effective transformations with the adoption of technology, thus potentially diluting their potential impact. Automation of tasks retains the threats where workers must be displaced by machines performing tasks previously done by human labour, including planting, harvesting and sorting. New jobs might be created in agri-tech support/analyst/maintenance services. However, it is possible that the skills required for these roles may not be transferable to all workers who will have roles threatened by a technology, and this may exacerbate rural unemployment and inequality.

Public sector agencies need to establish training programs, extension services to address the skills gaps in partnerships with local educational institutions. The skills development must not just be in technical abilities but must also ensure the establishment of greater digital literacy and an acceptance of risk taking and innovation within.

5.4 *Data Security and Privacy Concerns*

As farms become more interconnected and data-driven, data security and privacy will also play a role as farmers collect and store more sensitive data. There will be a number of IoT devices that will continually collect sensitive data (for example, everything around your farm/business operations or production levels, and proprietary genetics information). The requirements for data storage and collection could mean that, in most situations, data will move to a cloud-based platform, which could also be breached and represent huge risks for farms (i.e. a loss of a data set, a data breach, or a cloud account sharing private data without consent).

Privacy breaches related to information security could have a major financial impact on farms and could include threats like disclosing their trade secrets and their strategic competitive advantages. In addition to the potential risk to farmers who could lose their sensitive information, there are also data ownership concerns because the large technology companies continue to aggregate and mine the agricultural data. Farmers may be reluctant to utilize technologies if they believe they will not have control over how the data would be shared or used by others.

Constructive trust and transparency could assist with the responsible use of data and provide clarity in the roles of policymakers, technology companies, and users of that data, while developing equitable legislation and regulations surrounding data ownership and sharing and access issues. If farms strongly considered implementing protections to their farm data, especially in relation to cybersecurity protections like encryption, login credentials and a regular independent security audit the ownership or loss of ownership from breaches could become irrelevant. In addition, technology companies would have to demonstrate transparency about how they are managing and using farmers' data, particularly if they monetize their data.

5.5 Social and Ethical Considerations

The significant incorporation of AI and IoT into the agricultural industry now impacts other, broader social and ethical considerations. For instance, there is concern of potential inequality amongst large, wealthy agribusinesses and smallholder farmers. Generally speaking, large businesses are better equipped to invest in new technologies, gain access to technical assistance, and capitalize on economies of scale. A digital divide can further widen these inequities in such areas as income, productivity, and access to markets. This can undermine efforts to spur inclusive and sustainable agriculture development for all farmers and rural communities. Accessing the benefits of AI and IoT must be possible for all farmers no matter their size, socio-economic status, or location; this can be accomplished through policies, capacity building and cooperative models that enable shared access to technology.

Ethical dilemmas also arise with the use of AI in decision-making processes. There should be transparency, accountability, and oversight as algorithms are increasingly relied on to make choices that have impacts on all facets of farm management. Limited datasets can expose farmers to unintended consequences or, worse yet, bias in the training of a model. For example, if it recommends weeding based on a training dataset modelled with a crop or geography in mind with no relation to the farm, it creates the potential for inappropriate, failure and damage [5].

5.6 Environmental Risks and Unintended Consequences

AI and IoT hold promise for making agriculture more sustainable, but if poorly considered there is the possibility of unintended environmental consequences in the near future. Over-reliance on automation and data-based decision making can increase monocultures, continue loss of biodiversity, and increase vulnerability to pests and diseases. The environmental consequences of production and disposal of electronic equipment, including e-waste and rare earth metals, introduce numerous challenges to sustainable farming options.

These issues can be minimized by developing agroecological principles of AI and IoT; and by changing to a circular economy focus of no waste and reduced use of resources. In all cases of digital agriculture strategies and business models, environmental monitoring and impact assessments need to be a priority.

5.7 Regulatory and Policy Challenges

The swift advancement of AI and IoT in agriculture is moving faster than the development of an appropriate regulatory framework. Policymakers need to ensure that innovation does not come at the expense of safeguarding farmers, consumers and the environment. Important issues relating to data ownership, liability for automated decisions, intellectual property rights and cross-border data flows urgently require regulatory focus and clarity so that policy can become consistent.

International cooperation is essential in agrifood, given the increasingly globalized agricultural supply chains, and that harmonization of standards enables the safe and effective adoption of different digital technologies across farmland, regions and markets [8].

6 Future Prospects and Innovations

The future of agriculture is being shaped by the rapid development of Artificial Intelligence (AI) and Internet of Things (IoT), creating an unprecedented paradigm shift. As we head further into the 2020s, these technologies are contributing to solving long-standing agricultural problems and opening new avenues of productivity, sustainability, and resilience. This section looks to identify emerging trends, innovations, and forthcoming opportunities that are laid down to define the next stage of smart agriculture [2].

Figure 5 depicts the AI-guided decision loop in smart farming. Sensor devices are operating in the environment, and the data they collect is being added to a collection system. The AI analyses the data and makes a decision to drive automated actions, such as irrigation or fertilization. The action then alters the environment which is monitored by the feedback through sensor devices. This feedback process develops a closed loop system, allowing for a reactive and adaptive approach to managing farm activities, filling capacity gaps and improving the efficiency and resilience of agricultural production when external conditions are continuously changing.

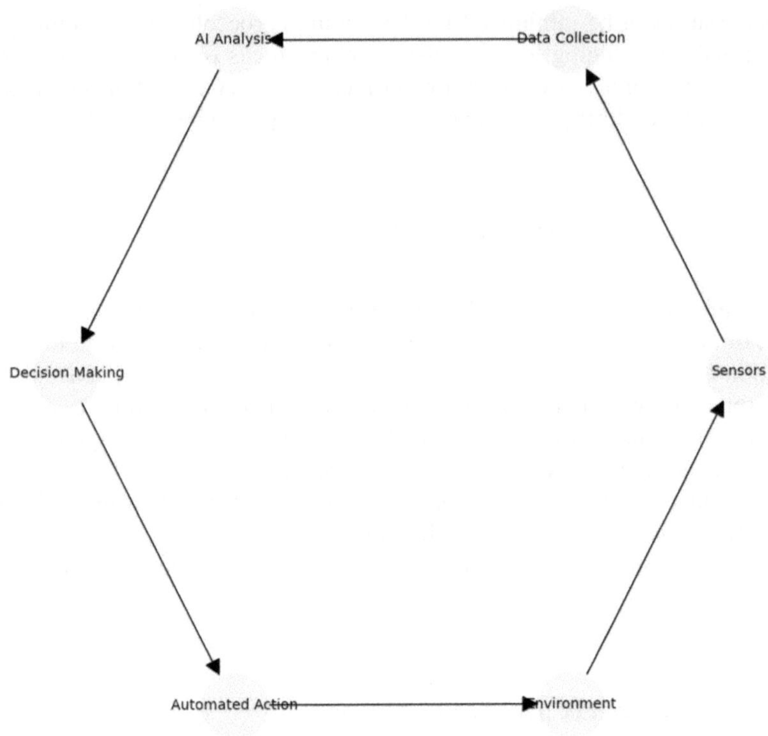

Fig. 5 AI-driven decision loop in smart farming

6.1 AI-Driven Predictive Analytics and Decision Support

The most important near-term trend is the growing use of AI-fueled predictive analytics. By 2025, intelligent platforms will turn to weather forecasts, soil health, historical conditions, and other data to inform decisions about crop planting and timing of harvest that is more productive and profitable for the farm [8–10]. These systems can manage large and complex datasets, streamlining the analysis of multiple factors, leveraging satellite imagery, Internet of things (IoT) data, soil analysis, market and/or commodity price data, and past historical conditions to generate the most accurate, actionable recommendations. For instance, predictive models can indicate yields based on AI and help farmers allocate inputs efficiently, as well as warnings about potential diseases and pests via computer vision systems.

AI is also re-shaping the way farmers will manage their farms in the coming years. For instance, generative AI will serve as a virtual agronomist and can assist farmers in determining the best crop rotations, the best irrigation schedule, and approaches to soil management in the coming weeks, months, or season/year. Democratisation of expert knowledge is essential for farmers, as all farmers will have access to the same information and decision-making resources, regardless of the size farm business diversities can close the knowledge gap between large agribusiness and family farms.

6.2 *The Internet of Farming: Hyperconnected, Real-Time Agriculture*

The Internet of Farming (IoF), which integrates IoT, AI, cloud, and big data analytic capabilities is changing the way we grow crops, manage livestock, and steward forests. By 2025, nearly 75 million—IoT devices will be deployed in agriculture, with the ability to collect and analyze data like never before. The IoT devices, including sensors that coats soil, cameras on drones, and automated tractors will many factors affecting environmental conditions, crop health, and farm machinery performance.

This hyperconnectivity allows for precision management at either the field level, the crop level, or even the individual plant and/or animal level. For example- smart irrigation systems provide dynamic scheduling of water for irrigation based on natural weather, soil moisture, and forecasts, allowing farmers to conserve water by as much as 60%. AI powered drones and cameras can survey fields and remote networks to look for early signs of disease or nutrient deficiency, enabling farmers to react quickly and precisely to improve yields and decrease chemical impacts.

6.3 *Robotics and Automation: The Next Agricultural Workforce*

Robotics and automation continue to develop at an accelerating pace and are proving to be the answer to the labor shortage, and are also improving efficiencies on the farm. For example, autonomous machinery is now able to tackle complicated tasks, such as planting, transplanting, weeding, and harvesting without the need for human touch—driverless tractors, robotic harvesters, and entire fleets of drones are able to perform complicated, skilled work with precision and efficiency. These autonomous systems promote constant output and quality with minimal labor costs and decrease variability at scale.

The third-generation agricultural robots will also be advanced and adaptable, versatile and skilled enough to perform multiple different tasks. The next generation of agricultural robots could very well be capable of performing anything during a growing season from soil characterization and land preparation to planting, crop protection, yield estimation and everything in between basically fully automating the entire agricultural life cycle.

6.4 Advanced Monitoring and Computer Vision

Emerging computer vision and enhanced monitoring systems have become key components of future agriculture. These technologies provide high-resolution monitoring of soil, crops and livestock, providing farmers with real-time feedback on the health of their plant populations, insect issues, or soil nutrients. With the application of AI, cameras or sensors can pick up small differences in either leaf colour or plant structure often before the symptoms are apparent to the human eye and provide the opportunity for early intervention and fine scale management.

AI monitoring will expand through satellite driven support to provide whole of farm specific knowledge in the future to support improved decision making for both farmers and agribusinesses. Such technologies will be particularly applicable to assess distant or large monitored agricultural areas where manual scouting is impossible.

6.5 Climate Adaptation and Resilience

As climate change progresses, the ability to adapt to the variability of climate and extreme events has never been more urgent. An important advance in climate adaptation is the application of AI and IoT in the development of climate adaptation modeling. Using multi-compartment simulation capabilities, AI can model and simulate various climate scenarios to help farmers develop resilient production systems and mitigate risk. Crops engineered by AI that are drought-tolerant, pest resistant, and heat tolerant, are moving from the research laboratory to commercial fields; thus, reducing reliance on chemical solutions and protecting food security.

AI-powered predictive analytics are helping us to not only predict disease outbreaks and model and forecast yield losses from weather events, but can advise adaptive management strategies to manage risk, creating proactive risk management strategies instead of reactive farming strategies.

6.6 Blockchain and Supply Chain Transparency

The future of agriculture will become increasingly transparent and traceable through the use of blockchain technology along with AI and the internet of things. You are not only developing immutable records of every step in the supply chain from planting to harvest, processing and lastly to distribution, thus providing consumer confidence but also providing access to premium markets by showing how to conform to sustainability and quality standards.

When these improvements are made using AI-powered supply chain management, farming will be even more efficient, allowing for demand forecasting and market prediction, while reducing overproduction and waste. These processes are especially crucial, as food production will need to increase to feed a growing global population while minimizing environmental impact.

6.7 Economic Growth and Market Expansion

There are great opportunities for the use of AI and IoT technology for agricultural use. The global AgriTech market is expected to more than double between 2023 and 2029. AI in agriculture is projected to grow from $1.7 billion to $4.7 billion by 2028. Furthermore, precision agriculture and the digital agriculture movement could contribute $450 billion to the annual agricultural GDP of low and middle-income countries. This demonstrates the transformation potential of agriculture from these technologies.

In the future, as more advanced tools become increasingly cheaper and readily accessible, it is anticipated that smallholder farmers will engage with digital agriculture and participate in the digital revolution at increased levels. This will facilitate more inclusive growth and rural development. New business models could be developed such as subscription-based advisory service, equipment sharing, and data-driven marketplace, that will enable access to technology and expertise for all.

6.8 Challenges and the Path Forward

While these developments are positive, challenges must be addressed to realize the full potential of AI and IoT in the agricultural future. These include:

1. Digital divide: Access to technology and connectivity remains equitable across urban and rural locations, or developed and less developed states [7].
2. Data privacy and protection: Farm-level data is sensitive in nature and susceptible to cyber threats. Efforts to organize data governance must also be transparent [8].
3. Improving skills and training: Farmers and agricultural workers need to be skilled to develop digital literacy or exchange digital skills and competencies [7].
4. Sustainable agricultural practices: Technological advancements need to be weighed against our environmental responsibility to reduce e-waste and assign value to circular economy, or material use coherence [7].

In response to these challenges, we will need a collaborative approach from governance, industry, academic, and civil society; as well as investment needs including, infrastructure, education, and policy frameworks that encourage responsible innovation.

6.9 The Road Ahead: Toward Intelligent, Sustainable Agriculture

Looking ahead, increasing AI, IoT, robotics, and advanced analytics will continue to turn what is possible in agriculture on its head. Farms of the future will be smart, connected ecosystems that can self-monitor, self-optimize, and even self-heal. [artisans] will increasingly turn to a network of digital advisors, autonomous machines, and insights from near-real-time data flows in order to make decisions that are better respected not only for the profitability but also the sustainability and resilience.

At this point of maturation, the impact of these technologies will change the agriculture landscape, but will also deliver a more expansive role for clues about food security, environmental impacts of agriculture or the economy of opportunity address agricultural viability of rural communities. The future is not just going to be growing enough food, but growing in a smarter, more sustainable, and more inclusive way than we've ever done before.

7 Conclusion

The marriage between Artificial Intelligence (AI) and the Internet of Things (IoT) in agriculture is a watershed moment in the history of food production around the globe. In this chapter, we have explored the fundamental concepts, transformative impacts, types of applications, barriers, and future opportunities for these technologies in agriculture. In conclusion, AI and IoT are not just advancements (one after another), they are catapulting agriculture to the next level of smarter, better and more sustainable food production.

AI and IoT have brought us to precision agriculture where real-time data processed with intelligent analytics are guiding decisions in every aspect of agriculture—soil preparation, planting, harvesting and supply chain. AI and IoT give farmers nimbleness to minimize resource use, maximize yield, lower impact on the environment, and proactively address challenges, such as climate change, resource constraints, and rising global food demands. Overall, the use of AI and IoT provides farmers with real-time crop and livestock monitoring, disease and pest prediction, mitigation via automation, automated redundant labor work, and supply chain and food transparency. Agriculture is operating on a different footing.

Nevertheless, this transition holds challenges. To make sure that people have equal and uninhibited access to competitive advantages from the tech/science commercial sector, we must assuage technical/infrastructure limitations, initial capital requirements, skill gaps, and critiques about data privacy and inequity. It is equally another challenge to ensure that technology adoption considers a complex nexus of ethical, social, and environmental factors, which, when carefully curated, support inclusivity, sustainability, and resilience.

The incoming view of agri-tech, thus far, remains positive. The advancement of artificial intelligence, the Internet of Things (IoT), robots and robots process driven on blockchain, and so on, are excitingly experiencing/and facing sustained performance (efficiency, sustainability, productivity, etc.) as well and they would only continue develop to be more accessible, affordable—democratically accessed knowledge and considered tools—and creating a new generation of digitally informal tech empowered rural poor, farmers and fisherfolk underdeveloped from a lot of development funding stakeholders.

In conclusion, the adoption of artificial intelligence AI and Internet of Things IoT would primarily define the trajectory of agriculture in the years to come. Therefore, having learnt from the past it accepted, was there as a sector to fulfill the demand to feed a growing global population, while also protecting the global environment, and locals invested in poverty the sector can begin its new journey towards intelligent and environmental, socially sustainable agriculture: what was early in engagement of that journey, and whether on 27 social media done, fully engrossed, this journey is just beginning; and hoping it is iterative along the way; with cooperative spirit in collaboration, innovation, in a shared responsibility, as a technologically based sector, not consuming, sustainably /environmentally commercially, standing by rural farmers and fishing family businesses towards an equal anticipating operating food and agricultural question about feeding nine billion people on the planet while making nine billion people are accommodated and have opportunities, are all contributing a more progressive and resilient society for other beings as human guides to the experience of sustainability and accept opportunity.

References

1. Ibrahim, A. S., Mohsen, S., Selim, I. M., Alroobaea, R., Alsafyani, M., Baqasah, A. M., & Eassa, M. (2025). AI-IoT based smart agriculture pivot for plant diseases detection and treatment. *Scientific Reports, 15*(1), 1–16.
2. Mohamed, E. S., Belal, A. A., Abd-Elmabod, S. K., El-Shirbeny, M. A., Gad, A., & Zahran, M. B. (2021). Smart farming for improving agricultural management. *The Egyptian Journal of Remote Sensing and Space Science, 24*(3), 971–981.
3. Avalekar, U., Patil, D. J., Patil, D. S., Khot, P., & Prathapan, P. (2024). Optimizing agricultural efficiency: A fusion of Iot, AI, cloud computing, and wireless sensor network. *Prof.(Dr.) Kesava, Optimizing agricultural efficiency: A fusion of IoT, AI, cloud computing, and wireless sensor network.*
4. Zhang, F., Zhang, Y., Lu, W., Gao, Y., Gong, Y., & Cao, J. (2022). 6G-enabled smart agriculture: A review and prospect. *Electronics, 11*(18), 2845.
5. Alreshidi, E. (2019). Smart sustainable agriculture (SSA) solution underpinned by internet of things (IoT) and artificial intelligence (AI). arXiv preprint arXiv:1906.03106
6. Dubey, A. K., Kumar, A., Kumar, S. R., Gayathri, N., & Das, P. (Eds.). (2021). *AI and IoT-based intelligent automation in robotics*. John Wiley & Sons.
7. Deepa, B., Swapna, B., Divya, B. S., & Chate, P. J. (2023). Internet of things (IoT) assisted smart agriculture monitoring and summarization system using NodeMCU and efficient sensor unit. In *2023 9th international conference on Smart Structures and systems (ICSSS)* (pp. 1–6). IEEE.

8. Sharma, K., & Shivandu, S. K. (2024). Integrating artificial intelligence and internet of things (IoT) for enhanced crop monitoring and management in precision agriculture. *Sensors International*, 100292.
9. Bhola, A., Sharma, H., Sagar, A. K., & Kumar, P. (2024). Pre-harvest to post-harvest: A review of AI and IoT applications in smart agriculture and the prospects of 6G-enabled IoT framework. In *2024 27th international symposium on wireless personal multimedia communications (WPMC)* (pp. 1–6). IEEE.
10. Miller, T., Mikiciuk, G., Durlik, I., Mikiciuk, M., Łobodzińska, A., & Śnieg, M. (2025). The IoT and AI in agriculture: The time is now—a systematic review of smart sensing technologies. *Sensors (Basel, Switzerland), 25*(12), 3583.
11. Jadon, J. K. S., & Singh, R. (2022). Challenges and opportunities of Internet of Things in smart agriculture: A review. *Electronic Systems and Intelligent Computing: Proceedings of ESIC, 2021*, 653–662.

Sensor Solutions: Cultivating Smart Agriculture

Indu Bala, Suverna Sengar, Komal Arora, Maad M. Mijwil, Karan Aggarwal, Jyoti Kedia, Deepika Kukreja, and Sapna Jarial

Abstract The integration of sensor technologies into agriculture is revolutionizing traditional farming practices, driving the development of smart agriculture. This chapter explores the critical role of sensor solutions in cultivating smart agriculture, focusing on how these technologies enable precision farming, resource optimization, and sustainable agricultural practices. By providing real-time data on soil conditions, crop health, environmental factors, and livestock monitoring, sensors empower farmers to make informed decisions, automate processes, and enhance productivity. The chapter delves into various types of sensors, including soil, climate, crop health, and livestock sensors, highlighting their applications in irrigation management, pest detection, yield forecasting, and animal welfare. Further, the chapter discusses the benefits of adopting sensor technologies, such as improved efficiency, cost savings, and sustainability, while addressing challenges like implementation costs, data management, and connectivity issues. Future trends, such as AI integration, IoT advancements, and blockchain for data integrity, are also explored, emphasizing the potential of sensor solutions to further transform agriculture and meet the growing global demand for food production.

I. Bala (✉) · S. Sengar
SEEE, Lovely Professional University, Punjab, India
e-mail: i.rana80@gmail.com

K. Arora
ECE Department, Chandigarh Group of Colleges, Landran, Mohali, India

M. M. Mijwil
Baghdad College of Economic Sciences University, Baghdad, Iraq

K. Aggarwal
ECE Department, Maharishi Markandeshwar University, Haryana, India

J. Kedia
Punjab Engineering College (Deemed to Be University), Chandigarh, India

D. Kukreja
Netaji Subash University of Technology, New Delhi, India

S. Jarial
Department of Agriculture, Lovely Professional University, Punjab, India

S. R. Kumar et al. (eds.), *AI-Driven IoT Solutions in Agriculture: New Insights*, Studies in Computational Intelligence 1236, https://doi.org/10.1007/978-3-032-05603-0_2

Keywords Precision agriculture · Sensor technologies · Smart farming · Irrigation management · Crop monitoring · Sustainable agriculture

1 Introduction to Smart Agriculture

Smart agriculture, also known as precision agriculture or digital farming, represents a modern approach to farming that leverages advanced technologies to improve efficiency, productivity, and sustainability. As the global population continues to grow and the demand for food increases, the agriculture sector faces significant challenges, including the need to produce more with fewer resources, mitigate the impact of climate change, and ensure environmental sustainability. Smart agriculture addresses these challenges by integrating technologies like sensors, Internet of Things (IoT), artificial intelligence (AI), and data analytics into traditional farming practices [1].

As shown in Fig. 1, the concept of smart agriculture is built on the ability to gather real-time data from various aspects of the farm, such as soil conditions, crop health, environmental factors, and livestock behavior. This data is used to optimize farming practices, enabling farmers to make informed decisions that enhance crop yields, reduce waste, conserve water, and minimize the use of pesticides and fertilizers. By applying data-driven insights, farmers can achieve precision in planting, watering, fertilizing, and harvesting, leading to more efficient resource use and reduced environmental impact [2].

One of the key drivers behind the growth of smart agriculture is the rapid development and affordability of sensor technologies. Sensors are small, reliable devices

Fig. 1 Smart agriculture with precision technology

that can monitor and measure a wide range of parameters on the farm. From soil moisture and nutrient levels to weather conditions and livestock movement, these sensors provide farmers with valuable information that can be accessed remotely and in real-time. The integration of these technologies into farming not only boosts productivity but also promotes sustainable agricultural practices by minimizing the overuse of resources [3].

In addition to sensors, the Internet of Things (IoT) plays a crucial role in smart agriculture by creating a network of interconnected devices that communicate with each other to automate farming processes. For instance, IoT-enabled irrigation systems can automatically adjust water levels based on soil moisture data, while AI-powered drones can assess crop health and detect pests early. These innovations reduce manual labor and improve the accuracy of farm operations.

As smart agriculture evolves, it has the potential to revolutionize the way food is produced globally. By optimizing resource use, reducing environmental impact, and increasing yields, it offers a promising solution to some of the most pressing issues in modern agriculture [4]. Moreover, smart agriculture technologies provide farmers with the tools to adapt to the effects of climate change, improve food security, and meet the growing demand for sustainably produced food.

2 Role of Sensor Technologies in Precision Farming

Sensor technologies play a pivotal role in precision farming, transforming traditional agricultural practices into data-driven, highly efficient systems. Precision farming, also known as precision agriculture, is a management strategy that utilizes detailed, site-specific data to make informed decisions on crop and resource management. The integration of sensor technologies allows farmers to monitor critical parameters, analyze real-time data, and optimize inputs like water, fertilizers, and pesticides. This leads to higher crop yields, lower costs, and more sustainable farming practices [5].

2.1 Importance of Sensors in Precision Agriculture

Sensors serve as the foundational technology for precision farming by collecting crucial information about soil conditions, plant health, climate, and livestock behavior. Unlike conventional farming methods that rely on uniform treatments across entire fields, sensor technologies enable farmers to adopt a targeted approach, applying resources where they are needed most. This not only conserves resources but also improves the overall efficiency and effectiveness of farming operations [6].

For instance, soil moisture sensors can determine the precise water needs of crops, ensuring that irrigation systems supply just the right amount of water, reducing both water wastage and the risk of overwatering. Similarly, nutrient sensors can detect soil deficiencies, allowing for targeted fertilization, which promotes healthier plant growth while preventing the excessive use of chemical inputs. The ability to monitor and control these variables in real time is what gives precision farming its edge over traditional methods.

2.2 *Types of Sensors in Precision Farming*

Different types of sensors are used in precision farming, each designed to monitor specific conditions and parameters. Some of the most widely used sensors include:

i. *Soil Sensors*: These measure factors like soil moisture, temperature, nutrient levels, and pH, providing farmers with a detailed understanding of soil health. This data is critical for optimizing irrigation, fertilization, and planting strategies (Fig. 2).
ii. *Crop Health Sensors*: Using multispectral or hyperspectral imaging, crop health sensors monitor plant vitality by analyzing factors such as chlorophyll levels, leaf area, and stress indicators. These sensors help in early detection of diseases, pest infestations, and nutrient deficiencies (Fig. 3).

(a) (b) (c)

Fig. 2 Soil sensors **a** pH sensor, **b** moisture sensor, **c** temperature sensor

(a) (b) (c)

Fig. 3 Crop health sensors, **a** chlorophyll levels sensor, **b** leaf area index meter, **c** plant nutrient deficiencies sensor

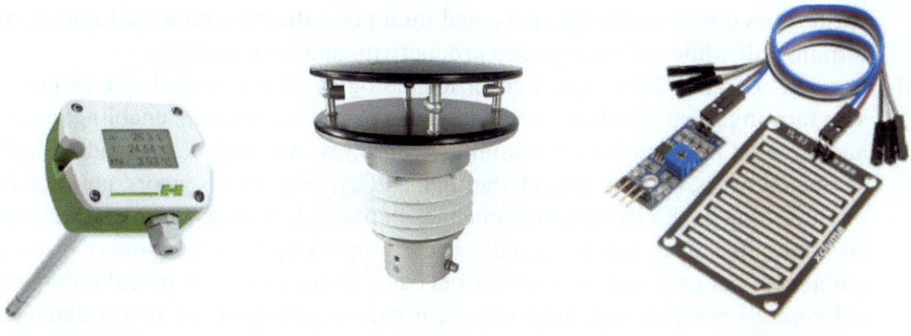

(a) (b) (c)

Fig. 4 Climate sensors: **a** temperature and humidity sensor, **b** wind speed sensor, **c** rain fall sensor

iii. *Climate Sensors*: Monitoring environmental variables such as temperature, humidity, wind speed, and rainfall, climate sensors provide vital information for optimizing planting schedules and managing crop growth (Fig. 4).
iv. *Livestock Sensors*: *In* animal farming, sensors are used to track animal behavior, monitor health conditions, and optimize feeding schedules. These technologies allow farmers to enhance livestock welfare and productivity [7].

2.3 *Applications of Sensor Technologies in Precision Farming*

Sensor technologies have a wide array of applications in precision farming. They contribute to various aspects of farm management, including:

i. *Irrigation Management*: Sensor technologies play a crucial role in precision farming, especially in irrigation management, by enabling farmers to optimize water usage and improve crop yields. Soil moisture sensors, for example, can measure the moisture content in the soil in real-time, allowing farmers to apply the exact amount of water needed for different crops and soil types, preventing overwatering or underwatering [7]. This not only conserves water but also ensures that plants receive the necessary hydration for optimal growth. These sensors can be integrated into automated irrigation systems, where they communicate with controllers to deliver water only when soil moisture drops below a certain threshold. Additionally, weather sensors and temperature monitors can provide valuable data on environmental conditions, helping farmers adjust irrigation schedules based on weather patterns, such as rainfall or evaporation rates. These sensor networks, when combined with GPS and remote sensing technologies, allow for site-specific irrigation, where different areas of a field receive tailored water amounts based on their unique needs. This precision reduces water waste, cuts down on energy costs, and minimizes the environmental impact of farming, all while enhancing crop productivity and sustainability.

ii. *Pest and Disease Detection*: Sensor technologies play a crucial role in precision farming, particularly in the detection of pests and diseases, enabling timely and targeted interventions that minimize crop loss and reduce pesticide usage. Various sensors, such as optical, thermal, and hyperspectral sensors, are used to monitor crop health by detecting changes in plant physiology, temperature, and spectral reflectance, which can indicate early signs of pest infestations or disease outbreaks. For example, optical sensors can capture variations in leaf color and chlorophyll content, signaling stress caused by pathogens or insect damage. Thermal sensors detect heat differences that can arise from localized plant infections, while hyperspectral imaging offers detailed spectral information to identify specific diseases or pest attacks by analyzing multiple wavelengths of light [8]. These sensors are often integrated with drones or ground-based systems to cover large agricultural fields, providing real-time data that helps farmers monitor crop health at a granular level. Additionally, data from these sensors can be fed into machine learning models and decision support systems to predict the spread of diseases or pests, optimizing the timing and application of treatments. By using sensor technologies for pest and disease detection, farmers can adopt more sustainable practices, reducing chemical inputs, improving crop yields, and ultimately enhancing food security [9].

iii. *Yield Optimization*: Sensor technologies are integral to yield optimization in precision farming, helping farmers maximize crop production while minimizing resource inputs. By providing real-time, site-specific data on soil conditions, plant health, and environmental factors, sensors enable farmers to make informed decisions that improve crop growth and productivity. Soil moisture sensors, for instance, allow for precise irrigation control by measuring water levels at various depths, ensuring crops receive optimal hydration without water wastage [10]. Nutrient sensors assess soil nutrient content, helping farmers apply fertilizers more efficiently and only where needed, preventing over-fertilization and reducing environmental impacts. Additionally, sensors like those based on weather and climate monitoring can predict environmental stressors such as drought, frost, or heatwaves, allowing farmers to take preventive measures that protect crop yields. In conjunction with GPS and remote sensing technologies, drones equipped with multi-spectral and infrared sensors can monitor plant growth patterns and detect areas of the field that require attention, such as regions with poor nutrient uptake or pest infestations. These insights lead to the implementation of variable-rate technology (VRT), where inputs like seeds, water, and fertilizers are applied in precise amounts to different parts of the field. The data-driven approach facilitated by sensor technologies not only enhances yield but also promotes sustainable agricultural practices by optimizing input usage, reducing waste, and improving overall crop management efficiency [11].

iv. *Livestock Monitoring*: Sensor technologies have revolutionized livestock monitoring in precision farming, providing farmers with real-time data that enhances animal health, productivity, and overall farm efficiency. Wearable sensors such as collars, ear tags, and leg bands equipped with GPS, accelerometers, and biometric sensors track key metrics like location, movement, heart rate, temperature, and behavior of individual animals [12]. These sensors help farmers detect early signs of illness, stress, or abnormal behavior, allowing for timely interventions that improve animal welfare and reduce losses. For instance, a sudden decrease in movement or changes in eating patterns could indicate health issues such as lameness or digestive problems. Additionally, temperature sensors can monitor heat stress in animals, especially in dairy cattle, where maintaining optimal conditions is critical for milk production [13]. These technologies also play a significant role in reproductive management, with sensors identifying the best time for breeding through behavioral and physiological indicators. By using data analytics and predictive modeling, farmers can optimize feeding regimens, track weight gain, and ensure optimal grazing patterns, reducing overgrazing and improving pasture management. Furthermore, integrated GPS tracking enables better management of livestock in large pastures, ensuring animals are safely located, reducing the risk of loss, and promoting better resource allocation [14]. With these advances, sensor technologies not only boost productivity and efficiency in livestock farming but also promote sustainable and humane practices.

2.4 Benefits of Sensor Technologies in Precision Farming

The integration of sensor technologies in precision farming offers numerous benefits, both economically and environmentally. Some of the key advantages include:

i. *Resource Efficiency*: Sensor technologies significantly enhance resource efficiency in precision farming by enabling the precise application of water, fertilizers, pesticides, and other inputs, which leads to reduced waste and optimized resource use [15]. By leveraging soil moisture sensors, farmers can monitor water levels at various depths and apply irrigation only when and where it is needed, preventing over-irrigation and reducing water consumption. This targeted approach to watering helps conserve a critical resource, particularly in regions facing water scarcity. Similarly, nutrient sensors assess the soil's nutrient composition, allowing for site-specific fertilizer application that avoids over-fertilization, minimizes runoff, and reduces the environmental impact of farming [16]. These sensors also contribute to more efficient pesticide use by identifying areas where pests are present, enabling farmers to apply pest control measures only in affected regions rather than across the entire field. Additionally, weather and climate sensors provide data on temperature, humidity, and other environmental factors, helping farmers predict and respond to changing conditions, which further reduces the need for excessive inputs. The integration of GPS and remote sensing technologies allows for precise mapping and monitoring of fields, facilitating the application of inputs based on the exact needs of specific zones. This level of precision improves crop growth while minimizing resource use, ultimately lowering production costs. By promoting the judicious use of resources, sensor technologies help farmers achieve higher yields with fewer inputs, contributing to more sustainable and economically viable agricultural practices.

ii. *Increased Productivity*: Sensor technologies play a pivotal role in boosting productivity in precision farming by providing farmers with real-time, accurate data that helps optimize every aspect of crop and livestock management [17]. With tools such as soil moisture, nutrient, and weather sensors, farmers can make data-driven decisions that improve crop health and growth. For example, by monitoring soil moisture levels, farmers can apply precise amounts of water to crops, ensuring they are neither under- nor over-irrigated, which promotes healthier plants and higher yields. Nutrient sensors allow for the accurate application of fertilizers, ensuring that crops receive the right nutrients at the right time, thus maximizing growth potential while reducing wastage. Moreover, drones and satellite-based remote sensors provide a comprehensive view of large fields, enabling farmers to detect areas of underperforming crops or early signs of stress due to disease, pest infestations, or nutrient deficiencies [18]. These insights allow farmers to intervene promptly, preventing crop loss and boosting overall productivity. In livestock farming, wearable sensors track animal health and behavior, allowing for early detection of illness, optimal feeding, and reproductive management, all of which contribute to higher productivity in terms of milk, meat, or other animal-based outputs. The ability to automate and optimize key farming activities such as planting, fertilization, irrigation, and pest control, based on real-time data, leads to better resource allocation, reduced input costs, and increased output. Overall, sensor technologies enable a more efficient, precise, and productive farming system, ultimately contributing to higher yields and profitability for farmers [19].

iii. *Cost Savings*: Sensor technologies in precision farming provide significant cost savings by optimizing resource use and reducing waste. By using real-time data on soil moisture, nutrients, and weather, farmers can apply water, fertilizers, and pesticides only when needed, cutting input costs [20]. For example, irrigation sensors help conserve water and lower energy costs, while soil sensors prevent over-fertilizing, reducing both expenses and environmental impact. Additionally, sensors enable early detection of crop stress, disease, or pests, allowing farmers to address problems before they escalate, thus minimizing crop losses and expensive interventions. Automated sensor-driven equipment, like drones, also reduces labor costs by streamlining tasks such as planting, monitoring, and harvesting. Long-term, sensors protect farmland and equipment, extending their useful life and preventing costly repairs. As sensor technology becomes more affordable, these cost-saving benefits will be accessible to a broader range of farmers, making agriculture more efficient and sustainable [21].

iv. *Sustainability*: Sensor technologies in precision farming enhance sustainability by optimizing resource use and reducing environmental impact. Soil moisture sensors ensure precise irrigation, conserving water and preventing over-irrigation, which can lead to erosion and nutrient runoff. Nutrient sensors allow targeted fertilizer application, minimizing chemical overuse and pollution [22]. Pest detection sensors reduce pesticide use by enabling selective spraying, preserving beneficial organisms and reducing chemical runoff. Automated farming equipment, powered by sensor data, cuts fuel consumption and lowers carbon emissions by optimizing machinery use. By maintaining soil health and promoting biodiversity, sensor technologies support long-term ecological balance, helping farms become more resilient to climate change while minimizing their environmental footprint. This sustainable approach improves resource efficiency, protects ecosystems, and promotes long-term agricultural productivity [23].

2.5 *Challenges in Adopting Sensor Technologies in Agriculture Sector*

Adopting sensor technologies in the agriculture sector presents a range of challenges despite their potential to transform farming practices. One of the primary obstacles is the high cost of sensor deployment and maintenance, particularly for small-scale farmers with limited financial resources. The initial investment in sophisticated sensors, coupled with ongoing maintenance and calibration, can be prohibitive. Additionally, lack of technical expertise poses a significant hurdle. Many farmers are unfamiliar with the operation of advanced sensor systems, and without adequate training or support, they may struggle to integrate these technologies into their farming practices effectively [24]. Another challenge lies in the connectivity and infrastructure required for sensor data to be transmitted and analyzed in real-time. Rural areas often lack reliable internet access, which can impede the functioning of IoT-enabled agricultural systems that rely on constant connectivity for optimal performance. Moreover, the sheer volume of data generated by these sensors necessitates advanced data analytics tools and platforms, adding another layer of complexity for farmers who may lack the resources to invest in sophisticated software solutions.

Data security and privacy concerns also arise as farmers collect sensitive information about their land, crops, and practices. Ensuring that this data is protected from cyber threats and unauthorized access is crucial but often overlooked. Additionally, the integration of sensor technologies into existing farming systems may face resistance due to traditional practices and cultural barriers, with some farmers hesitant to adopt new methods that deviate from conventional farming techniques. Moreover, environmental factors such as harsh weather conditions can affect the durability and functionality of sensors, leading to inaccuracies in data collection and increased costs for replacements or repairs. Addressing these challenges requires a collaborative effort between technology providers, government bodies, and agricultural stakeholders to create solutions that are affordable, accessible, and easy to implement across diverse farming environments [25].

2.6 The Future of Sensors in Precision Farming

As sensor technologies continue to evolve, their role in precision farming will become even more prominent. Future advancements, such as the integration of artificial intelligence (AI) and machine learning with sensor data, will enable predictive analytics and smarter decision-making. Additionally, the growing adoption of the Internet of Things (IoT) in agriculture will lead to more interconnected, autonomous farming systems that can operate with minimal human intervention [26].

The future of sensors in precision farming holds immense promise, as advancements in technology continue to reshape agricultural practices, making them more efficient, sustainable, and productive. As the global population grows and the demand for food increases, precision farming, driven by sensor technology, is set to play a pivotal role in enhancing crop yields while minimizing resource use. Miniaturization of sensors and the development of more cost-effective, robust, and energy-efficient devices are expected to accelerate adoption across farms of all sizes. These sensors, embedded in fields, crops, or machinery, will be able to monitor real-time data on soil moisture, nutrient levels, temperature, humidity, crop health, and pest infestations. The future will likely see the integration of sensors with artificial intelligence (AI) and machine learning algorithms, enabling predictive analytics that help farmers make proactive decisions, such as optimizing irrigation schedules or applying fertilizers only when needed, which conserves water and reduces chemical usage. Another critical development is the rise of autonomous farming equipment, such as drones and robots equipped with sensors that can perform tasks like planting, weeding, and harvesting with unparalleled precision and efficiency [27]. These machines, guided by sensor data, can significantly reduce labor costs and improve operational efficiency. Furthermore, the integration of sensor data with satellite imagery and remote sensing technologies will allow for large-scale monitoring of agricultural areas, enabling farmers to get a comprehensive view of their fields and make data-driven decisions on a macro scale.

In the coming years, Internet of Things (IoT) networks will become more widespread in rural and agricultural areas, improving connectivity and allowing seamless communication between sensors, devices, and platforms. This will lead to the creation of smart farming ecosystems, where every aspect of farming, from planting to harvest, is interconnected and optimized through data. Additionally, the future of sensors in precision farming will focus on sustainability, with technologies designed to minimize environmental impact. For instance, sensors can help reduce the overuse of fertilizers and pesticides, thus promoting healthier soils and ecosystems. The future will also address current challenges in sensor durability, data privacy, and ease of use, ensuring that farmers can adopt these technologies on a larger scale. Government initiatives, coupled with private sector investments, will likely drive research and innovation, making sensor technologies more accessible and affordable [28]. As these advancements continue to evolve, the future of sensors in precision farming will not only revolutionize agricultural practices but also play a crucial role in addressing global food security challenges while ensuring the sustainability of the planet's natural resources.

3 Real-Time Data in Agriculture

Real-time data is at the heart of modern agriculture, allowing farmers to make informed decisions based on current conditions in the field. By using sensor technologies to gather immediate data on soil, crops, climate, and livestock, farmers can respond quickly to changing conditions, optimize resource use, and enhance productivity. This approach minimizes guesswork and helps achieve precision in farming activities, such as irrigation, fertilization, and disease control. The integration of real-time data in agriculture is transforming traditional farming practices by enabling timely interventions that maximize efficiency and sustainability.

3.1 Soil Condition Monitoring

Soil condition monitoring is one of the most critical applications of real-time data in agriculture. Sensors that measure soil moisture, temperature, nutrient levels, and pH provide farmers with a detailed understanding of the soil's health and readiness for planting or irrigation. This data allows for precise water management, preventing overwatering or drought stress on crops. By continuously monitoring soil conditions, farmers can also adjust fertilization and planting schedules to optimize growth and improve crop yields, reducing unnecessary resource usage and environmental impact.

3.2 Crop Health Monitoring

Crop health monitoring involves using sensors to assess the vitality of plants through indicators such as chlorophyll content, leaf area, and stress levels. This real-time data allows farmers to detect early signs of nutrient deficiencies, diseases, or pest infestations, facilitating timely interventions before significant damage occurs. For example, multispectral sensors can analyze crop canopies to detect areas of stress, enabling targeted treatments rather than blanket applications of pesticides or fertilizers. Crop health monitoring ensures that resources are used efficiently, reducing input costs while maintaining healthy, high-yield crops [29].

3.3 Environmental and Climate Monitoring

Real-time monitoring of environmental and climate conditions, including temperature, humidity, wind speed, and rainfall, is essential for optimizing crop management strategies. Climate sensors provide farmers with up-to-the-minute data on weather conditions, allowing them to make informed decisions about planting, harvesting, and protecting crops from extreme weather events. This data also helps in adjusting irrigation schedules and understanding the microclimate around the farm, which can significantly impact crop growth. By leveraging real-time climate data, farmers can mitigate risks associated with unpredictable weather and improve overall crop resilience.

3.4 Livestock Monitoring

Livestock monitoring through real-time data helps farmers keep track of animal behavior, health, and well-being. Sensors attached to animals can monitor vital signs, movement patterns, and feeding habits, providing early warnings for health issues or abnormal behavior. This data enables farmers to take preventative measures, ensuring timely medical intervention and reducing the risk of diseases spreading among livestock [30]. Additionally, automated feeding and environmental control systems can be adjusted based on real-time data to optimize animal welfare and improve production efficiency in areas like milk yield or meat quality.

4 Agriculture Sensors

4.1 Soil Sensors

Soil sensors play a crucial role in modern agriculture by providing real-time data on various soil parameters. These sensors measure factors like soil moisture, temperature, pH, salinity, and nutrient levels, which are essential for optimal crop growth and resource management. By integrating soil sensors into precision agriculture systems, farmers can make informed decisions on irrigation scheduling, fertilization, and soil treatment. This data helps reduce water and chemical usage, increasing both the efficiency and sustainability of agricultural practices. Advances in wireless communication have enabled remote monitoring of soil conditions, further enhancing farm management.

4.2 Climate Sensors

Climate sensors are designed to monitor atmospheric conditions that directly affect crop growth, such as temperature, humidity, rainfall, wind speed, and solar radiation. These sensors are critical in predicting weather patterns, identifying optimal planting windows, and managing crop stress due to environmental changes. By integrating climate sensors into agricultural systems, farmers can adapt their strategies to current weather conditions, ensuring that crops are protected from potential climate-induced damage. Moreover, climate sensors contribute to long-term planning, helping farmers anticipate seasonal changes and adjust farming practices to mitigate the impact of climate variability [21].

4.3 Crop Health Sensors

Crop health sensors are essential for monitoring the physiological status of plants and detecting early signs of stress, diseases, or pest infestations. These sensors use technologies like multispectral and hyperspectral imaging, infrared sensors, and chlorophyll fluorescence to assess plant health. By identifying anomalies in crop health early, farmers can take targeted action to prevent yield losses and minimize the use of pesticides or herbicides. Crop health sensors are often integrated with drones or satellite systems, allowing large-scale monitoring of fields with high precision and reduced labor costs.

4.4 Livestock Sensors

Livestock sensors are increasingly used to monitor the well-being, behavior, and productivity of animals on farms. These sensors track various parameters, such as body temperature, movement, feeding habits, and heart rate, to assess the health and performance of livestock. Livestock sensors often come in the form of wearable devices, such as collars or ear tags, equipped with GPS and biosensors. This real-time data enables farmers to detect health issues, monitor reproductive cycles, and ensure proper nutrition, leading to improved animal welfare and farm efficiency. These sensors also help in managing large herds, reducing labor, and optimizing farm operations.

5 Applications of Sensors in Agriculture

Sensor technologies are transforming agriculture by enabling a wide range of applications that improve farm efficiency, productivity, and sustainability. These sensors collect real-time data on various aspects of the farm, such as soil moisture, crop health, and animal behavior, which can then be used to optimize resource use and decision-making processes. By integrating sensor systems into daily farming operations, farmers can automate tasks, reduce input costs, and minimize environmental impact. The growing use of sensors in agriculture is ushering in an era of precision farming, where data-driven insights help farmers respond to specific needs with greater accuracy and timing.

5.1 Irrigation Management

One of the most critical applications of sensors in agriculture is irrigation management. Sensors that measure soil moisture, temperature, and evapotranspiration rates provide real-time data that informs irrigation systems when and where water is needed. These sensors help farmers avoid overwatering, which can lead to water waste, soil erosion, and nutrient leaching, while also preventing underwatering, which can stress crops and reduce yields. By delivering water only where it is needed, automated sensor-driven irrigation systems optimize water use and improve crop health. For example, drip irrigation systems connected to soil moisture sensors can automatically adjust water levels to ensure plants receive just the right amount of hydration, reducing both water consumption and labor costs. This precise approach to irrigation is especially important in regions facing water scarcity, where efficient water use is critical for both economic and environmental sustainability.

5.2 Pest Detection

Pest detection is another key application of sensor technologies in agriculture, allowing farmers to address pest infestations before they cause significant damage. Traditional methods of pest management often rely on broad-spectrum pesticide applications, which can be costly, environmentally damaging, and ineffective in the long term. In contrast, sensors used for pest detection, such as cameras, traps, or acoustic devices, monitor pest activity in real time and alert farmers to early signs of infestation. These systems can also be integrated with predictive analytics tools to anticipate potential outbreaks based on environmental conditions like temperature and humidity [16]. With timely and accurate pest detection, farmers can apply targeted treatments to affected areas, reducing the use of pesticides and minimizing harm to beneficial insects and ecosystems. This leads to more sustainable pest management practices and better crop protection.

5.3 Yield Forecasting

Yield forecasting is a crucial application of sensors that enables farmers to predict harvest outcomes with greater accuracy. Sensors that monitor crop health, soil conditions, and climate variables provide continuous data on factors that influence crop growth. By analyzing this data, farmers can estimate potential yields early in the growing season, helping them make decisions on resource allocation, labor management, and market planning. For example, sensors that measure plant chlorophyll levels or track the development of fruit can offer insights into plant productivity, allowing farmers to anticipate how much produce they can expect at harvest time [27]. Additionally, advanced algorithms can use historical and real-time data to create predictive models that help farmers optimize inputs, such as fertilizers and irrigation, to maximize yields. Yield forecasting not only improves farm management but also helps reduce waste and increase profitability by aligning production with market demand.

5.4 Animal Welfare Monitoring

In livestock farming, sensor technologies play a crucial role in monitoring animal welfare, ensuring that animals are healthy, comfortable, and productive. Sensors attached to animals can track vital signs, behavior, and environmental conditions in real time, providing farmers with continuous updates on their livestock's health. For instance, sensors can detect changes in an animal's body temperature, heart rate, or movement patterns that may indicate illness or stress, allowing farmers to intervene before issues worsen. Automated systems can also monitor feeding and drinking behaviors, adjusting feed schedules or water supply as needed to ensure optimal nutrition. Environmental sensors in barns or pens can control temperature, humidity, and air quality, creating comfortable living conditions that enhance productivity, whether in terms of milk yield, meat quality, or reproductive health. By improving animal welfare, sensors contribute to more efficient and ethical farming practices, ultimately boosting both farm profitability and sustainability.

6 Benefits of Sensor Technologies in Agriculture

The integration of sensor technologies into agriculture offers a multitude of benefits that are reshaping traditional farming practices. By enabling real-time data collection and precise monitoring, sensors help farmers optimize various aspects of farm management, leading to significant improvements in efficiency, productivity, cost reduction, and sustainability. As the global demand for food production continues to rise, these technologies provide solutions that address both economic and environmental challenges. The ability to make data-driven decisions ensures that resources such as water, fertilizers, and labor are used effectively, contributing to the long-term sustainability of farming operations.

6.1 Improved Efficiency and Productivity

One of the most notable benefits of sensor technologies in agriculture is the dramatic improvement in efficiency and productivity. Sensors allow farmers to monitor soil conditions, crop health, and environmental factors with high precision, leading to optimized resource use. For example, by continuously tracking soil moisture and nutrient levels, sensors enable precision irrigation and fertilization, ensuring that crops receive exactly what they need for optimal growth. This reduces waste and ensures that plants are healthy and productive. Similarly, real-time monitoring of livestock health allows farmers to detect early signs of illness or stress, improving overall herd management and increasing productivity in dairy, meat, or egg production. The ability to automate many farming processes, from irrigation to pest control, also frees up labor and reduces human error, contributing to greater farm efficiency and higher yields [14].

6.2 Cost Savings

Cost savings are another significant advantage of sensor technologies in agriculture. By optimizing resource use, sensors help farmers reduce expenditures on inputs such as water, fertilizers, pesticides, and labor. For instance, automated irrigation systems connected to soil moisture sensors ensure that water is applied only when necessary, leading to lower water bills and reduced energy costs for pumping. Precision agriculture techniques, guided by real-time sensor data, also minimize the need for broad-spectrum pesticide and fertilizer applications, reducing both the cost of these inputs and the environmental damage they can cause [15]. Additionally, early detection of pest infestations or livestock health issues allows for targeted interventions, preventing large-scale losses and costly treatments. Over time, the initial investment in sensor technologies is often offset by the long-term savings achieved through reduced input costs, increased efficiency, and better management practices.

6.3 Sustainability in Farming Practices

Sensor technologies play a critical role in promoting sustainability in agriculture by minimizing waste and reducing the environmental impact of farming practices. With real-time data, farmers can adopt more sustainable approaches to managing water, soil, and energy resources. For example, precision irrigation systems reduce water wastage by ensuring that crops receive the exact amount of water they need, and soil condition monitoring helps prevent nutrient runoff by applying fertilizers in targeted, measured doses. Sensors also enable more precise application of pesticides and herbicides, reducing the risk of contamination to surrounding ecosystems and promoting biodiversity. Furthermore, the use of sensors in livestock farming contributes to animal welfare, improving both productivity and ethical standards in agriculture [10]. By enhancing the efficiency of resource use and reducing the environmental footprint, sensor technologies support the transition to more sustainable farming practices that can meet the growing demand for food while protecting natural resources for future generations.

7 Challenges in Implementing Sensor Solutions

Despite the numerous benefits of sensor technologies in agriculture, several challenges can hinder their widespread adoption. These challenges include high implementation costs, complexities in data management, and connectivity issues that can limit the effectiveness of sensor systems. Addressing these obstacles is essential for maximizing the potential of sensor technologies and ensuring that farmers can fully benefit from the advancements they offer. As the agriculture sector continues to evolve, finding solutions to these challenges will be crucial for the future of smart farming.

7.1 Implementation Costs

One of the primary barriers to adopting sensor technologies in agriculture is the high initial implementation costs. The expense associated with purchasing, installing, and maintaining sensor systems can be substantial, particularly for small-scale farmers or those operating on tight budgets. Costs can include not only the sensors themselves but also the necessary infrastructure, such as data management systems and software for analyzing the collected data. Additionally, training staff to effectively use and interpret sensor data adds to the overall expense. While the long-term benefits of increased efficiency and productivity can offset these costs, the upfront investment can be a significant deterrent for many farmers. Financial support, subsidies, or cooperative purchasing models may be needed to help lower the barriers to entry and facilitate the adoption of these technologies [2–5].

7.2 Data Management

The integration of sensor technologies generates vast amounts of data, leading to challenges in data management and analysis. Farmers must not only collect and store this data but also ensure that they can effectively interpret it to make informed decisions. Managing data from multiple sensors requires robust software systems and analytical tools, which can be complex and require specialized knowledge. Additionally, the need for data integration across different platforms can pose compatibility issues, making it difficult for farmers to gain a comprehensive understanding of their operations. Without effective data management strategies, the valuable insights that sensor data can provide may go untapped, limiting the potential benefits of these technologies. Developing user-friendly data management solutions and offering training on data analysis will be essential to help farmers leverage the full potential of sensor technologies [5].

7.3 Connectivity Issues

Connectivity issues can significantly impact the effectiveness of sensor solutions in agriculture, particularly in rural areas where internet access may be limited or unreliable. Many sensor systems rely on wireless communication to transmit data to central management systems, and poor connectivity can lead to delays or disruptions in data flow. This can hinder farmers' ability to receive real-time insights, reducing the overall effectiveness of their sensor-driven practices. Moreover, connectivity challenges can complicate the integration of sensors with other technologies, such as automated machinery and cloud-based analytics platforms [6]. To overcome these issues, investments in rural broadband infrastructure and the development of offline data processing solutions are necessary. By improving connectivity, farmers can ensure that they have consistent access to the data they need to make timely decisions, ultimately enhancing the success of sensor technologies in agriculture.

8 Future Trends in Sensor Technologies for Agriculture

The agricultural landscape is undergoing a profound transformation driven by advancements in sensor technologies. As the global demand for food increases, the integration of sophisticated technologies becomes essential to enhance productivity, sustainability, and efficiency in farming practices. This chapter explores three significant trends shaping the future of sensor technologies in agriculture: AI integration, IoT advancements, and blockchain for data integrity.

8.1 AI Integration in Agriculture

Artificial Intelligence (AI) is poised to revolutionize agricultural practices by facilitating data-driven decision-making processes. The integration of AI into sensor technologies enables farmers to leverage vast amounts of data collected from various sources, including soil sensors, weather stations, and drones. Key developments in this area include:

- *Predictive Analytics*: AI integration in agriculture is transforming farming by making operations more efficient, sustainable, and data-driven. Traditionally reliant on manual observations, farmers now benefit from AI's ability to analyze vast data from sensors, drones, satellite imagery, weather forecasts, and historical records. AI enables real-time insights, predicting optimal planting and harvesting times by analyzing weather and soil conditions, which helps farmers avoid adverse events and maximize yields. It also aids in selecting the best crop varieties for specific environments. In disease and pest management, AI systems analyze images and sensor data to detect early signs of problems, allowing farmers to take preventive action before issues escalate. This early detection reduces crop losses and minimizes chemical use, promoting sustainability. AI-driven predictive analytics also optimize resource management, helping farmers use water, fertilizers, and pesticides more efficiently. By analyzing soil moisture and weather forecasts, AI systems suggest the best times for irrigation, reducing waste and ensuring healthy crop growth while cutting operational costs [13].

  AI is also revolutionizing farm machinery, with autonomous drones and tractors performing tasks with greater precision than human labor. These machines can monitor fields, apply treatments, and optimize routes, reducing labor needs, fuel consumption, and environmental impact. Furthermore, AI helps farmers identify trends in crop performance, weather, and market conditions, enabling them to anticipate demand and adjust strategies. By revealing patterns and correlations, AI provides valuable insights to enhance productivity and support long-term planning. Overall, AI integration is making agriculture more efficient, sustainable, and resilient.
- *Automation of Agricultural Operations*: AI-powered robotics and machinery can perform tasks such as planting, watering, and harvesting with precision.

For example, autonomous tractors equipped with AI can adapt to varying field conditions, ensuring optimal operation based on sensor data.
- *Enhanced Crop Management*: By utilizing AI to process data from remote sensing technologies, farmers can obtain insights into crop health, nutrient deficiencies, and irrigation needs. This level of detail enables precision agriculture, where interventions are tailored to specific areas of a field, maximizing yield and minimizing resource wastage.

8.2 IoT Advancements

The Internet of Things (IoT) is rapidly advancing in agriculture, connecting a myriad of devices and sensors to facilitate real-time monitoring and data sharing. Future trends in IoT within agriculture include:

- *Smart Agriculture Solutions*: IoT-enabled devices such as soil moisture sensors, weather stations, and drone technology allow farmers to monitor environmental conditions continuously. This real-time data helps farmers make informed decisions regarding irrigation, fertilization, and pest management.
- *Integrated Farm Management Systems*: The emergence of IoT platforms enables farmers to centralize data from various sources, providing a holistic view of farm operations. These systems can automate processes, send alerts for critical events (e.g., extreme weather), and enhance overall farm management efficiency.
- *Sustainable Resource Management*: IoT advancements facilitate better resource management by optimizing water usage and reducing chemical inputs. For instance, smart irrigation systems can adjust water delivery based on soil moisture levels, significantly reducing waste while ensuring crops receive the necessary hydration.

8.3 Blockchain for Data Integrity

Blockchain technology offers unique solutions for ensuring data integrity and transparency in agricultural practices. By utilizing a decentralized ledger, blockchain can enhance trust among stakeholders in the agricultural supply chain. Key applications include:

- *Traceability in Food Supply Chains*: Blockchain enables the tracking of agricultural products from farm to consumer, providing transparency about the origins and journey of food. This traceability helps build consumer confidence and supports the demand for sustainably sourced products.
- *Smart Contracts for Automated Transactions*: The implementation of smart contracts on blockchain networks can facilitate transactions between farmers and suppliers. These contracts automatically execute agreements when pre-defined

conditions are met, reducing the need for intermediaries and enhancing operational efficiency [15–20].
- *Data Security and Ownership*: Blockchain can securely store sensor data, ensuring that farmers retain ownership and control over their information. This security mitigates risks associated with data tampering, providing reliable data for decision-making.

9 Sustainability and Global Food Production

Sustainability in global food production is critical for addressing the challenges posed by a rapidly growing population and the environmental pressures of climate change. As the demand for food increases, sustainable agricultural practices must be prioritized to ensure that production systems remain viable without depleting natural resources or harming ecosystems. This involves adopting methods that enhance resource efficiency, conserve biodiversity, and improve climate resilience. Innovations such as precision agriculture, agroecology, and sustainable supply chains are key to achieving these goals. By integrating technology and traditional knowledge, we can create agricultural systems that not only meet the nutritional needs of a burgeoning population but also protect the planet for future generations. Emphasizing sustainability in food production is not just an environmental imperative; it is essential for social and economic stability, ensuring that all communities have access to safe, nutritious food while fostering resilient ecosystems.

10 Conclusion

The integration of sensor technologies into agriculture is reshaping traditional farming practices, facilitating the rise of smart agriculture that prioritizes efficiency, sustainability, and productivity. By leveraging real-time data from various sensor types—ranging from soil and climate to crop health and livestock monitoring—farmers are empowered to make informed decisions that optimize resource use and improve yields. This chapter has highlighted the critical role these technologies play in enabling precision farming and addressing key challenges in modern agriculture, such as resource scarcity and environmental degradation. The benefits of adopting sensor technologies are manifold, including enhanced operational efficiency, cost savings, and sustainable agricultural practices. However, challenges such as implementation costs, data management complexities, and connectivity issues remain to be addressed.

Looking ahead, the future of sensor technologies in agriculture is promising. Continued advancements in artificial intelligence and the Internet of Things (IoT) will likely further enhance the capabilities of sensors, enabling even more precise and automated farming practices. Additionally, the incorporation of blockchain technology can improve data integrity and traceability, fostering transparency in agricultural supply chains. As these technologies evolve, they will not only drive innovation in agricultural practices but also play a crucial role in meeting the growing global demand for food in a sustainable manner. The future scope includes ongoing research and development aimed at lowering costs, enhancing accessibility for smallholder farmers, and integrating diverse technologies to create holistic smart farming solutions. Ultimately, embracing these advancements will be vital for cultivating resilient agricultural systems that can thrive amid the challenges of a changing world.

References

1. Shaikh, F. K., Karim, S., Zeadally, S., & Nebhen, J. (2022). Recent trends in internet-of-things-enabled sensor technologies for smart agriculture. *IEEE Internet of Things Journal, 9*(23), 23583–23598.
2. Khoa, T. A., Man, M. M., Nguyen, T. Y., Nguyen, V., & Nam, N. H. (2019). Smart agriculture using IoT multi-sensors: A novel watering management system. *Journal of Sensor and Actuator Networks, 8*(3), 45.
3. Navarro, E., Costa, N., & Pereira, A. (2020). A systematic review of IoT solutions for smart farming. *Sensors, 20*(15), 4231.
4. Ayaz, M., Ammad-Uddin, M., Sharif, Z., Mansour, A., & Aggoune, E. H. M. (2019). Internet-of-Things (IoT)-based smart agriculture: Toward making the fields talk. *IEEE access, 7*, 129551–129583.
5. Antonacci, A., Arduini, F., Moscone, D., Palleschi, G., & Scognamiglio, V. (2018). Nanostructured (Bio) sensors for smart agriculture. *TrAC Trends in Analytical Chemistry, 98*, 95–103.
6. Lakhiar, I. A., Jianmin, G., Syed, T. N., Chandio, F. A., Buttar, N. A., & Qureshi, W. A. (2018). Monitoring and control systems in agriculture using intelligent sensor techniques: A review of the aeroponic system. *Journal of sensors, 2018*(1), 8672769.
7. Ray, P. P. (2017). Internet of things for smart agriculture: Technologies, practices and future direction. *Journal of Ambient Intelligence and Smart Environments, 9*(4), 395–420.
8. Dhanaraju, M., Chenniappan, P., Ramalingam, K., Pazhanivelan, S., & Kaliaperumal, R. (2022). Smart farming: Internet of Things (IoT)-based sustainable agriculture. *Agriculture, 12*(10), 1745.
9. de Araujo Zanella, A. R., da Silva, E., & Albini, L. C. P. (2020). Security challenges to smart agriculture: Current state, key issues, and future directions. *Array, 8*, Article 100048.
10. Ragaveena, S., Shirly Edward, A., & Surendran, U. (2021). Smart controlled environment agriculture methods: A holistic review. *Reviews in Environmental Science and Bio/Technology, 20*(4), 887–913.
11. Kaewmard, N., & Saiyod, S. (2014). Sensor data collection and irrigation control on vegetable crop using smart phone and wireless sensor networks for smart farm. In *2014 IEEE conference on wireless sensors (ICWiSE)* (pp. 106–112). IEEE.
12. Dagar, R., Som, S., & Khatri, S. K. (2018). Smart farming–IoT in agriculture. In *2018 international conference on inventive research in computing applications (ICIRCA)* (pp. 1052–1056). IEEE.
13. Mehta, K., Arora, P., Arora, N., & Aayushi, A. (2022). Enhancement of smart agriculture using internet of things. *ECS Transactions, 107*(1), 7047.
14. Goel, R. K., Yadav, C. S., Vishnoi, S., & Rastogi, R. (2021). Smart agriculture–urgent need of the day in developing countries. *Sustainable Computing: Informatics and Systems, 30*, Article 100512.

15. Lavanya, G., Rani, C., & GaneshKumar, P. (2020). An automated low cost IoT based Fertilizer Intimation System for smart agriculture. *Sustainable Computing: Informatics and Systems, 28*, Article 100300.
16. Yin, H., Cao, Y., Marelli, B., Zeng, X., Mason, A. J., & Cao, C. (2021). Soil sensors and plant wearables for smart and precision agriculture. *Advanced Materials, 33*(20), 2007764.
17. Mekonnen, Y., Namuduri, S., Burton, L., Sarwat, A., & Bhansali, S. (2019). Machine learning techniques in wireless sensor network based precision agriculture. *Journal of the Electrochemical Society, 167*(3), Article 037522.
18. Suma, V. (2021). Internet-of-Things (IoT) based smart agriculture in India-an overview. *Journal of ISMAC, 3*(01), 1–15.
19. Yang, J., Sharma, A., & Kumar, R. (2021). IoT-based framework for smart agriculture. *International Journal of Agricultural and Environmental Information Systems (IJAEIS), 12*(2), 1–14.
20. Mahbub, M. (2020). A smart farming concept based on smart embedded electronics, internet of things and wireless sensor network. *Internet of Things, 9*, Article 100161.
21. Dhruva, A. D., Prasad, B., Kamepalli, S., & Kunisetti, S. (2023). An efficient mechanism using IoT and wireless communication for smart farming. *Materials Today: Proceedings, 80*, 3691–3696.
22. Boursianis, A. D., Papadopoulou, M. S., Diamantoulakis, P., Liopa-Tsakalidi, A., Barouchas, P., Salahas, G., & Goudos, S. K. (2022). Internet of things (IoT) and agricultural unmanned aerial vehicles (UAVs) in smart farming: A comprehensive review. *Internet of Things, 18*, Article 100187.
23. Muangprathub, J., Boonnam, N., Kajornkasirat, S., Lekbangpong, N., Wanichsombat, A., & Nillaor, P. (2019). IoT and agriculture data analysis for smart farm. *Computers and Electronics in Agriculture, 156*, 467–474.
24. Ferrández-Pastor, F. J., García-Chamizo, J. M., Nieto-Hidalgo, M., Mora-Pascual, J., & Mora-Martínez, J. (2016). Developing ubiquitous sensor network platform using internet of things: Application in precision agriculture. *Sensors, 16*(7), 1141.
25. Mathi, S., Akshaya, R., & Sreejith, K. (2023). An internet of things-based efficient solution for smart farming. *Procedia Computer Science, 218*, 2806–2819.
26. Triantafyllou, A., Tsouros, D. C., Sarigiannidis, P., & Bibi, S. (2019). An architecture model for smart farming. In *2019 15th international conference on distributed computing in sensor systems (DCOSS)* (pp. 385–392). IEEE.
27. Mohapatra, K., Singh, D., & Biswal, A. K. (2023). IoAT-based smart farming technology. In *2023 IEEE 2nd international conference on industrial electronics: developments & applications (ICIDeA)* (pp. 585–590). IEEE.
28. Alex, N., Sobin, C. C., & Ali, J. (2023). A comprehensive study on smart agriculture applications in India. *Wireless Personal Communications, 129*(4), 2345–2385.
29. Kamilaris, A., Gao, F., Prenafeta-Boldu, F. X., & Ali, M. I. (2016, December). Agri-IoT: A semantic framework for Internet of Things-enabled smart farming applications. In *2016 IEEE 3rd world forum on internet of things (WF-IoT)* (pp. 442–447). IEEE.
30. Jararweh, Y., Fatima, S., Jarrah, M., & AlZu'bi, S. (2023). Smart and sustainable agriculture: Fundamentals, enabling technologies, and future directions. *Computers and Electrical Engineering, 110*, 108799.

Connecting IoT for Precision Farming: Driving Efficiency and Sustainability in Agriculture

Aditya Vardhan and Sagar Sharma

Abstract Precision farming, powered by the Internet of Things (IoT), is revolutionising modern agriculture by enabling unprecedented levels of efficiency and productivity. Through the use of smart sensors, GPS technologies, and data analytics, IoT allows farmers to monitor and control various agricultural processes in real time. This targeted approach ensures optimal resource utilisation, such as water, fertilisers, and pesticides, while minimising waste and environmental impact. By leveraging IoT, precision farming enhances crop management through data-driven insights, improving yield forecasting, pest control, and soil health monitoring. This paper explores the key components of IoT-enabled precision farming, highlighting its role in maximizing efficiency, reducing costs, and promoting sustainability. Additionally, it presents case studies of successful implementations, demonstrating how IoT is transforming agriculture into a more precise, adaptive, and environmentally conscious practice.

Keywords Data analytics · Resource optimization · Smart agriculture · Agriculture 5.0 · IoT · Sustainability · Precision farming

1 Introduction to IoT-Enabled Precision Farming

Precision breeding is a transformative approach to agriculture that uses the Internet of Things (IoT) to provide more accurate, efficient and sustainable agricultural practices. In contrast to traditional farming methods based on uniform resource applications and manual observation, IoT-enabled precision breeding uses digital technology to collect, analyze and respond to data in real-time. This revolution allows farmers to make decisions that have been properly discovered, reduce waste, and improve crop

A. Vardhan (✉)
Department of Physics and Computer Science, Dayalbagh Educational Institute, Agra, India
e-mail: vardhana3098@gmail.com

S. Sharma
George Brown College, Toronto, ON, Canada
e-mail: er.sagar0594@gmail.com

S. R. Kumar et al. (eds.), *AI-Driven IoT Solutions in Agriculture: New Insights*, Studies in Computational Intelligence 1236, https://doi.org/10.1007/978-3-032-05603-0_3

yields. This ultimately contributes to nutritional safety and environmental protection [1, 2].

1.1 The Evolution from Traditional to Smart Agriculture

Planting, irrigation, fertilizing, and harvesting have all been guided by human experience, seasonal cycles, and manual effort for ages. Although somewhat successful, this conventional method frequently resulted in inefficiencies, excessive use of resources, and uneven results. The first significant technological advancement during the Green Revolution was the introduction of mechanization, which was followed by automation and the digitalization of farming through the use of Geographic Information Systems (GIS). But the move to smart agriculture is a more profound change, one in which automation, data, and connection come together to make farming responsive and intelligent. IoT, AI, robots, and cloud computing are all integrated in smart agriculture to enable site-specific farming that dynamically adjusts to crop and environmental conditions. These days, farmers employ mobile dashboards, sensor networks, and drone photography to keep an eye on field conditions, forecast illnesses, and maximize input utilization. By improving sustainability and profitability, this shift strengthens agriculture's ability to withstand global issues, including labour shortages, soil degradation, and climatic variability. A significant step toward effective, data-driven agricultural systems has been taken with the development of smart farming [3].

1.2 Role of IoT in the Agriculture 5.0 Landscape

According to Agriculture 5.0, farming will be completely automated, highly linked, and ecologically sensitive in the future. The Internet of Things (IoT), which enables real-time data gathering, communication, and control across all farming components, is at the centre of this revolution. An intelligent agricultural ecosystem is produced by the interaction of Internet of Things (IoT) devices, including weather stations, automated tractors, and soil sensors, via wireless networks, cloud services, and machine learning algorithms. Predictive analytics, precise input control, and ongoing condition adaptation are made possible by this degree of integration. With previously unheard-of precision, farmers may automate irrigation schedules, remotely monitor crop development, and get notifications about abnormalities [4].

2 Core Technologies Underpinning IoT in Agriculture

To monitor, automate, and enhance agricultural processes, IoT-based precision farming uses a mix of hardware, software, and communication protocols. These technologies serve as the cornerstone of smart agriculture, facilitating remote farm equipment operation, real-time field condition tracking, and data analysis for improved decision-making. Key components include sensors, actuators, GPS, embedded systems, and remote sensing instruments. A networked environment with smooth machine and system interaction is produced by their integration. Farmers will be able to enhance productivity, conserve resources, and decrease waste thanks to this combination of technologies. To fully appreciate how IoT turns conventional farming into an effective, intelligent, and sustainable ecosystem, it is essential to comprehend four fundamental elements shown in Fig. 1 [5, 6].

2.1 Sensor Networks, Actuators, and Embedded Systems

Precision farming relies heavily on sensor networks, which provide real-time information on crop health, temperature, humidity, and soil moisture. To record local variances and changes in real time, these sensors are dispersed throughout fields. This information is used by actuators to carry out automated tasks like pesticide spraying and irrigation. Sensor-actuator communication is controlled by embedded systems, which frequently process data locally via edge computing. With this configuration, rapid reactions are possible without depending solely on cloud servers. Together, they guarantee quicker decision-making, less manual effort, and increased productivity. Even small farms can profit from automation and data-driven interventions as these technologies become more affordable and widely available [7, 8].

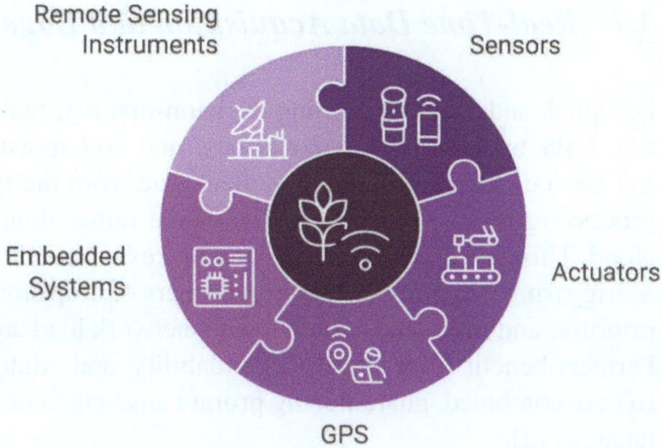

Fig. 1 Components of IoT-based precision farming

2.2 *Integration of GPS, Drones, and Remote Sensing Tools*

Precision farming accuracy is greatly increased by geospatial technologies. Farmers may use GPS to map fields, control a self-driving tractor, and keep an eye on equipment in real time. Aerial footage from drones can be used to detect problems such as nutritional deficits, water stress, or pest infestations. Drones and satellites with thermal and multispectral cameras provide comprehensive information about the health of plants. Remote sensing eliminates the need for physical field inspections by enabling rapid examination of wide areas. Resource optimization and prompt interventions are supported by this integration. Drones, remote sensing, and GPS work together to give farmers a potent arsenal for better land management, analysis, and visualization [9].

3 Data-Driven Decision Making in Precision Farming

IoT-based precision farming relies heavily on data. Large volumes of data are gathered about weather, soil, crops, and machinery using smart devices and sensors. Farmers can use this information to make well-informed decisions rather than depending on hunches or set schedules. Farmers can respond quickly to issues like disease outbreaks, irrigation requirements, or equipment malfunctions when they have access to real-time statistics. Predictive modeling can also be used with data to predict yields or spot potential dangers. Farmers can improve productivity and efficiency by examining trends and patterns. A more responsive, lucrative, and sustainable agricultural system where decisions are supported by facts rather than conjecture is the end outcome shown in Fig. 2 [10, 11].

3.1 *Real-Time Data Acquisition and Edge Computing*

For quick and efficient farming decision-making, real-time data collection is essential. GPS trackers, weather stations, and soil moisture sensors are examples of IoT devices that continuously collect data from the field. Edge computing enables processing to take place near the source rather than sending all data to a central cloud. This cuts down on delays and makes it possible to take faster responses, such as triggering irrigation or warning farmers of crop stress. Additionally, edge devices prioritise and filter data, which lowers network load and boosts system performance. Farmers benefit from speed, dependability, and control when edge computing and IoT are combined, guaranteeing prompt answers to field conditions without needless latency [12].

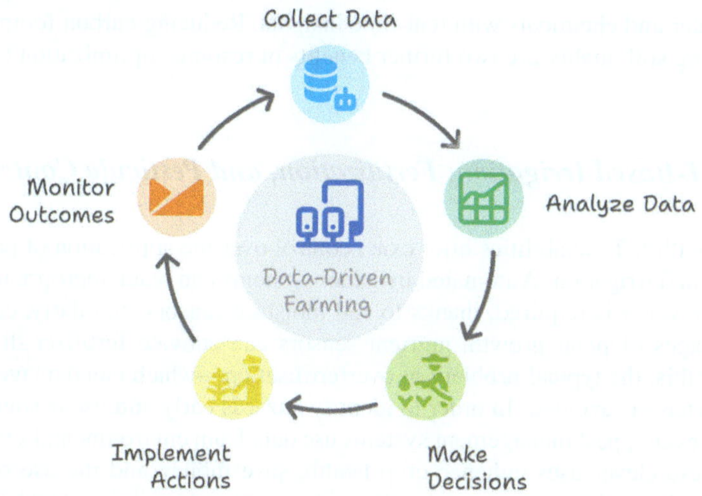

Fig. 2 Data-driven precision farming cycle

3.2 *Big Data Analytics for Predictive and Prescriptive Agriculture*

Big data analytics turns unprocessed agricultural data into insightful forecasts and suggestions. Large datasets are analyzed by precision farming methods like AI and machine learning to anticipate crop production, forecast weather impacts, and identify early disease and insect symptoms. Prescriptive analytics makes recommendations for what should be done, whereas predictive analytics foresees potential outcomes. For instance, using both historical and present data, data models might suggest the best times to plant or when to fertilize. These realizations enhance planning, cut expenses, and increase output. Big data analytics, in conjunction with IoT, enables farmers to take proactive rather than reactive action [13].

4 Smart Resource Management and Optimization

One of the main benefits of IoT in precision farming is effective resource management. Due to a lack of precise management, traditional farming frequently uses excessive amounts of energy, fertilizer, and water. Smart monitoring and automated resource distribution are made possible by IoT, guaranteeing that resources are used only when and where they are required. Sensors keep an eye on crop requirements, weather, and soil conditions, and intelligent controls modify fertilizer supply and irrigation as necessary. This specific use minimizes environmental harm, cuts expenses, and decreases waste. Farmers may improve sustainability by avoiding excessive use of water and chemicals with real-time insights. Reducing carbon footprints and maintaining soil quality are two further benefits of resource optimization [14].

4.1 IoT-Based Irrigation, Fertilization, and Pesticide Control

Systems with IoT capabilities offer exact control over the application of pesticides, fertiliser, and irrigation. Automated irrigation systems can react appropriately when and where water is required, thanks to soil moisture sensors. Similarly, depending on the stages of plant growth, nutrient sensors can activate fertilizer dispensers. By doing this, the typical problem of overfertilisation—which can damage soil and groundwater—is avoided. In order to identify hazards early and use pesticides only when necessary, pest management systems use data from environmental sensors and traps. These clever uses enhance crop health, save money and the use of chemicals, and safeguard important species. Better harvests and environmental balance are ultimately the results of focused treatments [15, 16].

4.2 Energy and Water Use Optimization for Sustainable Outputs

IoT aids in optimizing agriculture's use of energy and water, two essential components of sustainability. In order to minimize water loss, smart irrigation systems keep an eye on evapotranspiration rates and weather predictions. IoT devices that run on solar power, such as sensors and pumps, help reduce reliance on non-renewable energy. Tasks like planting and spraying are completed with the least amount of energy waste thanks to automated equipment that is controlled by GPS and real-time data. Farmers can monitor their electricity use, spot irregularities, and modify their plans to use less power during off-peak hours. These tactics encourage environmental stewardship while reducing operating expenses. Precision farming supports long-term agricultural resilience and sustainability by preserving energy and water [17].

5 Enhancing Crop Monitoring and Disease Prediction

Healthy yields and the early identification of issues depend on accurate crop monitoring. Proactive farming is made possible by IoT technology, which provide real-time visibility into plant status. Farmers can identify nutrient deficiencies, growth problems, and stress indicators before they become more serious by using instruments like multispectral photography, leaf sensors, and climate monitors. Without

Fig. 3 Enhancing crop management with technology

frequent field trips, these tools provide ongoing crop status monitoring. Farmers can reduce damage and stop disease outbreaks by acting quickly to spot patterns and abnormalities early. In addition to increasing production, IoT-based crop monitoring lessens the need for manual scouting. This results in more consistent harvest quality, better resource use, and better timing of treatments shown in Fig. 3 [18, 19].

5.1 Remote Crop Health Surveillance Using IoT and AI

Artificial intelligence and Internet of Things technologies allow for remarkably accurate remote crop health observation. Data on leaf colour, moisture content, chlorophyll levels, and canopy development are recorded in real time by drones and field sensors. AI examines this data to find anomalies like pest activity or nutrient stress. These technologies are able to identify trouble spots and provide immediate recommendations for solutions. Mobile apps are used to deliver alerts to farmers, enabling prompt action. Remote surveillance improves diagnosis accuracy while lowering the requirement for in-person field inspections. By ensuring that each plant receives the proper care at the appropriate time it enhances crop quality and output [20].

5.2 Pest Detection and Mitigation with Predictive Analytics

Pests are a major threat to crops, but IoT can detect them early and prevent outbreaks. Sensors placed around fields monitor temperature, humidity, and plant signals associated with pest presence. Cameras and pheromone traps relay visual and behavioural data. Predictive models analyze this information to forecast pest activity and recommend targeted treatment. Instead of spraying entire fields, farmers can treat only affected areas, saving cost and reducing environmental harm. Automated systems can also activate repellents or mechanical controls when threats are detected. This precise, data-driven approach improves pest management, protects beneficial insects, and supports sustainable agricultural practices [21, 22].

6 Soil Intelligence and Environmental Monitoring

Environmental factors and soil health are important determinants of agricultural productivity. Farmers may better understand nutrient levels, pH balance, compaction, and moisture content in various field zones with the use of IoT technology, which provides comprehensive insights into soil attributes. These realizations make it possible to apply water, fertilizer, and amendments more precisely, which enhances crop quality and soil sustainability. Furthermore, environmental sensors monitor wind patterns, light intensity, CO_2 levels, and local weather, enabling more intelligent farm activity planning. Risks from overfarming and climate variability are decreased by real-time soil and climate intelligence. Long-term productivity is increased, land degradation is avoided, and ecologically friendly farming methods are supported by this focused monitoring [23, 24].

6.1 IoT Systems for Soil Nutrient and Moisture Analysis

In-ground sensors are used by smart soil monitoring systems to measure important elements such as moisture, phosphorus, potassium, and nitrogen. These readings are transmitted to farmer dashboards or cloud platforms and are updated on a regular basis. Farmers can prevent both overuse and undernourishment by using this data to choose when and where to fertilize or irrigate. Plants are healthier, and input expenses are decreased as a result. Automated fertigation, in which nutrient-rich water is supplied in response to real-time soil readings, is even supported by certain sophisticated systems. IoT-based analysis is essential to maintaining soil ecosystems and guaranteeing long-term farm output since it maintains soil fertility and stops leaching or runoff [25, 26].

6.2 Climate-Smart Practices and Environmental Impact Reduction

IoT technologies help farmers adopt climate-smart practices by enabling better adaptation and resource conservation. Weather stations and environmental sensors monitor temperature, rainfall, humidity, and solar radiation. These data points help in scheduling irrigation, planting, and harvesting to align with favourable conditions.

Predictive models can also forecast extreme weather events, allowing for timely preparations. Moreover, IoT supports reduced greenhouse gas emissions by optimizing machinery usage and limiting excess fertilizer application. Smart monitoring of inputs and field conditions contributes to cleaner water, healthier air, and reduced chemical runoff. Altogether, IoT fosters an eco-friendly approach to farming that supports climate resilience and biodiversity [27].

7 Case Studies and Real-World Applications

IoT applications in agriculture demonstrate how technology may transform conventional farming practices. IoT technologies are being adopted by both large and small farms worldwide in order to manage irrigation, monitor crops, and make data-driven choices. These case studies show increases in sustainability, yield, and resource efficiency. IoT is versatile across crops and geographical areas, as evidenced by the use of soil sensors in Italian vineyards and weather-linked irrigation systems in Indian rice farms. Examining these effective models yields useful information for increasing the uptake of technology. They also emphasize how labour reduction, cost reductions, and productivity gains may all coexist. Future investments and regulations in digital agriculture are guided in part by lessons learned from practical applications [28].

7.1 Global Success Stories of IoT-Driven Precision Agriculture

In the Netherlands, high-tech greenhouses use IoT systems to control light, humidity, and nutrients, achieving high yields with minimal resource input. Real-time monitoring of planting and harvesting activities is made possible in the United States by John Deere's networked equipment. Similar to this, government-supported projects in India have installed Internet of Things irrigation systems in areas that are prone to drought, which has resulted in a 30% reduction in water usage. These illustrations demonstrate how IoT adapts farming to local difficulties. Smartphone-based services that transmit weather forecasts and sensor data are helping even smallholder farmers. These success examples demonstrate that IoT is a scalable solution that helps farms of all sizes, not just a luxury [29, 30].

7.2 Comparative Analysis of Outcomes: Cost, Yield, and Sustainability

IoT-enabled farms outperform conventional ones in several quantifiable ways. Because precision agriculture improves input management, output improvements of 15–25% are common. Fertilizer expenditures can be lowered by up to 30%, while water use can decrease by 20–40%. Smart equipment scheduling also frequently results in energy savings. Additionally, IoT promotes environmentally sustainable farming by lowering carbon emissions and chemical runoff. According to case study statistics, return on investment (ROI), particularly for high-value crops, is usually positive within two to three seasons. These advantages show how implementing IoT can boost profitability while supporting sustainability objectives and the ideas of climate-smart agriculture [28].

8 Challenges, Future Trends, and Policy Implications

IoT in precision farming has a lot of promise, but there are a number of obstacles to overcome. Widespread adoption is hampered by low internet availability in rural areas, expensive setup fees, concerns about data privacy, and farmers' lack of technological expertise. Another issue is interoperability across various IoT platforms and devices. Despite these challenges, smart agriculture appears to have a bright future thanks to quick developments in blockchain, artificial intelligence, and wireless communication. To close the digital divide, strategic policies and farmer education are crucial. To guarantee infrastructure, data security, and cost, the public and private sectors must work together. Precision agriculture will scale effectively and sustainably in the years to come if these issues are resolved [31, 32].

8.1 Technical, Ethical, and Infrastructural Barriers

Inadequate network coverage, unstable electricity, and expensive equipment are just a few of the infrastructure issues that frequently hinder the adoption of IoT in agriculture. Technically speaking, a lot of farms don't have standardized platforms to link different devices, which results in disjointed systems. Ethical concerns are also important, particularly those about privacy and data ownership. Farmers frequently worry about how service providers will use, store, and share their data. Adoption is further hampered by rural farmers' low levels of internet literacy. Strict regulations, user-friendly interfaces, reasonably priced gear, and a strong digital infrastructure are all necessary to address these problems. To fully profit from smart farming technologies, these issues must be resolved [33, 34].

8.2 *Future Outlook: AI Integration, Blockchain, and Policy Frameworks*

Deeper integration with cutting-edge technologies will influence the direction of IoT in agriculture. Artificial intelligence will improve crop recommendations, automate decision-making, and improve forecasting skills. Blockchain will increase stakeholder trust by ensuring safe data interchange and traceability in food supply chains. Even in remote locations, low-power networks like 5G and LoRaWAN will enable large-scale, real-time deployments. In terms of policy, governments need to provide educational initiatives, smallholder subsidies, and frameworks that encourage innovation while guaranteeing the ethical use of data. Precision agriculture will become more intelligent, accessible, and in line with international sustainability and food security objectives thanks to these developments [35, 36].

9 Conclusion

An important development in contemporary agriculture is IoT-driven precision farming, which offers more intelligent, effective, and sustainable methods of food production. Farmers can correctly monitor crop health, manage resources, and react proactively to environmental changes by combining real-time data collection, intelligent automation, and advanced analytics. There is no denying the advantage of greater yields, lower prices, and a reduced environmental impact, despite obstacles such as infrastructural constraints and data privacy issues. As AI, connectivity, and policy backing continue to advance, IoT will become increasingly widely available and influential across all farming scales. Building a resilient, successful, and ecologically conscious agricultural future requires embracing this technology.

References

1. Agrawal, H., Dhall, R., Iyer, K. S. S., & Chetlapalli, V. (2020). An improved energy efficient system for IoT enabled precision agriculture. *Journal of ambient intelligence and humanized computing, 11*, 2337–2348.
2. Atalla, S., Tarapiah, S., Gawanmeh, A., Daradkeh, M., Mukhtar, H., Himeur, Y., Mansoor, W., Hashim, K. F. B., & Daadoo, M. (2023). IoT-enabled precision agriculture: Developing an ecosystem for optimized crop management. *Information, 14*(4), 205.
3. Stamatescu, G., Drăgana, C., Stamatescu, I., Ichim, L., & Popescu, D. (2019). Iot-enabled distributed data processing for precision agriculture. In *2019 27th mediterranean conference on control and automation (MED)* (pp. 286–291). IEEE.
4. Phupattanasilp, P., & Tong, S.-R. (2019). Augmented reality in the integrative internet of things (AR-IoT): Application for precision farming. *Sustainability, 11*(9), 2658.
5. Brewster, C., Roussaki, I., Kalatzis, N., Doolin, K., & Ellis, K. (2017). IoT in agriculture: Designing a Europe-wide large-scale pilot. *IEEE Communications Magazine, 55*(9), 26–33.

6. Singh, R. K., Berkvens, R., & Weyn, M. (2021). AgriFusion: An architecture for IoT and emerging technologies based on a precision agriculture survey. *IEEE Access, 9*, 136253–136283.
7. Rajabzadeh, M., & Fatorachian, H. (2023). Modelling factors influencing IoT adoption: With a focus on agricultural logistics operations. *Smart cities, 6*(6), 3266–3296.
8. El-Basioni, B. M. M., & Abd El-Kader, S. M. (2020). Laying the foundations for an IoT reference architecture for agricultural application domain. *IEEE Access, 8*, 190194–190230.
9. Babar, A. Z., & Akan, O. B. (2024). Sustainable and precision agriculture with the internet of everything (IoE). arXiv preprint arXiv:2404.06341
10. Alreshidi, E. (2019). Smart sustainable agriculture (SSA) solution underpinned by internet of things (IoT) and artificial intelligence (AI). arXiv preprint arXiv:1906.03106
11. Caro, M. P., Ali, M. S., Vecchio, M., & Giaffreda, R. (2018). Blockchain-based traceability in Agri-Food supply chain management: A practical implementation. In *2018 IoT vertical and topical summit on agriculture-Tuscany (IOT Tuscany)* (pp. 1–4). IEEE.
12. Mishra, H., & Mishra, D. (2024). AI for data-driven decision-making in smart agriculture: from field to farm management. In *Artificial intelligence techniques in smart agriculture* (pp. 173–193). Springer Nature Singapore.
13. Linaza, M. T., Posada, J., Bund, J., Eisert, P., Quartulli, M., Döllner, J., Pagani, A., et al. (2021). Data-driven artificial intelligence applications for sustainable precision agriculture. *Agronomy, 11*(6), 1227.
14. Elsherbiny, O., Zhou, L., He, Y., & Qiu, Z. (2025). Precision farming: Using an IoT multimodal data-driven deep network to optimize irrigation in wheat crops. *Expert Systems with Applications* 128583.
15. van Klompenburg, T., & Kassahun, A. (2022). Data-driven decision making in pig farming: A review of the literature. *Livestock Science, 261*, 104961.
16. Ding, Y., Decker, C., Vassileva, I., Wallin, F., Beigl, M. (2011). A smart energy system: distributed resource management, control and optimization. In *2011 2nd IEEE PES international conference and exhibition on innovative smart grid technologies* (pp. 1–8). IEEE.
17. Sodhro, A. H., Pirbhulal, S., Luo, Z., & De Albuquerque, V. H. C. (2019). Towards an optimal resource management for IoT based Green and sustainable smart cities. *Journal of Cleaner Production, 220*, 1167–1179.
18. Kwon, H., Do, T. N., & Kim, J. (2022). Optimization-based integrated decision model for smart resource management in the petrochemical industry. *Journal of Industrial and Engineering Chemistry, 113*, 232–246.
19. Rahman, A., Islam, M. J., Montieri, A., Nasir, M. K., Reza, M. M., Band, S. S., Pescape, A., Hasan, M., Sookhak, M., & Mosavi, A. (2021). Smartblock-sdn: An optimized blockchain-sdn framework for resource management in iot. *IEEE Access, 9*, 28361–28376.
20. Li, J. (2020). Resource optimization scheduling and allocation for hierarchical distributed cloud service system in smart city. *Future generation computer systems, 107*, 247–256.
21. Vale, Z. A., Morais, H., Khodr, H., Canizes, B., & Soares, J. (2010). Technical and economic resources management in smart grids using heuristic optimization methods. In *IEEE PES general meeting* (pp. 1–7). IEEE.
22. Alotaibi, E., & Nassif, N. (2024). Artificial intelligence in environmental monitoring: In-depth analysis. *Discover Artificial Intelligence, 4*(1), 84.
23. Alzoubi, I., Almaliki, S., & Mirzaei, F. (2019). Prediction of environmental indicators in land leveling using artificial intelligence techniques. *Chemical and Biological Technologies in Agriculture, 6*, 1–19.
24. Himeur, Y., Rimal, B., Tiwary, A., & Amira, A. (2022). Using artificial intelligence and data fusion for environmental monitoring: A review and future perspectives. *Information Fusion, 86*, 44–75.
25. Stipanicev, D., Bodrozic, L., & Stula, M. (2007). Environmental intelligence based on advanced sensor networks. In *2007 14th international workshop on systems, signals and image processing and 6th EURASIP conference focused on speech and image processing, multimedia communications and services* (pp. 209–212). IEEE.

26. Teng, Y., Wu, J., Lu, S., Wang, Y., Jiao, X., & Song, L. (2014). Soil and soil environmental quality monitoring in China: a review. *Environment international, 69*, 177–199.
27. Gautam, K., Sharma, P., Dwivedi, S., Singh, A., Gaur, V. K., Varjani, S., Srivastava, J. K., Pandey, A., Chang, J.-S., & Ngo, H. H. (2023). A review on control and abatement of soil pollution by heavy metals: Emphasis on artificial intelligence in recovery of contaminated soil. *Environmental Research,* 225, 115592.
28. Ahmad, S., Kaushik, R., Ghatuary, R., Kotiyal, A., Jarial, S., & Kumar, R. (2025). Utilizing IoT and AI for soil health monitoring and enhancement in sustainable agriculture. In *IoT and advanced intelligence computation for smart agriculture* (pp. 110–125). CRC Press.
29. Kalantzopoulos, G., Paraskevopoulos, P., Domalis, G., Liopa-Tsakalidi, A., Tsesmelis, D. E., & Barouchas, P. E. (2024). The Western Greece soil information system (WESIS)—a soil health design supported by the internet of things, soil databases, and artificial intelligence technologies in Western Greece. *Sustainability, 16*(8), 3478.
30. Senanayake, S., Pradhan, B., & Wedathanthirige, H. (2024). Spatial intelligence in AI applications for assessing soil health to monitor farming systems and associated ESG risk. In *Intelligence systems for earth, environmental and planetary sciences* (pp. 81–111). Elsevier.
31. Wani, A. K., Rahayu, F., Amor, I. B., Quadir, M., Murianingrum, M., Parnidi, P., Ayub, A., et al. (2024). Environmental resilience through artificial intelligence: Innovations in monitoring and management. *Environmental Science and Pollution Research, 31*(12), 18379–18395.
32. Meenakshi, B., Sumathy, B., Haariharan, N. C., Krishnakanth, L., & Soumya, M. (2023). Artificial intelligence enhanced-plant and soil monitoring systems. In *2023 intelligent computing and control for engineering and business systems (ICCEBS)* (pp. 1–6). IEEE.
33. Nayar, P., & Dubey, A. K. (2025). Artificial intelligence based sensors for environmental monitoring. In *Emerging sensors for environmental monitoring* (pp. 187–199). Elsevier.
34. Zhao, J., & Wan, S. (2023). Artificial intelligence and hyperspectral modeling for soil management. In *Innovation for environmentally-friendly food production and food safety in China* (pp. 67–91). Springer Nature Singapore.
35. Wei, Y., Han, C., & Yu, Z. (2023). An environment safety monitoring system for agricultural production based on artificial intelligence, cloud computing and big data networks. *Journal of Cloud Computing, 12*(1), 83.
36. Abutaleb, K. A., Abdelsalam, A. A., & Khaled, M. A. (2025). Unleashing environmental intelligence through AI, image processing, and big data: Paving the path to a sustainable future. In *Modelling and advanced earth observation technologies for coastal zone management* (pp. 315–354). Springer Nature Switzerland.

Livestock Management Monitoring with AI

Kanakaprabha and K. Shanthi Latha

Abstract Artificial Intelligence (AI) in livestock management has completely changed how farmers and the agricultural sector track and maximize resource use, animal welfare, and output. By means of real-time analysis, predictive modeling, and automated data collecting, AI-driven solutions are significantly improving livestock monitoring systems. The many AI technologies utilized in livestock management are examined in this chapter, including computer vision, machine learning, and IoT-enabled devices that monitor behavior, environmental factors, and vital signs. Artificial Intelligence (AI) enhances productivity and sustainability by facilitating early disease detection, efficient breeding, and precision feeding through the provision of actionable insights from data. The future potential of AI in revolutionizing the cattle business is also highlighted, along with ethical considerations and problems like data protection and the use of AI in rural areas.

Keywords Artificial intelligence · Computer vision · Disease detection · Livestock management · Machine learning · Precision feeding

1 Introduction to Livestock Management and AI

Livestock Management is a major part of the agricultural sector, which includes for breed care, feeding and maintaining a check on the health of animals for nutrition, fibre, and other purposes. In traditionally, the farming of livestock has always depended on the manual processes, which are time-consuming, labor- intensive, and frequently to human error. Furthermore, Automation is has grown into an important tool in shifting how livestock are managed due to the increasing demand for efficiency, sustainability, and higher yields. The agriculture is for effective livestock

Kanakaprabha (✉) · K. S. Latha
Department of Artificial Intelligence and Machine Learning, Malla Reddy College of Engineering, Hyderabad, India
e-mail: skanakaprabha@gmail.com

K. S. Latha
e-mail: rshanthilatha@gmail.com

S. R. Kumar et al. (eds.), *AI-Driven IoT Solutions in Agriculture: New Insights*, Studies in Computational Intelligence 1236, https://doi.org/10.1007/978-3-032-05603-0_4

management both economic activity and the production of the food. The managing livestock is included keeping an eye on the general welfare, feed efficiency and health of the animals. The growth of artificial intelligence of livestock management of the new stage. The huge amounts of data to the sensors, cameras, and internet of things to the powered by artificial intelligence systems, real-time insights into the monitor, health and environment conditions of their animals. These advanced increase to the animal welfare and other optimize farm operations is to be improved productivity, feeding plans and other spotting early disease symptoms. The use of AI in livestock system is to identify the global need for the ethical and environmentally farming practices.

1.1 Overview of Livestock Management

The strategies and techniques are used for the care and maintain domesticated animals such as cattle, sheep, goats and pigs and poultry are known as livestock management. It is maintaining agricultural practices such as high-quality production and animal health are the main goals of livestock system. The several factors such as healthcare, breed and nutrition and caring animals are the managing livestock. The farmers are mainly conducting for animals that are effectively in the hygienic and secure the surroundings of illness and pain. In addition, preserving accurate methods and following rules are essential for the efficient operation of farms and livestock. The advancement technology is to be completely in the traditional handling livestock. These methods are to be identified in the range of livestock monitoring is to be possible development of many techniques, data mining. Furthermore, farmers can be identified various resources such as feed and water these instruments are to be identified in the sexual, growth, and overall health of their livestock in real time. The profitability and improving animal welfare, and handling livestock of animals is to improve the food chain and growing the need for ethical and environmentally agricultural methods.

1.2 Key Challenges in Livestock Farming

The numerous major challenges are more sustainable and production is mainly for the livestock farming sector. The prevention of disease is another main problem of cattle were at high risk at the several issues of disorders is to be spread more quickly and cause of huge losses. The early illness and identification are more prevention of ailments to be identify the farmers, especially those with big operations. These can be major benefits of nutrition and feed efficiency is another major concern is the direct impact of health and output of animals. This can be more problems in growing prices for high-quality feed and the changes in feed inventory. Furthermore, sustainable cattle farming is hindered by more issues like greenhouse gas emissions, soil degradation, water scarcity. These emissions of the methane from cattle animal husbandry are main important factor of the climate change. Finally, two challenges are to be facing the managing labor shortages of the technology changes. These automation and artificial intelligence are the novel innovations is to be identify the problems more effectively, are the farmers of lack the resources of implementing them.

1.3 Role of Technology in Livestock Monitoring

The technology is completely altered in the way of livestock monitored in the farmers in the innovative ways of the boost output better care of the animals, and improved for farm management. The real time livestock management of the easier to do more accurate because to advance in sensors, artificial intelligent and internet of things devices. The farmers may recognize early symptoms of disorder, other stress of odd behaviors is to the wearable technology is signs of body temperature, heart rate and other patterns. These devices included in the sensors to build in the potential of disease pandemics of detection improves animal's general health. Furthermore, AI-driven systems of the huge collections of data collected from the diverse sources to offer useful insights into environmental factors, breeding cycles and other dietary habits (Fig. 1).

The animals receive the huge amount of nutrients is to minimize of the waste, and these devices in the maximization of feed efficiency. In addition, automatic cameras and robots are being used for the remotely monitor conditions and livestock behavior. These data-driven decisions, save operating costs, and enhance the long-lasting of the agricultural methods through the technology of the livestock monitoring.

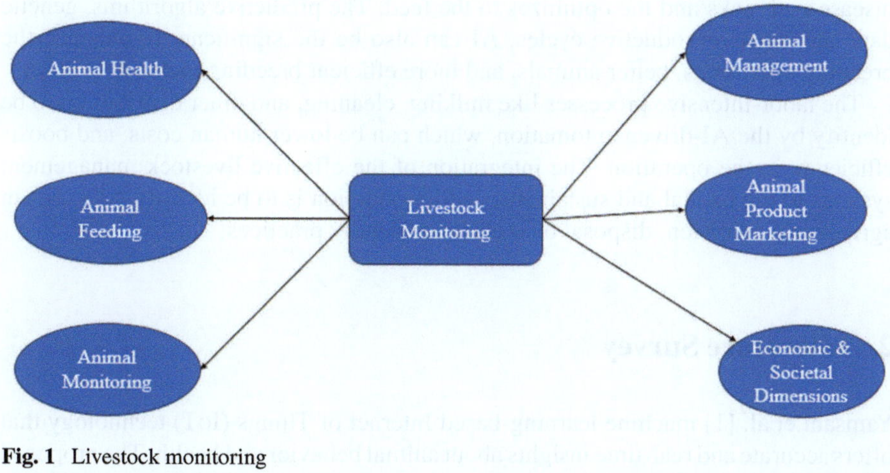

Fig. 1 Livestock monitoring

1.4 Evolution of Artificial Intelligence in Agriculture

The introduction of modern tools and system that can be improve the productivity, sustainability, and decision-making, artificial intelligence in the agriculture is entirely in the industry. The applications of artificial intelligence in the automation models like irrigation and machinery control. The artificial intelligence is identifying irrigation and machinery control. These evaluate can be massive datasets in the crop yields, and optimize utilization of resources. These are able to detect soil conditions, rainfall, pest control, planting, and harvest with the use of solutions in powered by AI. The AI is transformed into livestock monitoring is the sensors and prediction algorithms to identify the monitor health actions and wellbeing of cattle in real time. The AI-based system can be identifying the performance is too lowering the costs by improving mating procedures, feeding plans and identify disorder very early. In furthermore, farm activities are to AI integration with robotics and drones to improve agricultural techniques and diminish labor charge. The new era in farming integration is use to be identify significant charges such as solving issues like climate change and food safety.

1.5 The Scope of AI in Livestock Management

Artificial Intelligence is several types of applications in the livestock monitoring to improve farm efficiency, productivity, and health of animals. Artificial Intelligence is to identify the systems are able to track the behaviors of animals, identify early signs of disease, and provide real time data is identify the proactive treatment of disease. The patterns of motion, feeding habits and the other conditions is identifying the cameras and sensors. This farmer to improve the overall welfare of the animals, disease outbreaks and the optimizes to the feed. The predictive algorithms, genetic data and other reproductive cycles, AI can also be the significant to enhance the breeding operations, better animals, and more efficient breeding methods (Fig. 2).

The labor-intensive processes like milking, cleaning, and other distribution to be identify by the AI-driven automation, which can be lower human costs, and boosts efficiency in the operation. The integration of the effective livestock management systems to the ethical and sustainable food production is to be identify AI's uses in agricultural precision, disposal of waste and friendly practices.

2 Literature Survey

Yamsani et al. [1] machine learning-based Internet of Things (IoT) technology that offers accurate and real-time insights about animal behavior and health. The proposed system's cattle area is equipped with a network of Internet of Things (IoT) sensors

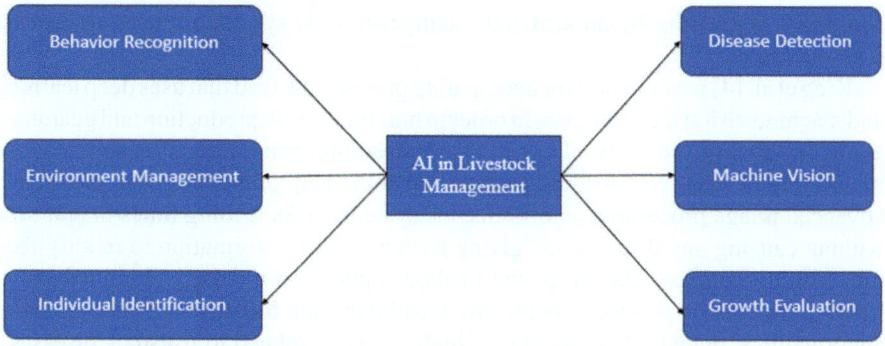

Fig. 2 AI in livestock management

that are positioned strategically to monitor a range of characteristics, such as temperature, humidity, grazing habits, and movements. These sensors collect data, which is wirelessly transferred to a central location where algorithms trained on experience in the field analyze it. Both supervised and unsupervised machine learning algorithms are used to glean valuable insights from collected data. The system keeps an eye on farmers or administrators in real time and instantly notifies them of any unusual occurrences pertaining to the behavior or wellbeing of the animals.

Arafat et al. [2] strong livestock management is necessary for efficient cattle farming in order to achieve the best possible health and output. This study presents a thorough framework for early disease identification and real-time livestock monitoring. It uses a mobile application, machine learning techniques, and LoRa communication IoT devices. Cattle equipped with LoRa-based sensors gather health data in real time. Long-distance transmission of this data is flawless. For processing, it is routed to a cloud platform. The suggested framework makes use of machine learning techniques. The Subspace k-Nearest Neighbors (KNN) classifier is the main importance. A quantitative investigation of ten classifiers has confirmed this. Subspace KNN-based is always better than the others. Early disease classification is accurate. Livestock managers are notified by sophisticated early detection algorithms via a mobile application. This makes immediate action possible. This makes it possible to take preventative healthcare actions. The suggested framework has several advantages. Proactive disease control, better livestock care, lower financial losses, and increased agricultural productivity are a few of these. The smartphone app offers warnings and an easy-to-use representation. This gives farmers useful information. This system provides a scalable and efficient solution for real-time livestock monitoring through the utilization of the Subspace KNN classifier, cloud computing mobile application, and LoRa IoT.

Camacho et al. [3] livestock farming accounting for the agricultural GDP, it is a significant source of revenue. However, a number of problems must be overcome for this industry to develop properly, including rustling, or the theft of cattle. With this study, we want to solve this issue by suggesting a technique for identifying and tracking animals through surveillance flights. Drone-captured photos that undergo additional processing by an artificial intelligence (AI) system are used to provide such outcomes.

King et al. [4] new method for anticipating precise cow feed that uses deep learning and machine vision technologies. In order to maximize milk production and guarantee animal wellbeing, the study addresses the vital requirement for accurate and effective monitoring of dairy cow feeding patterns. With the help of the YOLOv8 model and advanced image processing algorithms, the system tracks feeding times in real time without causing any disturbance, giving farmers useful information to modify feed mixes and schedules. The suggested method supports in the early identification of health problems by precisely measuring feed intake and identifying feeding pattern abnormalities, potentially reducing the high expenses related to livestock diseases.

Qazi et al. [5] a multimodal vision framework that uses the Grounding DINO, HQSAM, and ViTPose models to optimize precision livestock farming. Without the need for intrusive animal tagging, this unified package allows for thorough behavioral analytics using video data. HQSAM separates individual animals inside the precise bounding boxes that Grounding DINO creates around livestock. ViTPose facilitates movement and pose analysis by estimating important body points. Our technique, which has been shown on a sheep dataset that includes grazing, running, sitting, standing, and walking activities, extracts important insights such as broad position evaluations, interaction dynamics, and patterns of activity and grazing.

Sujitha et al. [6] the agricultural environment is changing, requiring creative ways to boost output while maintaining animal welfare. A cutting-edge Automated Animal Detection and Monitoring System from Farm guard is set to transform conventional farm management techniques. Farm guard provides real-time, non-invasive monitoring of animal movement, behavior, and health on farm grounds by utilizing state-of-the-art sensor technology, computer vision, and machine learning algorithms. Using a network of precisely placed sensors and cameras, this system tracks and identifies individual animals with simplicity.

Shankar et al. [7] applications for digital twins are found in many different industries, including manufacturing, healthcare, agriculture, automotive, aviation, construction, and real estate. Digital Twins, which are made up of three essential parts—data modeling, data application, and data acquisition use IoT, AI, XR, and cloud technologies to succeed in various fields. Digital twins enable remote asset control, sophisticated simulations, real-time monitoring, and innovation catalysts in manufacturing. More effective treatments can be obtained in healthcare through personalized medicine, precision treatment planning, and customized medication features. Digital twins are used in agriculture to monitor animals, forecast climate change, and manage fields more effectively. Digital twins help the automotive and aviation sectors track assets, improve customer engagement, expedite maintenance, and identify flaws. Real-time data is essential for project tracking, resource management, and increased safety in the building and real estate sectors.

Warbhe et al. [8] the use of technologies like Exact Hybrid Precision Farming, AI in animal farming, AI-based resource supply chain applications, and others, this article seeks to provide a quick overview of the application of AI and robots in agriculture. The concepts of culture as a factor in addressing the difficulties impacting agriculture and the potential use of artificial intelligence in the process are compared with the thoughts of other authors in this literature review. Presenting literary evidence demonstrating the use of robotic harvesting, automated crop monitoring, decision support systems, and predictive analytics is the goal of the theoretical literature study of the application of AI and robotics in precision farming.

Gyamfi et al. [9] a tech-driven transformation in agriculture, smart farming—also referred to as agriculture 4.0 aims to increase industry efficiency and productivity. It is caused by four main trends: population changes, resource scarcity, climate change, and food waste. The use of new technologies is causing changes in the agriculture sector. Smart farming is revolutionizing international agriculture regulations and conventional production methods with the use of cutting-edge technologies like IoT, AI, and other sensors. The fourth industrial age has brought about an important shift in the agricultural industry by fusing advanced technology with traditional farming methods to boost effectiveness, sustainability, and productivity. Governments, businesses, and other stakeholders must work together to fully realize the promise of modern technology in the agriculture sector.

Molapo et al. [10] in farm operations, livestock management and monitoring systems are essential. In this research, a deep learning-based system for farm animal management and monitoring is proposed. Manual observation is used in traditional livestock monitoring techniques, but it can be laborious and unreliable. Despite the development of numerous methods, issues with livestock classification and counting persist, such as training, opacity, and livestock overlap. The proposed model accepts animals of various sizes for counting on each anchor frame and conducts feature extraction on the original image using the original YOLOv5 backbone network. This model also recognizes and follows specific animals. Since YOLOv5 mainly depends on data augmentation to enhance its tracking and detection capabilities and evaluate the suggested method, the real-time and included a variety of animals, is employed. The Roboflow application handles the splitting dataset's scaling, resizing, and manipulation.

Mendes Lima et al. [11] the goal is to give a methodical summary of the majority of ongoing research in the field of dairy cattle comfort monitoring, together with the related image recognition-based methods and tools. Techniques: First, a systematic review was planned, and among other things, goals, research questions, and search terms were established. In order to provide an easily readable visual source of information (tables and graphics), a comprehensive survey was then carried out to collect, analyze, and assemble the data. Findings: Data was taken from the papers that were reviewed. The information gathered from the publications includes the following: the font used in the study, the publication date, the locality, target behaviors, and cow body parts recognized using visual computation.

Wang et al. [12] the primary features of this system are quality traceability, RFID e-label identification, feed use monitoring and management, monitoring and predicting growth, and monitoring the animal husbandry environment. The current screen provides a thorough overview of the algorithms employed by the system's prediction function, growth tracking, and environment monitoring for animal husbandry. Animal farming cycles are minimized by this approach, which also increases the survival and off-take rates of animal products and the production efficiency of animal farming.

Singh et al. [13] the Internet of Things (IoT) platforms and Artificial Intelligence (AI) approaches, a thorough analysis of poultry health monitoring is provided. IoT devices using various sensor types, video/image processing and classification capabilities, and vocalization (sound) based poultry livestock analysis are being studied for the purpose of monitoring the health of the birds and poultry farms. There is an excellent case for using contemporary technology to continually monitor the massive farms with millions of birds and increase overall efficiency due to the availability of increasingly affordable computer resources, Internet of Things sensors, and basic algorithms.

Erden et al. [14] the livestock monitoring system (LMS) aims to keep an eye on all animal transactions and animal products, track milk and milk products nationwide using a square code system, keep an eye on the cattle, poultry, and ovine animals being slaughtered so they can safely reach consumers, and register fisheries treatments and fishermen. Monitoring of all animal diseases, transportation activities, annual consumption of milk and meat, medication use and expenses, routine vaccinations, and hunting gear in fisheries will be protected, and far more effective support and preventative measures will be offered. Additionally, this gadget will be able to keep an eye on both single-shank animals and other animals like dogs and cats. Intentional or intentional misreporting and the resulting harm to the public will be avoided since the payment is intended to be made over the identified animals in the system.

Parakkal et al. [15] in a zoo, a group of officials tend to and care for the animals. In addition, it will amuse travelers and anyone who enjoy animals while educating them about animal behavior. In addition to generating income for tourists, it serves educational functions. With the assistance of zoo directors, it cares for and serves a large number of wild animals. Since those zoo directors are directly interacting with wild animals, their job is particularly stressful. The suggest a creative and clever way to use a supervised machine learning algorithm to forecast an animal's health. Furthermore, our architecture includes real-time monitoring, automatic feeding, cage temperature management, animal health monitoring, and virus-infected animal identification. This helps to the zoo's minimal costs. Furthermore, in order to prevent the spread of the diseases in this pandemic era, animals infected with the virus must be kept quite from both other animals and humans.

Joshi et al. [16] many industries have seen significant advances since the development of digital technology. This study presents a thorough review of the literature with an emphasis on AI-enabled Internet of Things applications for managing the health and welfare of chicken. Given that current poultry husbandry faces challenges with accepted standards for welfare evaluation and reliable monitoring systems, particularly for young wellness and disease outbreak prevention, the primary focus of this study was chicken welfare. The use of modern technologies to continuously monitor large farms with millions of birds and boost total production has a strong case. Since eggs and poultry are the most significant sources of protein in the diet, the deployment of cutting-edge technological solutions should be given top priority by poultry farm oversight. The most advanced IoT systems with AI capabilities are thus carefully examined together with the latest advancements in the development of smart systems in this domain.

Alanezi et al. [17] drone technology is booming due to its ease of use and improvements, which are causing Unmanned Aerial Vehicles (UAVs) to be widely applied in a variety of fields. One of the most potential uses of UAVs is in livestock husbandry, where they help with a variety of tasks for effective animal care. However, a number of technological, strategic, financial, and environmental issues define the field. However, the guarantee of animal welfare when using the UAVs, combined with the application of modern technological methods like artificial intelligence (AI), the Internet of Things (IoT), machine learning (ML), deep learning (DL), advanced sensors, etc., can result in the broad adoption of drone technology among livestock farmers.

Ashmitha et al. [18] agriculture is the primary source of income. In rural living, animals like goats, sheep, buffalo, and cows are vital. They are a source of revenue. Therefore, raising animals becomes a major issue. As a result, keeping barns and cows for commercial use is a difficult process. The suggested model addresses the monitoring of cattle health. Arduino-based sensors are used to identify diseases. The vibrating energy harvester gathers energy and uses it as the sensor array's input power source. With this application, keeping an eye on the health of cows is easy rather than a laborious process. To start control, health metrics that are measured in real time are shown on a thing discuss cloud channel. Before beginning the control, the acquired data are compared to the ideal values. A vibrating energy harvester is used in this project to capture energy.

Anand et al. [19] to manage livestock farming, the farmer must monitor and track the health of the cattle. Improperly, it is difficult to keep an eye on animals every day, especially in large farms where employees don't have the time to examine every animal and search for signs of sickness. This idea provides an automated monitoring system designed to track the whereabouts and well-being of the cattle. It measures the cattle's body temperature, respiration, heart rate, humidity, and location. If any anomaly is found in the cattle, the farmer is notified by buzzer or SMS on their smartphone.

Jeong et al. [20] smart livestock farming systems that can automatically identify animal behavior have been developed as a result of the aging and decrease of the rural population. Cattle mounting behavior can happen at any time of day or night, however previous research has mostly concentrated on identifying behavior during the day. A continuous, round-the-clock, real-time IoT monitoring system for identifying cattle mounting behavior is proposed in this study. This system uses the YOLO-v8 model, which is based on convolutional neural networks, to distinguish between mounting actions and normal cattle behavior.

Jijesh et al. [21] for livestock or cattle farms, exterior gating is a long-term investment. Cattle are protected by physical fences and the most recent IoT sensors. Biosecurity stops dangerous infections from spreading to cattle farms. The agricultural sector has grown significantly as a result of IoT. The IoT sensors preserve a topographical safety zone for the cattle system. Modern technology has completely transformed livestock monitoring, which was formerly done manually with physical fences. Now, the animals can be followed and monitored automatically. Numerous advantages and security features offered by IoT cow tracking software can significantly enhance your livestock operations. The Global Positioning System (GPS) and navigation satellites make it simple to track the location of livestock.

Yousefi et al. [22] precision livestock farming (PLF), utilizing cutting-edge information technologies, is emerging to enhance farming production systems in light of the growing significance of dairy and meat production. The most recent advancements in automation, networking, and artificial intelligence create new opportunities for livestock monitoring in open, controlled, and pasture settings. This systematic study collects and summarizes the current deep learning (DL) strategies in PLF employing unmanned aerial vehicles (UAV) because of the importance of livestock monitoring and detection. The DL architectures that have been put into practice and examines the widely used evaluation metrics, characteristics, and databases. The majority of UAV livestock monitoring systems that use DL approaches fall into one of three categories: localization, classification, or detection.

wa Maina [23] the use of inexpensive sensors to track dairy cattle's movements in order to integrate them into a system that can identify significant occurrences, including when a cow is sick or giving birth. This is based on developments in human activity recognition (HAR), which uses wearable sensors to gather information and determine human activity. The Raspberry Pi CPU coupled with an accelerometer sensor forms the foundation of our sensor system. We demonstrate how our basic system can identify various animal activities including walking, standing, and feeding by examining the application of basic machine learning algorithms to infer activity from the data we gather. The system on the detection of human behaviors gathered in a controlled environment to demonstrate the system's possible applications. In our ideal internet of things (IoT) system, a group of cows is equipped with sensors that send data to servers via the internet.

Henny et al. [24] precisely track critical physiological indicators including body temperature, heart rate, and respiration rate, this study set out to design and deploy an Internet of Things-based health monitoring system for sheep. The system integrated non-invasive sensors, the Thing Speak platform, and the ESP32 microcontroller, among other IoT technologies. System design, implementation, and testing were among the research and development techniques that were applied. The system successfully offered precise and real-time monitoring of the desired physiological indicators, according to the results.

Curti et al. [25] the growth of research fields like Precision Livestock Farming (PLF), certain areas of animal production and reproduction have demonstrated significant technical advancements. PLF is a novel method that makes it possible to monitor animals by utilizing state-of-the-art technology that continuously gather data in real-time by fusing sophisticated algorithms with sensors to give farmers decision-making capabilities. Large datasets and computer science are combined in the subject of artificial intelligence (AI) to produce expert systems that can make predictions and categorize information in a manner akin to that of human intellect. Simplified, machine learning (ML) is a subfield of artificial intelligence (AI) that includes deep learning (DL, a neural network with at least three layers).

Reigones et al. [26] the use of electronics and information and communication technologies (ICTE) in agriculture has shown itself to be both revolutionary and highly effective. The competitiveness and productivity of agricultural activities are conducted in a more sustainable manner with the deployment of increasingly contemporary and efficient technologies. The creation of a comprehensive and precise livestock monitoring system is aided by the use of sensors, communication systems, geographic mapping, precise positioning, and geolocation. A microprocessor inside it is used to measure the animal's heart rate and identify any unusual movement. Many diseases and irregularities that result in financial losses for the producers can be found with the help of accurate evaluation of these two characteristics.

Armando et al. [27] an affordable remote livestock management monitoring system that combines software (Arduino IDE and dashboard) and hardware (motion sensor, jumpers, ESP32 camera, and TTL programmer). This study's cost-effective strategy, which was effectively applied for real-time animal monitoring with internet-based communication, is a notable strength. The Arduino IDE was used to create the programming code, and a cattle production corral with calves, cows, cows, and oxen was used for validation. We utilized the confusion matrix and the Kappa index to evaluate the model, and the results indicated good agreement. This confirmed that the system is easy to use, secure, scalable, and cost-effective, allowing for dashboard data storage and animal monitoring.

3 Fundamentals of AI and Machine Learning for Livestock Management

It becomes more and more difficult for pigs and farmers to rely on these conventional monitoring methods, especially as hog farming grows in size. Overall economic productivity and pig welfare are subsequently impacted by this stress. A key element in the field of intelligent pig farming is the integration of artificial intelligence (AI). In order to continuously monitor biometric data that includes pig vocalizations and behavior, advanced pig farming systems use AI. This information is converted into crucial markers of pig health and welfare using sensors like cameras and RF ID devices. By analyzing these crucial factors, it is possible to identify pig health problems early on and treat them quickly, which greatly improves pig farming's financial results and production efficiency [28]. As the main source of food and other basic resources, agriculture is regarded as the cornerstone of life. In terms of the nation's economic growth, it is essential. The suggested solution explains how irrigation can be made smarter by combining IOT and machine learning techniques. The suggested system saves time by avoiding issues like ongoing field surveillance through the use of IOT devices, crop prediction that uses machine learning techniques to help farmers grow crops that are appropriate based on soil parameters, and helping to keep wild animals and other intruders out of the field. By using sensors to automatically feed the plants or field with the least quantity of water based on their needs, it also helps in water conservation [29].

3.1 Machine Learning Algorithms for Livestock Data

The use of artificial intelligence and innovative sound technology in mammal sound classification, such as background noise delays. We propose a web-based, real-time livestock monitoring system that can evaluate cattle conditions in order to solve the problem [30]. The evaluation and interpretation of the huge amounts of data is produced in livestock management heavily depends on Machine Learning. The farmers are to production and animal welfare by the various algorithms to the random choices based on the data. These field uses a variety of machine learning algorithms such as supervised, unsupervised, and semi-supervised. For classification and regression applications, supervised learning algorithms like decision trees, support vector machines, and neural networks are frequently employed. Farmers can correctly identify animals at risk and carry out prompt interventions by training models using labeled datasets. Patterns in unlabeled data can be found using unsupervised learning methods such as hierarchical clustering and k-means clustering. When grouping up cattle according to characteristics like weight, age, and breeding ability, these methods are particularly useful. Through the identification of fundamental patterns in the data, farmers can modify their management approaches in order to meet the distinct requirements of various animal populations (Fig. 3).

Reinforcement learning has become popular for resource allocation and feeding plan optimization. These algorithms can suggest ideal feeding schedules that optimize growth rates while decreasing waste by modeling various situations and learning from the outcomes. The use of machine learning algorithms in livestock data

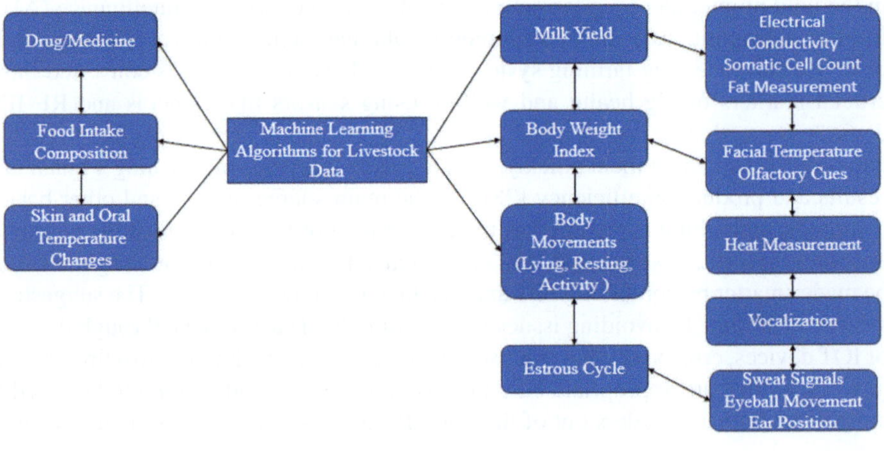

Fig. 3 Machine learning algorithms for livestock data

management enhances production and efficiency while also supporting the industry's long-term growth.

3.2 *Deep Learning and Neural Networks in Livestock Analysis*

A branch of machine learning called "deep learning" makes use of multi-layered neural networks to examine intricate patterns in big datasets. This method offers great potential in the analysis of cattle, especially in domains like health evaluation, behavior monitoring, and picture identification. Because of the way they consider unstructured input, neural networks are perfect for jobs combining both visual and audio information. Convolutional neural networks (CNNs) are frequently used in livestock management for image analysis tasks, including breed identification, injury detection, and body condition score monitoring via photos. CNNs can identify minor variations in animal appears with high accuracy by training on large datasets of annotated images, which helps with the early detection of medical conditions. In addition to provide the predictive analytics, deep learning models can be including data from multiple sources, including wearable sensors, environment factors and other records. The techniques are greatly can be enhance the productivity and health of cattle are made of the possible of the approach. The more accurate monitoring and management approaches is to supported by the long-term viability and efficiency of the livestock production, the use of deep learning in livestock evaluation is changing the agricultural industry.

3.3 *Role of Data Preprocessing and Feature Engineering*

The efficiency of the models is used for livestock management is greatly impacted by feature engineering and data preparation. The two phases in the machine learning pipeline. The noise, missing values, and the irrelevant information are frequently present in raw data is gathered from variety of sources such as sensors, imaging equipment and farm management systems. The data is clean and ready for the analysis, the effective preparation. These methods of the data preprocessing data transformation, normalization, and cleansing. For instance, standardizing measures, eliminating duplicates, and completing missing values all contribute to improving the dataset quality. The algorithms like neural networks are sensitive to the scale of input data, normalization guarantees are featuring similar scale. The categorical variables are efficiently used in machine learning models of the numerical representations. The model performance, feature engineering which involves choosing, changing, or producing novel characteristics from the current dataset. The animal age, weight feeding patterns and environment factors can all be significant components of livestock management. The farmers can provide models of the most useful inputs is improve the accuracy predictions of extracting new features, including the frequency of particular actions in the weekly gain in weight. In addition, more successful feature engineering is heavily relying on the domain knowledge. The health, productivity, and general well-being by having a thorough understanding of the distinctive traits and needs of the livestock. These models are in the real-world farming requirements in the addition of data-driven.

3.4 Supervised Versus Unsupervised Learning for Animal Health Predictions

The cattle management forecasting of animal health. The health outcomes or diseases are known as supervised learning- which is relies on the labeled datasets is commonly used. The recognizing trends in labeled data on the animal health variables like temperature, activity levels and vital signs, supervised algorithm such as decision trees, logistic regression, and neural network can forecast beginning of disorder. The models are preventing potential of the outbreaks and animal welfare by enabling proactive steps and early identification. The abnormalities and patterns to the new health problems is require data that is labeled. The animals are frequently grouped according to the behavioral or physiological features using algorithms such as Principal Component Analysis and K-means of clustering. These clusters might demonstrate trends be hard to spot in the small variations in food habits is indicate the start the medical condition. The farmers and physicians to obtain a complete set of tools for health monitoring by combining the two methods. The learning delivers a more general knowledge of new patterns and outliers to the specific insights. Those methods are allowed for more accurate and health forecasts, boosting data-driven management strategies is to improve efficiency and overall health.

4 AI-Driven Monitoring of Livestock Health

The real-time assessment of animal health and prompt detection of health problems of AI-driven monitoring is changing in the livestock health management. The large volumes of data on the feeding habits, movement patterns and early signs collected and processed through combination of sensors, cameras, and internet of things devices. The farmers can take the preventative steps to address potential illnesses to they develop by using AI algorithms and examine this data and detect trends and abnormalities. The AI-driven monitoring is to be analyzed in this chapter such as Machine Learning algorithms for the behavior analysis computer vision for image-based health assessments, predicting algorithms which likelihood of disease. The computers can detect changes from the conventional patterns, indicating possible of

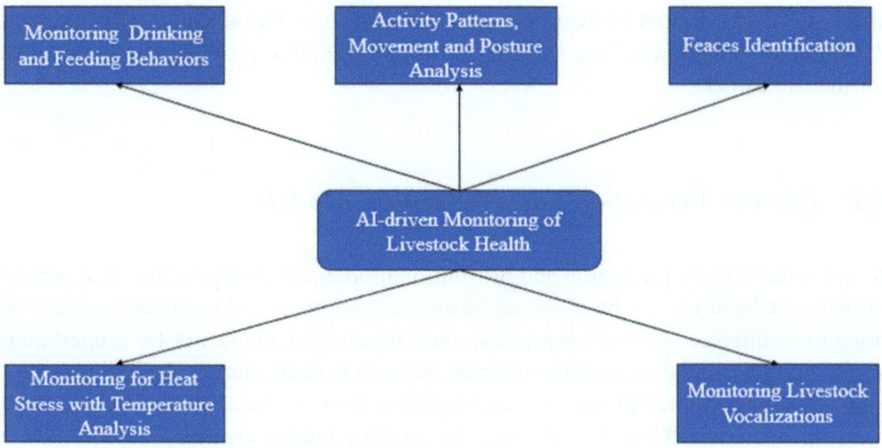

Fig. 4 AI-driven monitoring of livestock health

health problems in the using wearable sensors is measured heart rate and temperature. The computer vision is accurately identifying symptoms of impairment of position and movement (Fig. 4).

The demand for manual inspections, which may be laborious and inaccurate, by automating health monitoring. The personalized approach to livestock care is further backed by AI-driven monitoring, each animal's unique information is allowing for the specific actions is to improve wellbeing and productivity. This chapter is highlights to how AI-driven monitoring can be enhancing livestock health management by providing a smarter, data-driven techniques is to both farmers and animals.

4.1 Remote Monitoring of Livestock with AI

The remote livestock monitoring is to be changed animal health and welfare by farming a continuous, precise picture of their animals wellbeing without needing in-person checks. The variables like body temperature, movement, patterns and other factors is continuously collected and sent to AI-driven platforms in the internet of Things devices, sensors, and cameras. The real time processing and evaluation of information of these platforms of any unexpected trends is linked stress or medical conditions. The early illness or injury detection of the possible AI-powered systems such as computer vision and sensor fusion. The computer vision algorithms, for example, monitor an animal's posture, locomotion, and expressions in order to spot in the any signals of stress or behavior. The physical indicators are monitored by wearable sensors, which allows AI models to identify deviations from predefined health. The speed of response to health issues and reducing the time and labor needed to perform manual checks, this remote monitoring method improves animal welfare and lower costs related to treatment illnesses and loss. The scalable, effective, and dependable wellness tracking system, farms can use AI to provide the greatest care for their livestock.

4.2 *Disease Prediction and Prevention Models*

AI-powered disease prediction and prevention models are changing the management of livestock health issues by allowing farmers to anticipate and mitigate disease risks prior to outbreaks. To find correlations and trends that might not be immediately apparent, these models use data collected through several sources, including behavioral data, environmental sensors, and medical records. AI algorithms can predict the risk of disease through analyzing these data patterns, typically days or weeks before symptoms that are apparent show up. The risk indicators including temperature, humidity, herd, density and recent medical history, Machine learning methods such as logistic regression, random forests, and other neural networks is predict the disease. For instance, the variations in movement and pulmonary rates, a disease prediction model may be able to detect early signs of respiratory disease in the cattle. These findings, producers may the prevent the spread of disease within livestock by implementing preventive measures like modifying the ventilation systems of the infected animals. In addition, the farmers, or physicians to elevated danger levels can be identified with these AI-driven models, allowing for prompt intervention. These encouraging proactive, data-driven management practices, disease identification and other models for promote healthier cattle and lower financial losses and the other cattle operations.

4.3 *Behavior Analysis for Early Illness Detection*

The AI-powered behavior analysis is an essential tool for the livestock disease detection, providing an efficient approach to preserve animal productivity and health. These AI models can spot the small variations in the start of the health problems by continuously observing an animal's behavior patterns, social interactions and feeding frequency. The early signs of illnesses or stress can be included for the reduction intake or shift interactions with withdrawal from the group. The algorithms for computer vision and machine learning analyze data gathered in the wearable technology sensors and other cameras. The symptoms are to be appear these models is to be identify departures from normal patterns and possible health issues. For instance, a system may identify lameness or the dairy calves is to detecting mobility or longer intervals. The early signs of these farmers to take prompt action, increasing likelihood of recovery and halting the disease among the cattle. These AI can be evaluating each animal's foundation behavior of the analysis also for make unique animal care and results in extremely tailored in health interventions. These farmers can be maintaining a healthier cattle's, reduce medical expenses, and ensure is to be more effective and livestock operation is to behavioral evaluation of illness recognition.

4.4 *AI for Reproductive Health Monitoring*

The AI-powered reproductive to the health monitoring to the cattle producers in the advanced resources to improve breeding results and other reproductive cycle intervention. The AI-systems can be monitor important to the reproductive markers like hormone levels, body temperature and other behaviors are linked to estrus or pregnancy using combination of sensors, wearable technology, and computer vision. These AI models can be identifying patterns in the changes such as activity levels or the behaviors in estrus of cattle. These signs, machine learning algorithms to forecast the most favorable times for breeding, improving the other successful conception and reducing time between calving cycles. These AI can be also tracking the pregnancy, spotting possible problems in early in the examining patterns in the vital signs. These farmers can be concerning significant occurrences like labor, automated alerts for the swift assistance and other types of risks. The farmers can be increases fertility rates, lower reproductive losses and optimize breeding schedules of health monitoring. This data-driven strategy is to be supports healthier and more productive cows by improving the sustainability and efficiency of animal husbandry.

5 AI and Precision Feeding Systems

In recent years, the integration of artificial intelligence (AI) into precision feeding systems has transformed the agricultural industry, notably cattle management. Precision feeding is customizing feed kinds and quantities to the specific requirements of individual animals or groups, hence optimizing nutrient intake, and increasing overall production. Precision feeding systems use data analytics and real-time monitoring to alter feed supply based on variables such as animal age, weight, health status, and environmental conditions. These systems, using AI algorithms, can analyses massive volumes of data acquired from sensors and other sources in order to make smart feeding decisions. AI technologies, including as machine learning and predictive analytics, are critical to enhancing precision feeding techniques. Key applications include AI can process and analyses data from a variety of sources, including feed composition, animal behavior, and health parameters, in order to optimize feeding techniques. Real-Time Monitoring: IoT devices provide continuous monitoring of cattle, allowing feeding schedules and quantities to be adjusted in response to immediate needs.

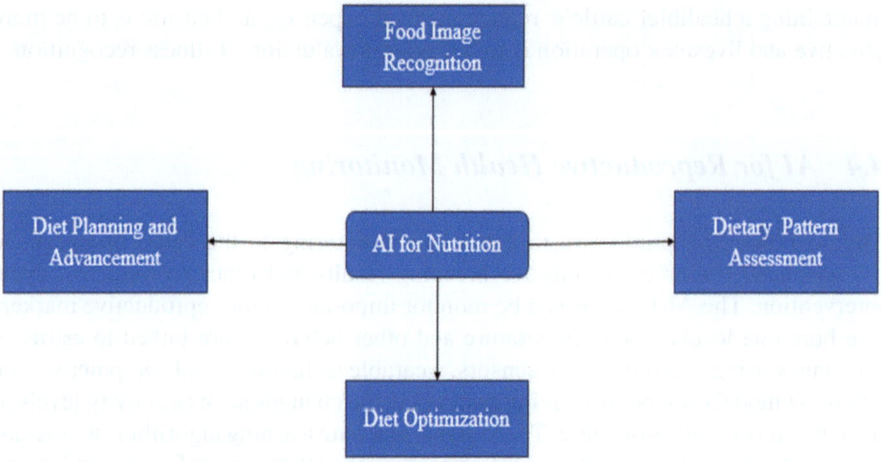

Fig. 5 AI for nutrition assessment

5.1 AI for Nutritional Needs Assessment

AI-based nutritional needs assessment is an important component of precision feeding systems, allowing farmers to optimize livestock feeds based on individual animal needs. This procedure begins with the gathering of information from numerous sources, such as animal weight, age, health status, and production objectives. AI systems use this information to calculate the best nutrient composition for each animal. For example, machine learning algorithms can detect patterns in growth rates and feed intake, allowing for personalized feeding plans that improve growth efficiency and health (Fig. 5).

AI-driven nutritional needs assessment can improve animal health and productivity by integrating real-time data from wearable sensors that monitor animal behavior and physiological parameters. Farmers can adjust diets based on these insights to ensure each animal receives the exact nutrients it requires, reducing waste.

5.2 Automated Feeding Systems for Livestock

An important development in livestock management is automated feeding systems, which combine robotics and artificial intelligence to expedite the feeding procedure. At predetermined intervals or in reaction to real-time data inputs, these systems use technology to precisely feed animals. Sensors that keep an eye on feed levels, animal behavior, and surrounding factors are essential components of automated feeding systems. By analyzing this data, AI systems optimize feed distribution, guaranteeing that animals get the proper nutrients at the appropriate times. By delivering timely and consistent nutrition, this can lower labor expenses, minimize feed waste, and improve the general health of the herd. Additionally, a lot of automated systems may be modified to work with various livestock species and feeding schedules. Because of this flexibility, farmers can put customized solutions into place that support animal welfare and achieve certain production targets. As these technologies continue to evolve, they offer promising solutions for increasing efficiency and sustainability in modern livestock farming.

5.3 Role of AI in Sustainable Feed Management

Furthermore, a variety of automated systems can be tailored to suit various livestock species and feeding techniques. Farmers can use customized solutions that satisfy particular production targets and advance animal welfare thanks to this flexibility. With their ongoing development, these technologies present viable ways to boost sustainability and efficiency in contemporary animal production. AI systems are able to suggest the best feeding practices by continuously monitoring and analyzing feed composition, animal behaviors, and growth patterns. This prevents overfeeding, which can result in waste and higher greenhouse gas emissions, and guarantees that animals get the nutrients they need. Furthermore, to further promote resource conservation, AI-driven insights can assist in identifying substitute feed ingredients, such as renewable sources like insects or algae or by-products from other industries. AI helps create a more resilient and environmentally friendly food system, which in turn supports the objectives of nutrition security and safeguarding the environment by increasing feed efficiency and lowering dependency on traditional feed sources.

5.4 Case Studies: AI for Feed Optimization in Dairy, Poultry, and Cattle Farms

Several case studies showcase the successful deployment of AI for feed optimization in several livestock sectors, demonstrating the technology's disruptive potential. Dairy Farms: In a large-scale dairy farm, AI systems were employed to track milk production and individual cow health. The technology optimized the ration composition through analysis of wearable and feed intake data, improving feed efficiency by 15% and increasing milk yield. Farmers reported lower feed costs and better cow health, proving AI's capacity to customize diets for peak performance.

Poultry Farms:

A chicken farm adopted an AI-powered feeding software that modified feed volumes depending on real-time growth data and environmental conditions. Using machine learning techniques, the system estimated flock nutritional needs, saving 20% on feed waste. The farm achieved faster growth rates and higher feed conversion ratios, resulting in a significant increase in total profitability.

Cattle Farms:

AI techniques were used to produce personalized feeding regimens for beef cow farms based on weight gain, feed intake, and health markers. The technology improved feed delivery schedules and ingredient compositions, resulting in a 10% rise in average daily weight gain. Furthermore, the farm observed a reduction in methane emissions per kilogram of beef produced, demonstrating AI's role in fostering economic and environmental sustainability. These case studies demonstrate how AI technology may improve feed optimization across a wide range of livestock operations, thereby increasing agricultural efficiency, profitability, and sustainability.

6 Real-Time Monitoring Systems Using IoT and AI

Real-time monitoring systems have become essential components in a variety of industries due to their capacity to track, analyses, and act on data in real time. These systems use advances in the Internet of Things (IoT) and Artificial Intelligence (AI) to enable seamless communication, automation, and smart decision-making in a variety of fields, including healthcare, manufacturing, environmental monitoring, and smart cities. The Internet of Things (IoT) is the connecting of physical things over the internet, enabling them to collect and exchange data. Sensors, actuators, and connected devices are integrated into systems, delivering real-time data on factors like as temperature, pressure, humidity, and more. This information is sent to centralized platforms for additional analysis. IoT systems rely on lightweight protocols like MQTT (Message Queuing Telemetry Transport) and CoAP (Constrained Application Protocol), which enable efficient data transmission over low-bandwidth networks. This improves the ability to monitor systems in real time, even from faraway locations.

Artificial intelligence (AI) is critical for analyzing the vast amounts of data generated by IoT devices. AI techniques such as machine learning (ML), deep learning, and natural language processing (NLP) are used to extract meaningful insights from raw data and perform predictive analytics. In healthcare, wearable sensors and AI-driven monitoring systems enable continuous patient monitoring and real-time notifications to healthcare personnel when vital signs deviate from normal. This improves patient outcomes, particularly in critical care settings. Sensors placed in natural surroundings collect data on air quality, water levels, and meteorological conditions. AI systems use this data to predict natural calamities such as floods and forest fires, allowing for prompt intervention.

6.1 *Integrating IoT with AI for Livestock Tracking*

The combination of IoT and AI in livestock tracking has transformed the agricultural economy by providing advanced technologies for monitoring the health, behavior, and movement of animals in real time. Traditional livestock management frequently requires manual tracking, which can be time-consuming and inefficient. With the introduction of IoT-enabled devices and AI-driven analytics, farmers and ranchers can now automate and optimise different parts of livestock management.

IoT for Livestock Tracking

IoT for Livestock Tracking GPS trackers, RFID tags, and biometric sensors are used to monitor animals' location, health, and activity levels. These gadgets continuously capture information on characteristics like as body temperature, heart rate, grazing patterns, and movement. GPS-enabled IoT devices allow farmers to pinpoint the exact location of each animal in huge pastures or farms. This helps to avoid livestock loss or theft while also ensuring efficient grazing management. Wearable IoT sensors can monitor animals' vital signs in real time, alerting farmers to any unusual behaviour or health issues, such as disease or damage. Early diagnosis of health concerns aids in disease prevention and animal welfare.AI improves the capabilities of IoT by analyzing large amounts of data generated from animals. Machine learning algorithms and predictive analytics tools use this data to deliver actionable insights, allowing farmers to make more educated decisions. AI systems use health-related data from IoT devices to forecast future diseases or health hazards. By finding trends in the data, AI can inform farmers before an animal exhibits apparent signs, allowing for timely intervention and lower veterinary expenditures. Artificial intelligence may detect changes in livestock behavior, such as decreased mobility or irregular grazing habits, which could signal stress or sickness. This allows farmers to change feeding schedules or environmental factors to optimise animal health.

6.2 *Sensor Networks for Environmental and Behavioral Monitoring*

Sensor networks play an important role in environmental and behavioral monitoring, offering real-time insights into natural ecosystems as well as animal, plant, and human behaviour. When integrated with IoT technology, sensor networks are made up of interconnected devices that gather, transmit, and analyses data from the environment and live beings. Integrating these networks with AI enables comprehensive analysis, including predictive insights, trend detection, and automated decision-making. Sensor networks are used to observe and track environmental characteristics such as air quality, soil moisture, temperature, humidity, and pollution levels. These networks can be built in urban, rural, or natural habitats to provide a complete picture of environmental conditions. Sensors that detect pollutants, particulate matter, and dangerous gases in the air give real-time information that is critical for urban planning and public health. Similarly, water quality sensors measure pH, salinity, and pollution levels in rivers, lakes, and water supply systems. Precision agriculture uses sensor networks to monitor soil moisture, nutrient levels, and weather variables in order to optimise irrigation, fertilizer application, and crop management. This improves yields, conserves resources, and promotes sustainable farming techniques. Behavioral monitoring is the process of tracking the activities, movements, and physiological signals of living species like animals or humans in order to analyses trends, health, and responses to environmental changes. Sensor networks with GPS and RFID technology can be fitted to animals to track their movement, feeding, and resting routines. This is especially useful in wildlife conservation and livestock management, where information on behaviour and migration can assist protect species and improve farming techniques. Wearable devices and ambient sensors can track human behaviors such as physical activity, sleep habits, and posture. This is useful in healthcare and wellness since monitoring can reveal early warning signs for illnesses such as heart disease or detect changes in mental health.

6.3 Edge Computing and Real-Time Data Processing in Farms

When paired with IoT devices and AI analytics, it forms a solid foundation for smart farming that optimizes resource consumption, increases production, and improves sustainability. In traditional cloud-based systems, data collected by IoT sensors on farms is transferred to remote data centers for processing and analysis. However, delays in uploading data to the cloud and receiving response might be problematic in instances demanding real-time action, such as changing irrigation or identifying equipment problems. Edge computing solves this problem by bringing data processing closer to the farm, allowing for real-time monitoring and decision-making. Edge computing offers real-time data processing, which is critical in precision agriculture. IoT sensors placed across the farm collect information on soil conditions, weather patterns, crop health, and mechanical performance. Edge devices interpret this data locally, offering real-time insights and automating activities like precision Irrigation, pest and Disease Detection, Real-time data from tractors, harvesters, and other farm machinery may be processed on the edge to improve fuel efficiency, detect faults, and schedule repairs. This enhances efficiency and lowers operating expenses.

7 Conclusion

The most significant development in livestock management is the use of AI, which gives farmers precise tools for proactive control of health, prediction, and monitoring. Farmers can obtain ongoing, real-time insights on the behavior, reproductive health, and welfare of animals by utilizing AI-driven technology like computer vision, machine learning algorithms, and Internet of Things-based collection of data. This knowledge makes it easier to identify diseases early, manage breeding, and provide individual treatment for animals, which leads to healthier sheep and more sustainable agricultural methods. AI makes it possible to go from reactive to proactive management, addressing problems before they become sufficiently serious to cause expensive health or productivity effects. Moreover, farmers can manage and profit from the vast volumes of data produced in contemporary livestock operations because to AI's ability to process data on a big scale. As a result, the livestock industry is more mature and resilient, able to satisfy the increasing demands for sustainability, animal care, and economy. The data-driven, ethical, and sustainable farming methods, AI-driven solutions are not only enhancing the daily management of livestock but also helping to the agricultural sector's future stability. These technologies will offer further opportunities to improve cattle productivity and health as they develop more thoroughly, fostering a healthy and conscientious agricultural the natural world.

References

1. Yamsani, N., Muthukumaran, K., Kumar, B. S., Singh, N., & Dhanraj, J. (2024). IoT-based livestock monitoring and management system using machine learning algorithms. In *2024 international conference on science technology engineering and management (ICSTEM)* (pp. 1–6). https://doi.org/10.1109/ICSTEM61137.2024.10560908
2. Arafat, S., Priya, C., Premkumar, R., & S.M., H. N. (2024). Real time cattle health monitoring and early disease detection using IoT and machine learning. In *2024 5th international conference on electronics and sustainable communication systems (ICESC)* (pp. 450–455). https://doi.org/10.1109/ICESC60852.2024.10690107
3. Camacho, S. A. A., Guerrero, A. F. A., Galán, J. A. M., Fajardo, C. A., & Perez, J. G. B. (2024). Transforming agriculture: Drones enhancing livestock management with AI precision. In *2024 XXIV symposium of image, signal processing, and artificial vision (STSIVA)* (pp. 1–5). https://doi.org/10.1109/STSIVA63281.2024.10637862
4. King, K., Thornton, C., & Roopaei, M. (2024). Smart feeding: Integrating deep learning into dairy farm practices. In *2024 IEEE world AI IoT congress (AIIoT)* (pp. 0151–0156). https://doi.org/10.1109/AIIoT61789.2024.10579035
5. Qazi, A., Razzaq, T., & Iqbal, A. (2024). AnimalFormer: Multimodal vision framework for behavior-based precision livestock farming. In *2024 IEEE/CVF conference on computer vision and pattern recognition workshops (CVPRW)* (pp. 7973–7982). https://doi.org/10.1109/CVPRW63382.2024.00795
6. Sujitha, S., Harshika, V., Hemavathi, V., Disha, M., & Nafiza, A. (2024). Implementation of Farmguard with automated animal detection and monitoring system using IoT. In *2024 ninth international conference on science technology engineering and mathematics (ICONSTEM)* (pp. 1–4). https://doi.org/10.1109/ICONSTEM60960.2024.10568785

7. Jyothi, A. P., Shankar, A., J R, A. N., Koliparthy, K., Badiger, V. N., & Shivakumar, V. (2024). Exploring the applications and challenges of digital twins in various industries. In *2024 IEEE 9th international conference for convergence in technology (I2CT)* (pp. 1–7). https://doi.org/10.1109/I2CT61223.2024.10543818
8. Warbhe, M. K., Verma, P., & Rewatkar, R. (2024). A review on the use of AI and robotics in agricultural practices. In *2024 second international conference on intelligent cyber physical systems and internet of things (ICoICI)* (pp. 1444–1450). https://doi.org/10.1109/ICoICI62503.2024.10696164
9. Gyamfi, E. K., ElSayed, Z., Kropczynski, J., Yakubu, M. A., & Elsayed, N. (2024). Agricultural 4.0 leveraging on technological solutions: study for smart farming sector. In *2024 IEEE 3rd international conference on computing and machine intelligence (ICMI)* (pp. 1–9). https://doi.org/10.1109/ICMI60790.2024.10585910
10. Molapo, M., Tu, C., Du Plessis, D., & Du, S. (2023). Management and monitoring of livestock in the farm using deep learning. In *2023 international conference on artificial intelligence, big data, computing and data communication systems (icABCD)* (pp. 1–6). https://doi.org/10.1109/icABCD59051.2023.10220556
11. Mendes Lima, L., Calebe Cavalcante, V., Guimarães de Sousa, M., Afonso Fleury, C., Oliveira, D., & de Andrade Freitas, E. N. (2021). Artificial intelligence in support of welfare monitoring of dairy cattle: A systematic literature review. In *2021 international conference on computational science and computational intelligence (CSCI)* (pp. 1708–1715). https://doi.org/10.1109/CSCI54926.2021.00324
12. Wang, Y., Yong, X., Chen, Z., Zheng, H., Zhuang, J., & Liu, J. (2018). The design of an intelligent livestock production monitoring and management system. In *2018 IEEE 7th data driven control and learning systems conference (DDCLS)* (pp. 944–948). https://doi.org/10.1109/DDCLS.2018.8516021
13. Singh, M., Kumar, R., Tandon, D., Sood, P., & Sharma, M. (2020). Artificial intelligence and IoT based monitoring of poultry health: A review. In *2020 IEEE international conference on communication, networks and satellite (Comnetsat)* (pp. 50–54). https://doi.org/10.1109/Comnetsat50391.2020.9328930
14. Erden, H., Kaya, İ. A., & Çamaşircioğlu, E. (2015). Livestock monitoring system. In *2015 fourth international conference on agro-geoinformatics (agro-geoinformatics)* (pp. 73–76). https://doi.org/10.1109/Agro-Geoinformatics.2015.7248081
15. Parakkal, K. S., Rahul, P. M., John, R., Madhavan, S., & Reshmi, S. (2021). Management system using internet of things and artificial intelligence. In *2021 8th international conference on smart computing and communications (ICSCC)* (pp. 215–218). https://doi.org/10.1109/ICSCC51209.2021.9528142
16. Joshi, K., Kumar, A., Anandaram, H., Kumar, A., Aggarwal, A., & Chhabra, G. (2023). Incorporating AI and IoT for smart poultry monitoring and health detection. In *2023 second international conference on augmented intelligence and sustainable systems (ICAISS)* (pp. 76–81). https://doi.org/10.1109/ICAISS58487.2023.10250564
17. Alanezi, M. A., Shahriar, M. S., Hasan, M. B., Ahmed, S., Sha'aban, Y. A., & Bouchekara, H. R. E. H. (2022). Livestock management with unmanned aerial vehicles: A review. *IEEE Access, 10*, 45001–45028. https://doi.org/10.1109/ACCESS.2022.3168295
18. Ashmitha, G., Manish Daniel, K., Saravanan, J., Ayyar, K., & Jaibhavani, K. S. (2023). IoT based sustainable live stock health monitoring system. In *2023 international conference on distributed computing and electrical circuits and electronics (ICDCECE)* (pp. 1–5). https://doi.org/10.1109/ICDCECE57866.2023.10151233
19. Anand, J. A., Pawan, K., Narendran, M., Johnson, M., & Rakesh, S. (2022). Real time livestock rearing. In *2022 international conference on computer, power and communications (ICCPC)* (pp. 603–606). https://doi.org/10.1109/ICCPC55978.2022.10072153
20. Jeong, K., et al. (2024). A monitoring system for cattle behavior detection using YOLO-v8 in IoT environments. In *2024 IEEE international conference on consumer electronics (ICCE)* (pp. 1–4). https://doi.org/10.1109/ICCE59016.2024.10444145

21. Jijesh, J. J., Niranjana, C., Vineeth Kumar, P. K., Divakara, M., & Sruthi, P. V. (2023). Design and implementation of smart livestock security system. In *2023 international conference on distributed computing and electrical circuits and electronics (ICDCECE)* (pp. 1–6). https://doi.org/10.1109/ICDCECE57866.2023.10150783
22. Yousefi, D. B. M., Rafie, A. S. M., Al-Haddad, S. A. R., & Azrad, S. (2022). A systematic literature review on the use of deep learning in precision livestock detection and localization using unmanned aerial vehicles. *IEEE Access, 10*, 80071–80091. https://doi.org/10.1109/ACCESS.2022.3194507
23. wa Maina, C. (2017). IoT at the grassroots—exploring the use of sensors for livestock monitoring. In *2017 IST-Africa week conference (IST-Africa)* (pp. 1–8). https://doi.org/10.23919/ISTAFRICA.2017.8102356
24. Henny, A. H. S. B., Anthara, I. M. A., Azharan, S. F., & Juanda, E. A. (2023). Design of an IoT-based sheep health monitoring system. In *2023 international conference on informatics engineering, science & technology (INCITEST)* (pp. 1–6). https://doi.org/10.1109/INCITEST59455.2023.10396935
25. Curti, P. F., Selli, A., Pinto, D. L., Merlos-Ruiz, A., Balieiro, J. C. C., & Ventura, R. V. (2023). Applications of livestock monitoring devices and machine learning algorithms in animal production and reproduction: An overview. *Animal Reproduction, 20*(2), Article e20230077. https://doi.org/10.1590/1984-3143-AR2023-0077
26. Reigones, A. R., & Gaspar, P. D. (2021). Real-time vital signs monitoring system towards livestock health furtherance. In *2021 6th international conference on inventive computation technologies (ICICT)* (pp. 753–758). https://doi.org/10.1109/ICICT50816.2021.9358658
27. Armando, E., Felicia, D. D., & Gatera, O. (2023). Remote monitoring system in real time for cattle production, through low-cost technologies. In *2023 first international conference on the advancements of artificial intelligence in African context (AAIAC)* (pp. 1–5). https://doi.org/10.1109/AAIAC60008.2023.10465273
28. Shukla, G., Gowri, K., & Sudhamsu, G. (2023). Fostering smart agriculture: Using vision-based AI for livestock managing. In *2023 international conference on advances in computation, communication and information technology (ICAICCIT)* (pp. 784–790). https://doi.org/10.1109/ICAICCIT60255.2023.10465732
29. Nikhil, R., Anisha, B. S., & Kumar, P. R. (2020). Real-time monitoring of agricultural land with crop prediction and animal intrusion prevention using internet of things and machine learning at edge. In *2020 IEEE international conference on electronics, computing and communication technologies (CONECCT)* (pp. 1–6). https://doi.org/10.1109/CONECCT50063.2020.9198508
30. Jung, D.-H., Kim, N. Y., Moon, S. H., Jhin, C., Kim, H.-J., Yang, J.-S., Kim, H. S., Lee, T. S., Lee, J. Y., & Park, S. H. (2021). Deep learning-based cattle vocal classification model and real-time livestock monitoring system with noise filtering. *Animals, 11*, 357. https://doi.org/10.3390/ani11020357

IoT-Driven Climate Smart Agriculture: Adapting to a Changing Climate for Sustainable Farming

Aditya Vardhan, Amarjeet Singh Chauhan, Sanjay Saini, and Sagar Sharma

Abstract Climate Smart Agriculture (CSA) characterizes a transformative method to farming that not only augments agricultural efficiency but also builds flexibility to the opposing effects of climate change. As global climate decorations become increasingly erratic, the integration of the Internet of Things (IoT) has emerged as a critical tool for adapting farming practices. IoT enables real-time monitoring of environmental conditions, precise control of water and nutrient levels, and predictive analytics to optimize crop yields while minimizing resource use. This paper explores how IoT-driven systems are reshaping traditional farming, providing actionable insights that empower agriculturalists to make data-driven decisions in response to changing climatic circumstances. By leveraging IoT, CSA systems can enhance sustainability, reduce greenhouse gas emissions, and improve food security. The conversation will cover key technologies such as smart sensors, autonomous systems, and cloud-based platforms that underpin IoT in agriculture, along with case studies demonstrating successful implementations of IoT in CSA. This integration of IoT into Climate Smart Agriculture offers a pathway to a more resilient and sustainable agricultural future.

Keywords Climate change · IoT · Green house · Autonomous systems · Sustainable agriculture

A. Vardhan (✉) · A. S. Chauhan · S. Saini
Department of Physics and Computer Science, Dayalbagh Educational Institute, Agra, India
e-mail: vardhana3098@gmail.com

A. S. Chauhan
e-mail: amarjeetsinghchauhan@dei.ac.in

S. Saini
e-mail: sanjay.saini@dei.ac.in

S. Sharma
George Brown College, Toronto, ON, Canada
e-mail: er.sagar0594@gmail.com

S. R. Kumar et al. (eds.), *AI-Driven IoT Solutions in Agriculture: New Insights*, Studies in Computational Intelligence 1236, https://doi.org/10.1007/978-3-032-05603-0_5

1 Introduction

Climate change is one of the most pressing global challenges, with profound effects on agriculture. Rising temperatures, altered rainfall patterns, increased frequency of extreme weather events, and shifting growing seasons are all contributing to reduced crop yields, degraded soil health, and disrupted ecosystems. These changes impact food security, threaten the livelihoods of farmers, and challenge the sustainability of agricultural systems.

As global temperatures rise, many regions are experiencing prolonged droughts, heatwaves, and unpredictable weather conditions that strain agricultural productivity. Farmers are also facing new pests and diseases that thrive in changing climates, while water scarcity is becoming an acute issue in many parts of the world. The cumulative impact of these factors requires a transformation in farming practices to ensure the long-term viability of agriculture [1, 2].

1.1 Importance of Climate Smart Agriculture

Climate Smart Agriculture (CSA) is a holistic approach designed to increase agricultural productivity, enhance resilience to climate change, and reduce greenhouse gas emissions where possible. It aims to create farming systems that are more sustainable and better adapted to climate variability, while ensuring food security for growing populations.

CSA Focus on three main Objectives:

- **Sustainably increasing agricultural productivity and incomes**—to ensure food security even under climate stress.
- **Building resilience and adapting to climate change**—helping farmers mitigate the effects of changing weather patterns, droughts, and floods.
- **Reducing greenhouse gas emissions**—by promoting sustainable farming practices, reducing waste, and improving resource efficiency.

The adoption of CSA practices is essential for enabling farmers to adapt to and overcome the challenges posed by climate change. Techniques such as conservation tillage, crop diversification, integrated pest management, and improved irrigation are key elements of CSA. These practices not only help in adapting to changing climates but also contribute to environmental sustainability and ecosystem health [3].

1.2 Role of IoT in Agriculture for Climate Adaption

The Internet of Things (IoT) has emerged as a game-changing technology in agriculture, offering new tools to help farmers adapt to climate change more effectively.

Fig. 1 Application of smart farming in climate smart agriculture

IoT refers to the network of connected sensors, devices, and systems that collect and share real-time data, enabling farmers to make informed decisions about their crops, soil, and water usage shown in the Fig. 1.

In the context of climate smart agriculture, IoT can play several vital roles:

- **Real-Time Monitoring**: IoT sensors can monitor soil moisture, temperature, and humidity levels, providing farmers with real-time data to make timely interventions, such as adjusting irrigation systems or applying fertilizers based on precise needs.
- **Predictive Analytics**: By combining IoT data with machine learning algorithms, farmers can predict weather patterns, identify potential risks such as droughts or frost, and take proactive measures to protect their crops.
- **Water and Resource Efficiency**: IoT enables the efficient use of water and other resources by automating irrigation systems and ensuring that water is applied only when and where it is needed, minimizing waste and enhancing drought resilience.
- **Remote Management**: Farmers can monitor and manage their fields remotely through IoT-enabled systems, allowing them to respond to climate challenges even when they are not physically present.

The integration of IoT in agriculture is not only improving productivity but also making farms more resilient to the uncertainties posed by climate change. By harnessing the power of IoT, farmers can better adapt to climate impacts, reduce resource usage, and implement precision farming techniques that are crucial for sustainable agricultural practices [4].

2 Key Concepts of Climate Smart Agriculture (CSA)

Climate Smart Agriculture (CSA) is an integrated approach aimed at transforming agricultural systems to support food security while addressing the challenges posed by climate change. The concept emphasizes the development of farming practices that not only increase productivity but also enhance resilience to climate impacts and contribute to the reduction of greenhouse gas emissions.

CSA is grounded in three key pillars:

- **Sustainably increasing agricultural productivity and incomes**: This involves improving the efficiency of agricultural systems to ensure that they can produce more food, feed, and fiber even under the constraints of changing environmental conditions. This pillar also focuses on improving the livelihoods of smallholder farmers who are often the most vulnerable to climate change.
- **Building resilience to climate change**: CSA promotes practices that help farmers adapt to climate variability, such as drought-resistant crops, efficient water management, and the use of predictive analytics to respond to extreme weather events. Enhancing resilience ensures that farmers can continue producing food even in adverse conditions.
- **Reducing greenhouse gas emissions**: CSA encourages the adoption of techniques that mitigate the impact of agriculture on the environment, such as carbon sequestration, agroforestry, and reduced reliance on chemical inputs. By reducing emissions, CSA helps limit the contribution of agriculture to climate change [5].

2.1 Benefits of Climate Smart Agriculture for Sustainable Farming

Climate Smart Agriculture offers numerous benefits for sustainable farming by helping farmers respond to the challenges of climate change while ensuring food production remains reliable and environmentally responsible. The major benefits include:

- **Increased Resilience to Climate Variability**: CSA equips farmers with tools and techniques to better manage risks associated with climate variability. Through practices such as crop diversification and conservation agriculture, CSA improves the ability of farming systems to withstand droughts, floods, and temperature fluctuations.
- **Improved Resource Efficiency**: CSA encourages the efficient use of water, energy, and inputs like fertilizers and pesticides. Precision farming techniques, supported by IoT devices, help in optimizing the application of resources, reducing waste, and lowering costs for farmers.
- **Enhanced Productivity and Food Security**: By promoting sustainable farming methods, CSA helps boost crop yields and agricultural productivity. This is crucial

for feeding a growing global population, especially in regions where climate change is exacerbating food insecurity.

- **Reduction in Greenhouse Gas Emissions**: Agriculture is a significant contributor to global greenhouse gas emissions, but CSA practices such as agroforestry, improved livestock management, and reduced tillage can help sequester carbon and lower emissions, contributing to global climate mitigation efforts.
- **Economic Benefits for Farmers**: CSA helps farmers optimize their operations and increase profitability. By improving productivity and reducing the need for costly inputs, CSA contributes to higher incomes for farmers, especially smallholders who are often disproportionately affected by climate change [6].

2.2 CSA in the Context of Global Climate Challenges

The world is facing a rapidly changing climate that poses significant risks to agriculture and food systems. In this context, Climate Smart Agriculture is not just an option but a necessity. As temperatures rise and weather patterns become increasingly unpredictable, traditional farming methods are becoming less reliable. CSA offers a framework for transforming agricultural practices to address these global climate challenges. In many developing countries, agriculture is the backbone of the economy, and farmers are particularly vulnerable to the impacts of climate change. CSA provides these farmers with strategies to adapt to changing conditions, ensuring that they can continue to produce food and support their livelihoods [7].

3 IoT-Enabled Climate Monitoring and Forecasting

The integration of IoT (Internet of Things) into agriculture has revolutionized the way farmers monitor and respond to climate conditions. IoT-enabled climate monitoring and forecasting systems provide farmers with real-time data, allowing them to make informed decisions and take timely actions to protect their crops and resources from the effects of changing weather patterns. By using advanced sensors and data analytics, IoT offers precise, actionable insights that enhance climate resilience in agricultural practices shown in the Fig. 2 [8].

3.1 Sensors for Monitoring Temperature, Humidity, and Rainfall

IoT devices equipped with sensors are at the core of climate monitoring systems in agriculture. These sensors are deployed across farms to continuously measure key environmental factors such as:

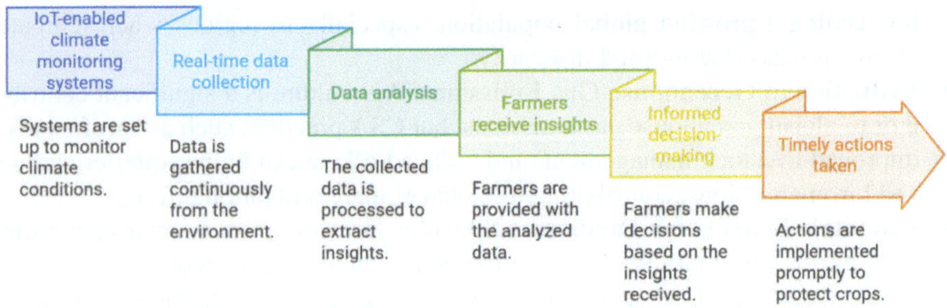

Fig. 2 IoT steps in agriculture

- **Temperature**: Sensors monitor air and soil temperatures, giving farmers insights into heat levels that affect crop growth stages, pest proliferation, and plant stress.
- **Humidity**: Humidity sensors track moisture levels in the air, which are critical for understanding conditions conducive to plant diseases, mold, and evapotranspiration rates.
- **Rainfall**: Rain gauges linked to IoT systems provide accurate measurements of rainfall, enabling farmers to adjust irrigation schedules and optimize water usage based on real-time precipitation data.

These sensors collect continuous data, which is transmitted to cloud platforms for further processing. The real-time tracking of such variables helps farmers stay ahead of environmental changes and adjust their practices as needed to prevent crop loss or damage.

4 IoT-Driven Precision Agriculture for Climate Adaption

Precision agriculture has emerged as a key strategy for optimizing farming practices, allowing farmers to use resources more efficiently and increase crop yields while reducing environmental impacts. As climate change increasingly affects agricultural productivity, IoT (Internet of Things) is playing a critical role in precision farming, enabling farmers to adapt to these changes with real-time data and advanced technologies. This section delves into how IoT drives precision agriculture, particularly in relation to climate adaptation shown in the Table 1 [9].

4.1 Role of IoT in Precision Farming Techniques

IoT-powered precision farming leverages a network of sensors, drones, automated systems, and cloud-based platforms to collect and analyze data on various farming conditions. This data helps farmers make informed decisions about when to plant,

Table 1 Application of related technologies

Technology	Application	Benefits	Examples
GPS and mapping	Field mapping for crop management	Enhances planting accuracy	Variable rate seeding
Variable rate technology	Adjusting inputs based on field variability	Reduces waste and increases yield	Precision fertilizer application
Remote sensing	Satellite imagery for large-scale analysis	Identifies areas needing attention	Crop health monitoring

irrigate, fertilize, and harvest crops, all while responding dynamically to changing climate conditions. Key IoT-driven techniques include:

- **Real-Time Data Collection**: Sensors deployed in fields continuously monitor soil moisture, nutrient levels, temperature, humidity, and plant health, transmitting the data to a central system for analysis. This real-time monitoring helps farmers adapt to immediate climate conditions and make timely decisions, improving the accuracy of their farming practices.
- **Automated Decision-Making**: With the data collected from IoT devices, machine learning algorithms can analyze patterns and predict future conditions, such as droughts, floods, or pests. Automated systems then adjust farming operations accordingly, such as initiating irrigation systems when soil moisture levels drop below a certain threshold.
- **Resource Optimization**: By using precise data, farmers can apply inputs like water, fertilizer, and pesticides only where they are needed and in the right amounts, minimizing waste and environmental impact. IoT facilitates this by enabling variable rate application and efficient use of resources, crucial for adapting to unpredictable climate scenarios.

4.2 Variable Rate Technology (VRT) and Resource Efficiency

Variable Rate Technology (VRT) is a cornerstone of precision farming that allows for the application of inputs—such as fertilizers, water, and pesticides—at different rates across a field. Rather than treating an entire field uniformly, VRT uses data from IoT devices to optimize resource distribution based on local conditions, enhancing both efficiency and productivity.

- **Targeted Input Application**: IoT sensors detect variations in soil quality, moisture levels, and crop health across a field. Based on these readings, VRT systems can apply precise amounts of water or nutrients in areas where they are needed the most, avoiding over-application in regions that don't require it. This is especially beneficial for adapting to climate-related variability in rainfall and temperature.

- **Water Conservation and Efficiency**: In regions facing water scarcity due to climate change, VRT combined with IoT-based irrigation systems ensures that crops receive the right amount of water at the right time. This prevents water wastage while maintaining crop health, making it an effective tool for drought-prone areas.
- **Energy and Cost Savings**: By applying inputs more efficiently, farmers can reduce their overall input costs, such as fertilizers and pesticides. This not only improves profitability but also contributes to more sustainable farming practices, a necessity in the face of fluctuating climate conditions.
- **Environmental Impact Reduction**: VRT helps minimize the environmental impact of farming by preventing overuse of fertilizers and chemicals, which can lead to runoff and pollution. This technology, driven by IoT, supports sustainable practices that align with climate smart agriculture [10].

4.3 Climate-Specific Crop Management and Precision Techniques

As climate change introduces new challenges such as irregular rainfall patterns, extreme temperatures, and increased pest pressures, IoT-enabled precision farming techniques become vital for managing crops effectively. These tools allow for climate-specific adjustments to be made in real time, helping farmers optimize their practices under varying environmental conditions.

- **Climate-Specific Crop Selection**: Using historical climate data and real-time weather forecasts from IoT systems, farmers can choose crop varieties that are more resilient to local climate conditions. For example, IoT data may suggest planting drought-tolerant or heat-resistant crop varieties in regions where climate models predict prolonged dry spells or heatwaves.
- **Precision Irrigation Techniques**: IoT-based precision irrigation systems ensure that crops receive the right amount of water based on the current climate conditions. For example, in arid regions, IoT sensors can monitor soil moisture and trigger drip irrigation systems to conserve water while maintaining optimal moisture levels for crop growth.
- **Climate-Specific Fertilization and Pest Management**: IoT sensors monitor soil fertility and detect early signs of pests or diseases. Based on the collected data, farmers can apply fertilizers and pest control measures more precisely and only when necessary, reducing input costs and environmental impact. Climate-specific variations, such as increased humidity or temperature, can also be accounted for in pest and disease management strategies.
- **Automated Greenhouse Management**: For farmers using greenhouses to protect crops from extreme climate conditions, IoT systems can automate the management of temperature, humidity, and light levels. Sensors continuously track the internal

environment, and automated systems adjust ventilation, heating, and irrigation as needed to maintain optimal growing conditions for climate-sensitive crops [11].

5 Water Management in CSA with IoT

Water management is a critical component of Climate Smart Agriculture (CSA), particularly in the context of climate change, which brings unpredictable rainfall, droughts, and water scarcity. The integration of the Internet of Things (IoT) in water management offers a solution that not only enhances water use efficiency but also supports sustainable agricultural practices. By leveraging IoT-driven smart irrigation systems and conservation strategies, farmers can optimize water usage, conserve resources, and improve crop resilience in water-scarce environments [12].

5.1 Smart Irrigation Systems Powered by IoT

Smart irrigation systems, powered by IoT, utilize real-time data from sensors placed in fields to automate and optimize water usage. These systems collect data on soil moisture, temperature, humidity, and even plant health, transmitting this information to cloud-based platforms. Advanced algorithms then analyze the data to determine the precise amount of water needed by crops at any given time shown in the Table 2.

Key components of IoT-powered smart irrigation systems include:

- **Soil Moisture Sensors**: These sensors monitor the soil's water content, ensuring crops receive water only, when necessary, thus preventing over- or under-watering.
- **Weather Stations**: Integrated weather stations track local climate conditions, such as temperature and rainfall, helping farmers make informed irrigation decisions.
- **Remote Control and Automation**: Farmers can remotely monitor and control irrigation systems through mobile apps or dashboards, making real-time adjustments based on the system's data analysis.

Table 2 Application of water managements in CSA with IoT

System	Description	Benefits	Examples
Smart irrigation systems	Automated irrigation based on sensor data	Conserver water and optimizes usage	Drip irrigation systems
Soil moisture sensor	Real-time soil moisture monitoring	Reduces over and under watering	Soil Moisture probes
Weather forecasting	Predict rainfall to adjust irrigation	Optimizes water usage	IoT-based weather stations

- **Drip and Sprinkler Systems**: IoT sensors are often paired with precision irrigation methods, like drip or sprinkler systems, ensuring water is delivered directly to the root zones of plants, minimizing wastage.

5.2 Water Conservation Strategies for Drought-Resistant Farming

Drought-resistant farming relies heavily on water conservation techniques, and IoT plays a significant role in implementing these strategies effectively. IoT-driven systems enable farmers to maximize water efficiency, even in regions facing severe water shortages. Below are key water conservation strategies enhanced by IoT:

- **Deficit Irrigation**: This strategy involves deliberately applying less water than what is required for optimal crop growth. IoT sensors monitor soil moisture levels and ensure that crops receive just enough water to survive drought conditions without compromising overall yield quality.
- **Rainwater Harvesting Systems**: IoT sensors can be used to monitor rainwater collection and distribution systems, ensuring water from natural sources is stored and used efficiently. Automated systems regulate water flow based on demand and availability, allowing farmers to conserve groundwater and rely more on stored rainwater.
- **Evapotranspiration Monitoring**: IoT sensors track the rate of evapotranspiration (the combination of evaporation from the soil and transpiration from plants). By analyzing this data, farmers can make more precise irrigation decisions, minimizing water wastage due to over-irrigation and evaporation.
- **Irrigation Scheduling**: Using IoT devices, farmers can create irrigation schedules that adapt to real-time environmental conditions. By delivering water during cooler parts of the day or when wind speeds are lower, they can reduce water loss through evaporation.
- **Crop-Specific Irrigation**: Different crops have different water requirements. IoT systems allow farmers to tailor irrigation schedules and water quantities to the specific needs of each crop, improving overall water use efficiency [13].

6 Soil Health and Fertility Management Using IoT

Soil health and fertility are critical components of sustainable agriculture. The integration of IoT technologies in soil monitoring and fertility management allows for real-time insights into soil conditions, ensuring optimal resource use and promoting environmentally friendly farming practices. This section explores how IoT-driven solutions can monitor and manage soil health, optimize fertilizer application, and ensure efficient nutrient and moisture levels to enhance crop yield and resilience to climate variability shown in the Fig. 3 [14].

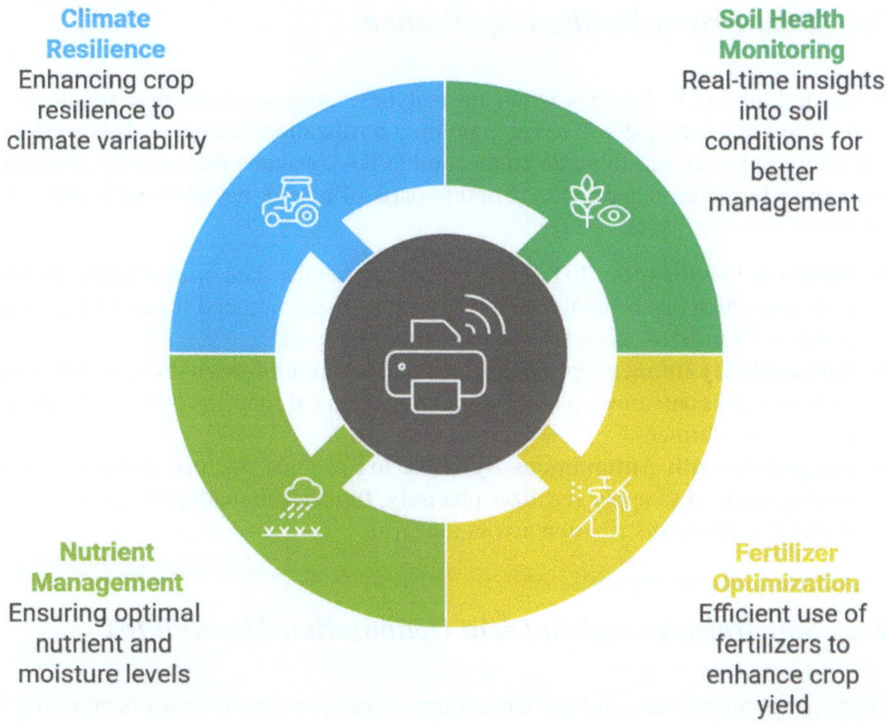

Fig. 3 Internet of things in soil management applications

7 Sensors for Real-Time Soil Monitoring

IoT-based soil sensors are revolutionizing the way farmers monitor and manage their soil. These sensors continuously collect data on various soil parameters such as pH, temperature, moisture levels, and nutrient concentrations. The use of real-time data ensures that farmers can make timely decisions regarding irrigation, fertilization, and soil treatment.

- **Types of Sensors**: Capacitive moisture sensors, electrochemical sensors for nutrient levels, and pH meters are some examples of IoT-enabled devices that provide precise data on soil conditions.
- **Benefits**: Real-time monitoring helps prevent under- or over-irrigation, reduces unnecessary fertilizer application, and enhances crop growth by maintaining an ideal soil environment.
- **Challenges**: Issues such as sensor calibration, data accuracy, and connectivity in remote areas can sometimes hinder the full effectiveness of IoT sensors [15].

7.1 Data-Driven Fertilizer Application

By integrating IoT technology with data analytics, farmers can use precise information on soil nutrient levels to develop tailored fertilization strategies. This process, often referred to as variable rate application (VRA), ensures that only the required amount of fertilizer is applied to different parts of a field, based on real-time data collected by IoT sensors.

- **Precision Fertilization**: IoT sensors provide granular data on soil nutrient deficiencies, which can be analyzed by algorithms to recommend the exact type and amount of fertilizer needed in specific areas of a farm.
- **Sustainability Impact**: Optimizing fertilizer use minimizes environmental harm, such as water contamination from excessive chemical runoff, while also reducing costs for the farmer.
- **Integration with Autonomous Systems**: IoT-enabled tractors and drones can be programmed to apply fertilizer precisely, further automating the process and ensuring uniform distribution across the farm.

7.2 Soil Moisture and Nutrient Optimization Through IoT

Optimizing soil moisture and nutrient content is crucial for maintaining healthy crops and preventing resource wastage. IoT solutions enable farmers to automate irrigation systems based on real-time moisture data, ensuring that crops receive the right amount of water. Additionally, nutrient levels can be optimized through IoT-based monitoring systems that track the soil's chemical composition.

- **Smart Irrigation Systems**: With moisture sensors connected to IoT networks, irrigation systems can automatically activate or deactivate based on the soil's current moisture level, ensuring water efficiency.
- **Nutrient Optimization**: IoT sensors can continuously monitor key nutrient levels, like nitrogen, potassium, and phosphorus. The data collected helps farmers adjust fertilization practices to keep the soil nutrient-rich without over-saturating it.
- **Long-Term Benefits**: By maintaining optimal soil moisture and nutrients, IoT systems improve crop yields, reduce the need for chemical inputs, and support long-term soil health, making farms more resilient to climate changes.

8 Smart Pest and Disease Control

Pest and disease management is one of the most significant challenges in agriculture. Traditional methods often rely on blanket chemical treatments, which can harm the environment, biodiversity, and even human health. IoT technologies offer a smarter, more precise approach to pest and disease control by enabling early detection,

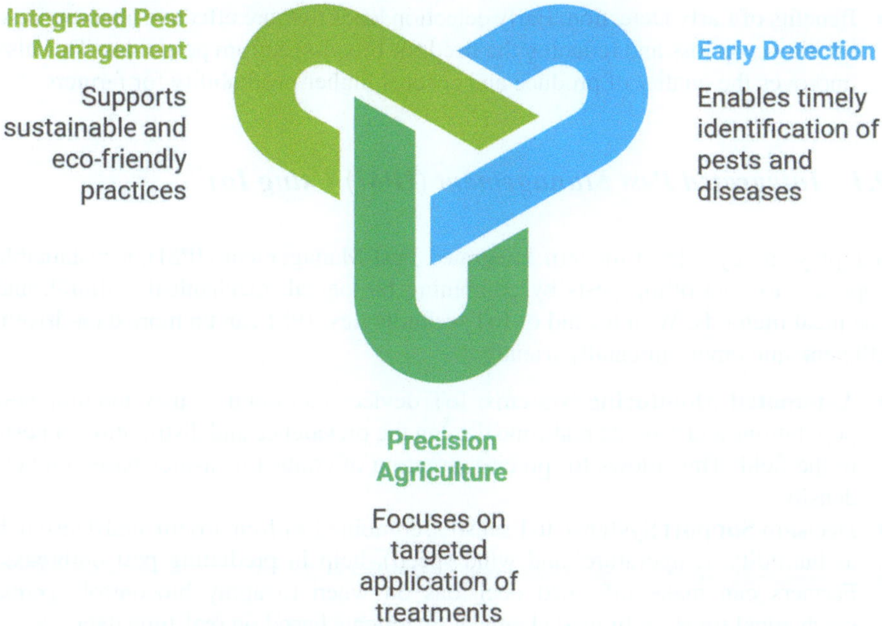

Fig. 4 IoT innovations in pest management

reducing unnecessary chemical use, and supporting Integrated Pest Management (IPM) systems. This section delves into how IoT is transforming pest and disease management in agriculture shown in the Fig. 4 [16].

9 IoT Solutions for Early Detection of Pests and Diseases

Early detection of pests and diseases is critical to prevent large-scale damage to crops. IoT solutions provide real-time monitoring systems that can identify the presence of pests or disease outbreaks at their earliest stages, allowing for immediate intervention before they spread [17].

- IoT Sensors and Networks: Sensors such as pheromone traps, camera-enabled drones, and environmental monitoring devices are connected through IoT networks to gather data on pest activity, humidity, temperature, and other environmental factors that can influence disease outbreaks.
- Machine Learning and Data Analysis: IoT systems often integrate machine learning algorithms to analyze sensor data and detect patterns indicating the presence of pests or early signs of disease. These predictive models can alert farmers to potential threats before they become widespread.

- Benefits of Early Detection: Early detection leads to more effective interventions, limiting crop loss and reducing the need for broad-spectrum pesticides. This also improves the quality of produce and ensures higher profitability for farmers.

9.1 Integrated Pest Management (IPM) Using IoT

IoT plays a key role in modern Integrated Pest Management (IPM), a sustainable approach to controlling pests by combining biological, mechanical, cultural, and chemical methods. With the aid of IoT technologies, IPM can be more data-driven, efficient, and environmentally friendly.

- **Automated Monitoring Systems**: IoT devices can continuously monitor pest populations and provide real-time data on the prevalence and distribution of pests in the field. This allows for precise targeting of control measures based on pest density.
- **Decision Support Systems**: IoT sensors, combined with environmental data (such as humidity, temperature, and wind speed), help in predicting pest outbreaks. Farmers can make informed decisions on when to apply biocontrol agents, mechanical traps, or limited chemical treatments based on real-time data.
- **Sustainability and Cost-Effectiveness**: By reducing the reliance on chemical treatments and utilizing biological or mechanical controls when feasible, IoT-enhanced IPM supports sustainable farming practices and reduces long-term costs for farmers.

10 Energy Efficiency and Automation in Climate-Smart Agriculture (CSA)

In the face of climate change, energy efficiency and automation have become vital components of climate-smart agriculture (CSA). IoT technologies, combined with automation and renewable energy, enable farmers to reduce energy consumption, enhance operational efficiency, and lower greenhouse gas emissions. This section explores how IoT-driven energy management, automation, and renewable energy integration can support sustainable farming practices and reduce environmental impact shown in the Fig. 5 [18].

10.1 IoT-Based Energy Management Systems

Energy management is critical in agricultural operations, as inefficient energy use can lead to increased costs and carbon emissions. IoT-based energy management

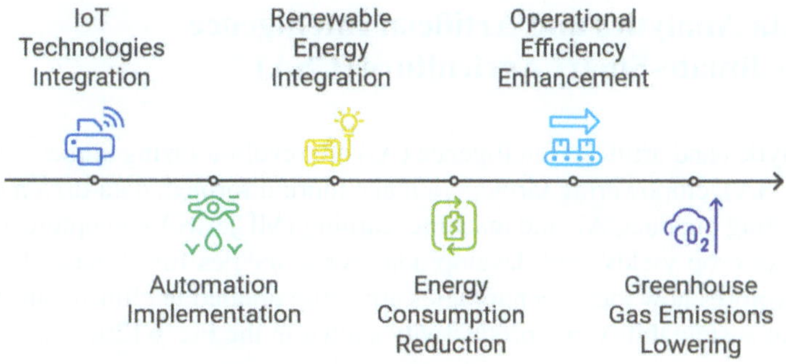

Fig. 5 IoT-driven sustainable farming practices

systems (EMS) provide real-time monitoring and control of energy consumption, enabling farmers to optimize their energy use across different farming processes.

- **Energy Monitoring Devices**: IoT-enabled sensors and smart meters can monitor energy usage in irrigation, lighting, climate control, and other farm operations. These devices provide real-time data on energy consumption, identifying inefficiencies and areas for optimization.
- **Energy Optimization Algorithms**: With the help of predictive analytics and AI, IoT systems can adjust energy usage automatically based on crop needs, environmental conditions, and power availability, ensuring that energy is used only when necessary.
- **Cost and Sustainability Benefits**: By reducing energy consumption, farmers can lower operational costs and minimize their carbon footprint, contributing to more sustainable farming practices and climate resilience.

10.2 Automation of Farm Equipment and Processes

Automation is transforming agricultural practices, enabling farmers to streamline labor-intensive tasks and improve overall efficiency. IoT technologies, combined with automation, enable precision control of farm equipment, reducing manual labor and ensuring timely interventions.

- **Autonomous Farm Machinery**: IoT-connected autonomous tractors, drones, and harvesters can perform tasks such as planting, spraying, and harvesting with minimal human intervention. These machines are equipped with sensors that allow them to adapt to real-time conditions, optimizing performance based on crop needs and environmental factors.
- **Automated Irrigation and Climate Control**: IoT-powered systems can automate irrigation, ventilation, and temperature control based on real-time data from soil and weather sensors. These systems ensure that crops receive optimal care without wasting resources [19].

11 Data Analytics and Artificial Intelligence in Climate-Smart Agriculture (CSA)

Data analytics and artificial intelligence (AI) are revolutionizing climate-smart agriculture (CSA), empowering farmers to make more informed, data-driven decisions. By leveraging big data, AI, and machine learning (ML), CSA can optimize resource use, predict crop yields, and develop adaptive strategies for climate change. This section explores how these technologies are being applied in climate-smart farming to enhance sustainability and productivity shown in the Fig. 6 [20].

11.1 Big Data for Climate and Farm Management

The use of big data in agriculture enables farmers to collect, store, and analyze vast amounts of information, leading to better farm management and climate adaptation strategies. IoT devices, remote sensors, drones, and satellite imagery continuously generate large volumes of data, which can be processed to extract actionable insights.

- **Data Sources**: IoT sensors collect data on soil health, crop growth, weather conditions, water levels, and pest activity. Combined with external data like weather forecasts and climate models, farmers can create a comprehensive understanding of their farm ecosystem.
- **Climate Resilience**: Through the analysis of historical and real-time data, farmers can identify trends in weather and climate that may affect crop performance, helping them adopt climate-resilient farming techniques.

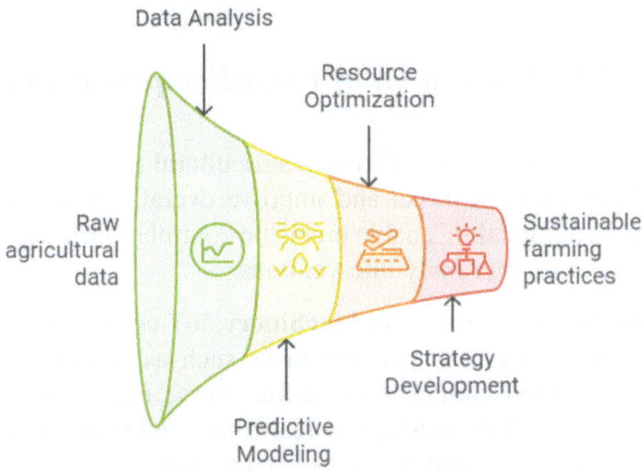

Fig. 6 Enhancing climate-smart agriculture with data and AI

11.2 AI-Driven Predictive Models for Crop Yield and Climate Adaptation

Artificial intelligence plays a crucial role in developing predictive models that can forecast crop yields, detect potential threats, and adapt to changing climate conditions. These AI models use real-time data inputs to provide accurate predictions, allowing farmers to plan ahead and mitigate risks.

- **Crop Yield Prediction**: AI algorithms analyze data on soil health, weather conditions, planting schedules, and crop variety to predict potential crop yields. This allows farmers to optimize planting strategies and make informed decisions about resource allocation.
- **Climate Adaptation Models**: AI-driven predictive models can simulate different climate scenarios, helping farmers identify the best crops and practices for adapting to future climate conditions. These models can forecast the impacts of extreme weather events, such as droughts or floods, on crop growth and suggest appropriate countermeasures.
- **Pest and Disease Forecasting**: AI models can also predict the likelihood of pest infestations or disease outbreaks by analyzing environmental factors and historical trends, allowing for early intervention and targeted control strategies.

11.3 Machine Learning for Decision Support Systems

Machine learning (ML) enhances decision support systems (DSS) in agriculture by continuously learning from new data and improving recommendations over time. ML-powered DSS provide farmers with real-time, actionable insights that help them optimize farm operations and respond to challenges proactively.

- **Precision Agriculture**: ML algorithms process data from multiple sources, such as satellite imagery, weather stations, and field sensors, to recommend precise actions like when to irrigate, fertilize, or apply pest control. These recommendations are based on the specific conditions of each farm, improving resource efficiency and crop health.
- **Automated Decision-Making**: By integrating ML into farm management software, IoT systems can automatically adjust irrigation schedules, temperature controls, and other farm operations in real-time based on the data analyzed, reducing the need for constant human intervention.
- **Improved Risk Management**: ML models can assess risk factors such as fluctuating market prices, weather uncertainty, and supply chain disruptions. By analyzing these variables, farmers can make more informed decisions on when to harvest, store, or sell their produce [21].

12 Blockchain Integration in Climate-Smart Agriculture

Blockchain technology offers powerful solutions for enhancing transparency, traceability, and efficiency in climate-smart agriculture (CSA). By providing a decentralized, secure, and immutable ledger, blockchain ensures that data across the agricultural supply chain is reliable and accessible to all stakeholders. This section explores how blockchain can improve supply chain transparency, enable smart contracts for climate insurance, and facilitate carbon credit trading for sustainable farming practices [22].

12.1 Traceability and Transparency in Agricultural Supply Chains

Blockchain technology provides a transparent and secure way to track the journey of agricultural products from farm to consumer. This transparency is critical for promoting sustainable and climate-smart practices in the agricultural industry.

- **Farm-to-Table Tracking**: With blockchain, every step in the supply chain—planting, harvesting, processing, distribution, and sale—is recorded in an immutable ledger. This allows consumers to trace the origin of their food, ensuring that it was produced using sustainable, climate-friendly methods.
- **Preventing Fraud and Ensuring Quality**: Blockchain can verify the authenticity of certifications such as organic or climate-smart labeling, preventing fraud and ensuring that consumers receive what they are promised. It also ensures that quality and safety standards are met, reducing the risk of contamination or spoilage.
- **Impact on Global Markets**: As consumers become more conscious of the environmental impact of their food choices, blockchain can provide a competitive advantage to farmers and producers who adopt sustainable and transparent practices. This transparency also supports fair trade and ethical farming initiatives [22].

12.2 Smart Contracts for Climate-Based Agricultural Insurance

Climate change increases the risks of extreme weather events, making agricultural insurance a critical tool for protecting farmers. Blockchain-based smart contracts can revolutionize the insurance process by automating payouts based on real-time data, offering faster, more reliable support to farmers in the event of crop failure due to climate-related disasters.

- **Automated Insurance Payouts**: Smart contracts can be programmed to trigger insurance payouts when specific conditions are met, such as drought, flooding, or severe temperature changes. IoT sensors and weather data integrated with blockchain can provide the necessary real-time information to automate these contracts.
- **Reduced Bureaucracy and Costs**: By eliminating intermediaries and automating processes, blockchain-based insurance reduces the time and cost involved in traditional insurance claims. This leads to faster payouts and greater accessibility for small-scale farmers who may struggle with complex claim procedures.
- **Increased Trust**: Blockchain's transparent nature ensures that all terms and conditions of the insurance policy are visible and immutable, building trust between insurers and farmers. This reduces the potential for disputes or delays in receiving compensation [23].

13 IoT-Driven Agroecological Zoning and Precision Landscape Management

Agroecological zoning and precision landscape management are essential for optimizing land use while ensuring sustainability and ecosystem preservation. IoT technologies, combined with geospatial data and sensors, can help farmers make informed decisions regarding land use, biodiversity conservation, and tailored farming practices. This section discusses how IoT enables data-driven zoning, supports biodiversity monitoring, and helps farmers adapt to the nuances of local microclimates shown in the Table 3 [24].

13.1 Land-Use Optimization Using IoT and Geospatial Data

Efficient land-use planning is crucial for sustainable agricultural development, especially in regions facing environmental stress due to climate change. IoT devices,

Table 3 IoT-driven agroecological zoning and precision landscape management

Zoning type	Description	Applications	Examples
Agroecological zones	Areas defined by ecological characteristics	Tailors farming practices to local conditions	Mapping zones for specific crops
Microclimate monitoring	Specific climate conditions within a field	Customizes crop management strategies	Sensors for local weather patterns
Biodiversity mapping	Tracks species and habitat changes	Supports conservation efforts	IoT systems for habitat monitoring

paired with geospatial data, enable farmers to analyze land characteristics in real time, ensuring that agricultural activities are aligned with ecological conditions.

- **Geospatial Mapping and IoT Sensors**: IoT-enabled sensors collect data on soil composition, moisture levels, topography, and land use history. Geospatial mapping systems like GPS, satellite imagery, and drones provide detailed information about land characteristics, helping to identify areas suitable for specific crops or conservation efforts.
- **Zoning for Sustainable Practices**: Using this data, farmers can create agroecological zones where specific farming practices are tailored to the land's unique attributes. This zoning helps to prevent land degradation, improve resource allocation, and optimize crop yields in areas best suited for cultivation.
- **Maximizing Land Potential**: IoT and geospatial data can help identify underutilized land and suggest the best use, whether for crop production, agroforestry, or conservation. This promotes efficient use of available resources while maintaining ecological balance.

13.2 IoT for Biodiversity Monitoring and Conservation

Biodiversity is critical for maintaining healthy ecosystems that support agriculture. IoT systems provide real-time data on species diversity, habitat health, and environmental changes, enabling farmers and conservationists to monitor and preserve biodiversity within agricultural landscapes.

- **Real-Time Biodiversity Monitoring**: IoT sensors, drones, and camera traps can monitor species presence, movements, and habitat conditions in real-time, offering valuable insights into the impact of farming practices on local ecosystems. This data helps to prevent biodiversity loss and supports conservation efforts.
- **Wildlife Protection**: IoT systems can alert farmers to the presence of wildlife in agricultural areas, reducing human-wildlife conflicts. For example, motion sensors can detect the movement of animals and trigger alerts or protective measures to safeguard both crops and local species.
- **Ecosystem Health Indicators**: By continuously monitoring environmental factors such as soil quality, water levels, and vegetation cover, IoT systems help track the overall health of an ecosystem. This allows farmers to implement corrective actions to maintain biodiversity and avoid harmful agricultural practices.

13.3 Customizing Farming Practices Based on Microclimates

Microclimates—localized climate variations within a region—play a significant role in determining the success of certain crops. IoT technology can help farmers track microclimate data, enabling them to tailor their farming practices to the specific conditions of different areas within their fields.

- **Microclimate Monitoring**: IoT sensors track temperature, humidity, soil moisture, wind speed, and solar radiation across different zones of a farm. This data provides a detailed picture of how microclimatic conditions vary within a farm, enabling more precise farming decisions.
- **Adapting Practices to Local Conditions**: By analyzing microclimate data, farmers can adjust planting schedules, irrigation levels, and crop selection to match the specific conditions of each area. This ensures that crops are grown in the optimal environment, reducing resource waste and increasing yields.
- **Precision Landscape Management**: IoT-based precision farming systems allow for the implementation of customized agricultural interventions in different microclimatic zones. For example, areas that receive less rainfall can be managed with more efficient irrigation systems, while regions with cooler temperatures may support different crop varieties [25].

14 Edge Computing and Decentralized IoT for Smart Agriculture

The integration of edge computing and decentralized IoT systems in smart agriculture is transforming how real-time data is processed and utilized on farms. By enabling faster data processing closer to the source, edge computing reduces latency and enhances the efficiency of IoT devices in making time-sensitive decisions. This section explores how edge computing improves real-time capabilities, reduces latency, and enables decentralized IoT networks, especially for small-scale and remote farms shown in the Fig. 7 [26].

14.1 Enhancing Real-Time Data Processing with Edge Computing

Edge computing brings data processing closer to where it is generated—on the edge of the network, such as IoT devices in the field. This reduces reliance on centralized cloud systems and enhances the ability to process and analyze data in real-time, making it highly useful for time-sensitive agricultural operations.

- **Local Data Processing**: Instead of sending raw data to distant cloud servers, IoT devices equipped with edge computing capabilities can process data on-site. This allows for immediate analysis and decision-making, improving the speed and efficiency of agricultural tasks such as irrigation, pest control, and soil monitoring.
- **Improved Response Time**: By processing data locally, edge computing minimizes the delay in receiving actionable insights, enabling farmers to make quicker

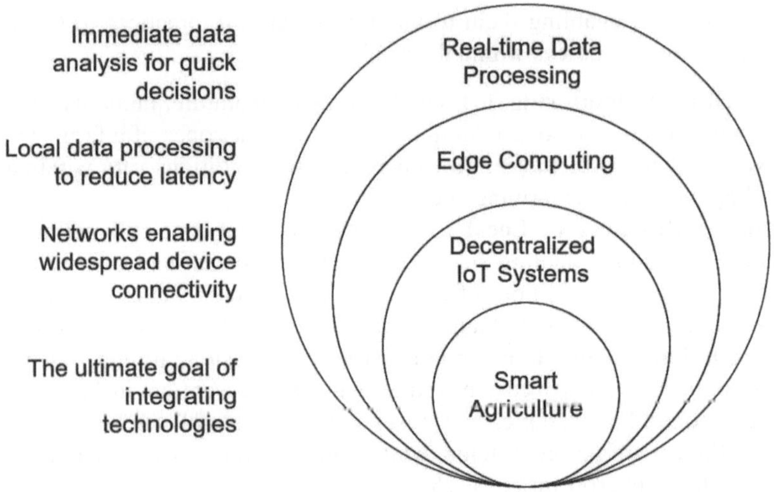

Fig. 7 Enhancing smart agriculture with edge computing

adjustments based on changing conditions in the field. For instance, irrigation systems can instantly react to soil moisture levels, preventing overwatering or drought stress.

- **Bandwidth Efficiency**: Edge computing reduces the need to constantly send large volumes of data to the cloud, freeing up bandwidth and reducing the cost associated with data transmission, particularly in areas with limited connectivity.

14.2 Reducing Latency in IoT Systems for Faster Decision Making

Latency—the delay between data generation and action—can significantly impact the effectiveness of IoT-driven agriculture, especially when immediate decisions are required. Edge computing reduces latency, enabling IoT systems to deliver faster responses, which is crucial for precision agriculture.

- **Real-Time IoT System Optimization**: In precision agriculture, decisions such as when to irrigate, apply fertilizer, or adjust environmental controls need to be made quickly. With edge computing, IoT devices can analyze data and trigger automated actions in real-time, without waiting for remote cloud servers to process information.
- **Immediate Responses to Environmental Changes**: When environmental parameters such as temperature, humidity, or soil moisture fluctuate, edge computing ensures that corrective actions (e.g., adjusting irrigation or shading systems) are taken almost instantaneously, preserving crop health and optimizing resource use.

- **Seamless Operation in Low Connectivity Areas**: For farms in remote locations where internet connectivity may be unreliable, edge computing allows IoT systems to function autonomously, as they are not dependent on constant cloud access. This autonomy ensures that critical farming operations are not disrupted by latency or connectivity issues [27].

14.3 Distributed IoT Networks for Small-Scale and Remote Farms

Decentralized IoT networks, powered by edge computing, enable small-scale and remote farms to harness the power of IoT without relying on centralized infrastructure. This opens up opportunities for farmers who might not have access to large-scale cloud computing resources or consistent internet connectivity.

- **Localized IoT Networks**: Small-scale farms can establish their own distributed IoT networks, where IoT devices, sensors, and actuators communicate with each other via edge computing nodes. This allows for the collection, processing, and sharing of data within the farm without relying on external servers.
- **Cost-Effective and Scalable Solutions**: Distributed IoT networks reduce the need for high-bandwidth internet connections, making them more cost-effective for small-scale and remote farmers. These networks can also be easily scaled up by adding more IoT devices and edge nodes as needed, supporting the growth of farming operations over time.
- **Enhanced Autonomy and Resilience**: Decentralized IoT systems enhance the resilience of smart farms by allowing them to function independently of centralized cloud servers. In the event of an internet outage or network failure, edge computing nodes can continue to process data and manage farm operations autonomously, ensuring continuous functionality.

15 Advanced IoT Sensors and Nanotechnology in Agriculture

The integration of advanced IoT sensors with nanotechnology is pushing the boundaries of precision agriculture. Nanotechnology enhances the sensitivity, efficiency, and capabilities of IoT devices, allowing for more detailed monitoring of soil, crop health, and resource management at an unprecedented scale. This section explores how nanotechnology-enhanced sensors, IoT nanobots, and nanoscale irrigation systems are revolutionizing agricultural practices [28].

15.1 Nanotechnology-Enhanced Sensors for Precision Monitoring

Nanotechnology has the potential to enhance the sensitivity and accuracy of IoT sensors, allowing for more detailed and precise monitoring of various environmental and biological factors in agriculture. These sensors can detect changes at the molecular level, providing deeper insights into soil and crop conditions.

- **Precision Soil and Crop Monitoring**: Nanotechnology-enabled sensors can detect minute changes in soil composition, moisture levels, pH, and nutrient availability. This allows farmers to make real-time adjustments to optimize soil health and improve crop yields. For example, nano-sensors can monitor nitrogen levels in soil to ensure that crops receive the appropriate amount of fertilizer.
- **Early Detection of Plant Diseases**: Nanotechnology-based sensors can detect pathogens, toxins, or stress signals in plants at early stages. This helps in the early diagnosis of diseases or pest infestations, enabling timely interventions that minimize crop losses and reduce reliance on chemical treatments.
- **Enhanced Sensitivity and Accuracy**: Due to their nanoscale size, these sensors are highly sensitive and can detect even the smallest fluctuations in environmental parameters. This leads to more accurate data collection and better decision-making in precision farming.

15.2 IoT Nanobots for Soil and Crop Health

IoT nanobots represent a cutting-edge innovation in agricultural technology, where tiny, nanotechnology-powered robots work in tandem with IoT systems to monitor and improve soil and crop health. These nanobots can perform highly specific tasks that enhance precision agriculture.

- **Soil Health Monitoring**: Nanobots can be deployed into the soil to monitor its structure, moisture content, nutrient levels, and microbial activity. They can move through the soil and report back data to IoT systems, helping farmers understand soil health at a much deeper level.
- **Nanobots for Targeted Crop Treatment**: These tiny robots can deliver nutrients, pesticides, or water directly to the plant roots, ensuring efficient resource use. For example, if a plant is identified as being deficient in a particular nutrient, nanobots can deliver the needed amount precisely, avoiding waste and runoff.
- **Self-Healing and Autonomous Actions**: Some nanobots are being designed with the capability to repair damaged plant tissues or remove harmful substances from the soil. By functioning autonomously, these nanobots can continuously improve crop health without requiring constant human intervention [29].

16 Challenges, Limitations, and Future Trends of IoT in Climate-Smart Agriculture (CSA)

While IoT holds tremendous potential in driving climate-smart agriculture (CSA), several challenges must be addressed to ensure its successful adoption. Simultaneously, emerging trends and innovations in IoT technology are poised to reshape the agricultural landscape, fostering sustainability and climate resilience. This section outlines the key technological, economic, and policy-related challenges as well as the future trends that will influence IoT's role in agriculture [30].

16.1 Technological Barriers, Connectivity, and Security Concerns

IoT deployment in CSA faces several technological challenges, primarily related to connectivity, infrastructure, and data security.

- **Connectivity Issues in Rural Areas**: Many farming operations take place in remote regions where reliable internet or cellular networks are limited or unavailable. This poses a significant barrier to the deployment of IoT systems that rely on real-time data transmission.
- **Interoperability and Infrastructure Limitations**: The lack of standardized protocols and infrastructure for IoT devices creates difficulties in integrating different systems across farms. Ensuring that various IoT devices can communicate seamlessly remains a challenge.
- **Data Privacy and Security**: With vast amounts of data being collected, the risk of cyber threats, data breaches, and misuse of sensitive farm-related information is a major concern. Farmers may hesitate to adopt IoT solutions if they perceive a lack of adequate data protection and cybersecurity measures.

16.2 Economic Viability and Farmer Adoption

Adopting IoT technologies in agriculture involves significant upfront costs, which can be prohibitive for many smallholder farmers. Moreover, widespread adoption requires greater awareness and education about IoT's benefits.

- **High Costs of IoT Devices and Infrastructure**: For many farmers, the investment in IoT sensors, equipment, and associated infrastructure may seem too costly, especially if the return on investment (ROI) is unclear or slow.
- **Lack of Technical Knowledge**: Farmers may be reluctant to adopt IoT technologies due to a lack of familiarity with how these systems work. Training and technical support are needed to improve farmer confidence in using IoT solutions.

- **Economic Disparities**: Smaller or resource-constrained farmers might find it difficult to invest in advanced IoT systems, potentially widening the gap between large, tech-savvy farms and smaller, traditional operations.

17 Conclusion

In summary, this chapter highlights the transformative potential of IoT in fostering climate-smart agriculture, emphasizing how real-time data, advanced technologies, and innovative practices can enhance resilience and sustainability in farming. Key findings underscore the critical role of IoT in optimizing resource management, improving soil health, and developing climate-resilient crop genetics, all while addressing significant challenges such as connectivity issues and economic viability. Looking ahead, the continued integration of IoT with emerging technologies, coupled with supportive policies and regulations, will pave the way for more effective and adaptive farming systems. Ultimately, embracing IoT solutions offers a pathway toward sustainable farming that not only meets the needs of a growing population but also addresses the pressing challenges posed by climate change.

References

1. Kumar, V., Sharma, K. V., Kedam, N., Patel, A., Kate, T. R., & Rathnayake, U. (2024). A comprehensive review on smart and sustainable agriculture using IoT technologies. *Smart Agricultural Technology, 8*, Article 100487. https://doi.org/10.1016/j.atech.2024.100487
2. Geetha, P., & Karthikeyan, R. (2024). Embracing IoT and precision agriculture for sustainable crop yields. In S. Balasubramanian, G. Natarajan, & P. R. Chelliah (Eds.), *Intelligent robots and drones for precision agriculture. Signals and communication technology* (pp. 139–158). Springer Nature Switzerland. https://doi.org/10.1007/978-3-031-51195-0_8
3. Ajatasatru, A., Prabhu, V., Pal, B. D., & Mukhopadhyay, K. (2024). Economy-wide impact of climate smart agriculture in India: A SAM framework. *Journal of Economic Structures, 13*(1), 4. https://doi.org/10.1186/s40008-023-00320-z
4. Piancharoenwong, A., & Badir, Y. F. (2024). IoT smart farming adoption intention under climate change: The gain and loss perspective. *Technological Forecasting and Social Change, 200*, Article 123192. https://doi.org/10.1016/j.techfore.2023.123192
5. Poudel, S., Thapa, R., & Mishra, B. (2024). A farmer-centric cost-benefit analysis of climate-smart agriculture in the Gandaki River Basin of Nepal. *Climate, 12*(9), 145. https://doi.org/10.3390/cli12090145
6. Mbanasor, J. A., et al. (2024). Climate smart agriculture practices by crop farmers: Evidence from south east Nigeria. *Smart Agricultural Technology, 8*, Article 100494. https://doi.org/10.1016/j.atech.2024.100494
7. Roy, T., et al. (2023). Climate crisis and adoption of climate-smart agriculture technologies. In U. Chatterjee, R. Shaw, S. Kumar, A. D. Raj, & S. Das (Eds.), *Climate crisis: Adaptive approaches and sustainability. Sustainable development goals series* (pp. 229–252). Springer Nature Switzerland. https://doi.org/10.1007/978-3-031-44397-8_13
8. Ansari, M., & Alam, M. (2024). A novel hybrid time-series approach for IoT-cloud-enabled environment monitoring. *The Journal of Supercomputing, 80*(7), 9019–9053. https://doi.org/10.1007/s11227-023-05782-3

9. Yesankar, P. R., Bandre, P., Verma, P., Gudadhe, A. G., Gourshettiwar, P., & Puri, C. (2024). A review on the role of IoT in smart agriculture with reference to efficiency, sustainability and precision farming. In *2024 5th international conference on electronics and sustainable communication systems (ICESC)* (pp. 533–537). IEEE. https://doi.org/10.1109/ICESC60852.2024.10689999
10. Papadopoulos, G., Arduini, S., Uyar, H., Psiroukis, V., Kasimati, A., & Fountas, S. (2024). Economic and environmental benefits of digital agricultural technologies in crop production: A review. *Smart Agricultural Technology, 8*, Article 100441. https://doi.org/10.1016/j.atech.2024.100441
11. Pandian, K., et al. (2024). Synergistic conservation approaches for nurturing soil, food security and human health towards sustainable development goals. *Journal of Hazardous Materials Advance, 16*, Article 100479. https://doi.org/10.1016/j.hazadv.2024.100479
12. Et-taibi, B., et al. (2024). Enhancing water management in smart agriculture: A cloud and IoT-Based smart irrigation system. *Results in Engineering, 22*, Article 102283. https://doi.org/10.1016/j.rineng.2024.102283
13. Zhang, G., et al. (2024). Combining water-saving and drought-resistant rice with plastic film mulching mitigates CH_4 emissions with higher net economic benefits. *Resources, Conservation and Recycling, 202*, Article 107372. https://doi.org/10.1016/j.resconrec.2023.107372
14. Meena, S. K., et al. (2024). Advanced and emerging techniques in soil health management. In R. K. Bhatia & A. Walia (Eds.), *Advancements in microbial biotechnology for soil health, microorganisms for sustainability* (Vol. 50, pp. 343–362). Springer Nature Singapore. https://doi.org/10.1007/978-981-99-9482-3_15
15. Zhang, J., Dyck, M., Quideau, S. A., & Norris, C. E. (2024). Assessment of soil health and identification of key soil health indicators for five long-term crop rotations with varying fertility management. *Geoderma, 443*, Article 116836. https://doi.org/10.1016/j.geoderma.2024.116836
16. Reddy, K. P., Roshni Thanka, M., Edwin, E. B., Ebenezer, V., & Joy, P. (2024). Farm_era: Precision farming with GIS, AI pest management, smart irrigation, data analytics, and optimized crop planning. In *2024 international conference on inventive computation technologies (ICICT)* (pp. 1134–1140). IEEE. https://doi.org/10.1109/ICICT60155.2024.10544687
17. Ali, Md. A., Sharma, A. K., & Dhanaraj, R. K. (2024). Heterogeneous features and deep learning networks fusion-based pest detection, prevention and controlling system using IoT and pest sound analytics in a vast agriculture system. *Computers & Electrical Engineering, 116*, Article 109146. https://doi.org/10.1016/j.compeleceng.2024.109146
18. Morchid, A., Marhoun, M., El Alami, R., & Boukili, B. (2024). Intelligent detection for sustainable agriculture: A review of IoT-based embedded systems, cloud platforms, DL, and ML for plant disease detection. *Multimedia Tools and Applications, 83*(28), 70961–71000. https://doi.org/10.1007/s11042-024-18392-9
19. Padhiary, M., Barbhuiya, J. A., Roy, D., & Roy, P. (2024). 3D printing applications in smart farming and food processing. *Smart Agricultural Technology, 9*, Article 100553. https://doi.org/10.1016/j.atech.2024.100553
20. Noma, F., & Babu, S. (2024). Predicting climate smart agriculture (CSA) practices using machine learning: A prime exploratory survey. *Climate Services, 34*, Article 100484. https://doi.org/10.1016/j.cliser.2024.100484
21. Goel, M., & Pandey, M. (2024). Crop yield prediction using AI: A review. In *2024 2nd international conference on disruptive technologies (ICDT)* (pp. 1547–1553). IEEE. https://doi.org/10.1109/ICDT61202.2024.10489432
22. Konfo, T. R. C., et al. (2024). Recent climate-smart innovations in agrifood to enhance producer incomes through sustainable solutions. *Journal of Agricultural Food Research, 15*, Article 100985. https://doi.org/10.1016/j.jafr.2024.100985
23. Thottadi, B. P., & Singh, S. P. (2024). Climate-smart agriculture (CSA) adaptation, adaptation determinants and extension services synergies: A systematic review. *Mitigation and Adaptation Strategies for Global Change, 29*(3), 22. https://doi.org/10.1007/s11027-024-10113-9

24. Chintamaneni, V., Janipalli, A., Rajaram, J., Devi, T. A., Krishna Vajjala, K. V., & Vivekanandhan, V. (2024). Revolutionizing farming: An analysis of IoT-based smart agriculture monitoring systems. In *2024 5th international conference on electronics and sustainable communication systems (ICESC)* (pp. 486–492) IEEE. https://doi.org/10.1109/ICESC60852.2024.10689990
25. Thwin, K. M. M., Horanont, T., & Phatrapornnant, T. (2024). Machine-learning microclimate forecasting for adaptive equipment control via web integration in open-ventilated greenhouses. *AgriEngineering, 6*(3), 2845–2869. https://doi.org/10.3390/agriengineering6030165
26. Sathya, D., Thangamani, R., & Balaji, B. S. (2024). The revolution of edge computing in smart farming. In S. Balasubramanian, G. Natarajan, & P. R. Chelliah (Eds.), *Intelligent robots and drones for precision agriculture, Signals and communication technology* (pp. 351–389). Springer Nature Switzerland. https://doi.org/10.1007/978-3-031-51195-0_17
27. Hasan, M. D. A., Balasubadra, K., Vadivel, G., Arunfred, N., Ishwarya, M. V., & Murugan, S. (2024). IoT-driven image recognition for microplastic analysis in water systems using convolutional neural networks. In *2024 2nd international conference on computer, communication and control (IC4)* (pp. 1–6). IEEE. https://doi.org/10.1109/IC457434.2024.10486490
28. Faqir, Y., Qayoom, A., Erasmus, E., Schutte-Smith, M., & Visser, H. G. (2024). A review on the application of advanced soil and plant sensors in the agriculture sector. *Computers and Electronics in Agriculture, 226*, Article 109385. https://doi.org/10.1016/j.compag.2024.109385
29. Bharti, A., Jain, U., & Chauhan, N. (2024). From lab to field: Nano-biosensors for real-time plant nutrient tracking. *Plant Nano Biology, 9*, Article 100079. https://doi.org/10.1016/j.plana.2024.100079
30. Raihan, A., Ridwan, M., & Rahman, M. S. (2024). An exploration of the latest developments, obstacles, and potential future pathways for climate-smart agriculture. *Climate Smart Agriculture, 1*(2), Article 100020. https://doi.org/10.1016/j.csag.2024.100020

Sustainable Practices: Balancing Efficiency and Environmental Impact

Chandra Prakash Choudhary, Ashwini Yadav, Sarita Kumari, and Priyanka Kumari

Abstract In an era of profound technological innovation, the agriculture sector faces the exigent dual challenge of enhancing production efficiency while mitigating its considerable environmental footprint. This chapter critically examines the role of Artificial Intelligence (AI) and Internet of Things (IoT) solutions in navigating the intricate balance between agricultural productivity and ecological stewardship. While the dominant narrative focuses on techno-optimism, we argue that the true potential of these technologies is realized only when their deployment is synergistically aligned with agroecological principles and equitable governance. By leveraging AI-driven data analytics and sophisticated IoT sensor networks, farmers can transition from prescriptive, input-heavy practices to predictive, knowledge-intensive systems. This facilitates optimized resource allocation, a significant reduction in agrochemical use, and minimized energy consumption. We analyze advanced applications—including AI-powered integrated pest management, dynamic digital twins for water management, and sensor-driven precision nutrient strategies—as core components of this new paradigm. Through a critical evaluation of real-world case studies, this chapter demonstrates how AI and IoT can foster not only economic efficiency but also promote long-term environmental sustainability. Ultimately, we contend that as the agricultural landscape evolves, the thoughtful and ethical adoption of these smart technologies is an indispensable pillar in the global pursuit of climate-smart agriculture and resilient food systems.

Keywords Sustainable agriculture · Artificial intelligence · Internet of things · Precision farming · Digital twin · Edge AI · Climate-smart agriculture · Food security · Agri-tech ethics

C. P. Choudhary (✉) · P. Kumari
Sri Karan Narendra Agriculture University, Jobner, Rajasthan, India
e-mail: cpchoudhary022@gmail.com

A. Yadav
Indian Council of Agricultural Research, New Delhi, India

S. Kumari
Maharana Pratap University of Agriculture and Technology, Udaipur, Rajasthan, India

S. R. Kumar et al. (eds.), *AI-Driven IoT Solutions in Agriculture: New Insights*, Studies in Computational Intelligence 1236, https://doi.org/10.1007/978-3-032-05603-0_6

1 Introduction: The Agri-Environmental Dilemma

The global agricultural sector is tethered to a paradox: it is both a fundamental pillar of human civilization and a primary driver of planetary-scale environmental change. The imperative to feed a global population projected to approach ten billion by 2050 has historically justified agricultural intensification, a model that has successfully boosted yields but at a severe ecological cost [1, 2]. Conventional farming systems, predicated on intensive tillage, homogeneous cropping, and the prophylactic application of agrochemicals, have precipitated widespread soil degradation, aquifer depletion, biodiversity collapse, and contributed significantly to anthropogenic greenhouse gas emissions [3]. This agri-environmental dilemma forms the central challenge of our time: how to decouple food production from environmental degradation.

The contemporary response to this challenge is increasingly articulated through the lens of a fourth agricultural revolution, powered by the confluence of the Internet of Things (IoT) and Artificial Intelligence (AI). IoT infrastructures, comprising vast networks of interconnected sensors, actuators, and devices, enable the capture of high-resolution, real-time data from the field, effectively giving the agricultural landscape a digital voice [4]. AI, particularly its sub-field of machine learning, provides the analytical power to interpret this complex data stream, discerning subtle patterns, forecasting future states, and automating intelligent decisions [5]. The synergy between these technologies underpins the concept of precision agriculture—a management philosophy that acknowledges and responds to in-field variability to optimize inputs and maximize outcomes.

This chapter advances a critical perspective on this technological revolution, deliberately moving beyond a purely techno-optimistic overview. We argue that the transformative power of AI and IoT is not an inherent property of the technologies themselves, but is contingent upon their deployment in service of proven agroecological principles. The ultimate objective should not be the mere optimization of an inherently flawed, input-intensive system—doing the wrong things more efficiently—but its fundamental reshaping into a system that is knowledge-intensive. This represents a paradigm shift from a model reliant on generalized, prophylactic applications of chemicals and water, to one based on a profound, data-driven understanding of biophysical processes at a micro-level. It involves moving from farming by prescription and calendar dates to farming by observation and prediction. By transitioning from broad-spectrum, uniform management to highly targeted, surgically precise interventions, farmers can achieve unprecedented levels of resource efficiency. This targeted methodology, which lies at the heart of precision agriculture, inherently reduces operational expenditures while simultaneously curtailing the pervasive negative externalities—nutrient runoff into waterways, atmospheric drift of pesticides, and the significant carbon emissions from machinery and fertilizer production—that have long defined industrial agriculture's environmental legacy [3, 5].

Herein, we will dissect the core technological frameworks, from multi-layered IoT architectures to the emerging paradigms of Edge AI and agricultural Digital Twins.

We will analyze, through meticulously chosen case studies, how these systems are being operationalized for sustainable pest, water, and soil management. Crucially, this analysis will extend to the socio-economic and ethical dimensions of this transition, addressing critical questions of data sovereignty, the digital divide, and the future of rural labor. The central thesis of this chapter is that the strategic, ethical, and context-aware implementation of AI and IoT is an essential, though not solely sufficient, condition for building the resilient, productive, and environmentally sustainable agricultural systems demanded by the twenty-first century.

2 The Sentient Farm: IoT Architectures and Data Acquisition

The foundation of any smart agriculture system is its ability to perceive and quantify the farm environment with high fidelity. The Internet of Things (IoT) provides this perceptual capability, creating a sentient digital nervous system that overlays the physical landscape. Agricultural IoT is an ecosystem of distributed devices—sensors, cameras, drones, and actuators—that collect and transmit data, offering an unprecedentedly granular view of farm dynamics and enabling a shift from reactive to predictive management [6].

2.1 A Multi-layered Architectural Framework

A robust agricultural IoT architecture is typically conceptualized in four distinct layers, each with a specific function:

1. **Perception Layer**: This foundational layer constitutes the physical interface with the farm environment, acting as the digital senses of the entire system. It is a diverse constellation of sensors tasked with quantifying key biophysical parameters. The quality, calibration, and strategic placement of these sensors are paramount, as they directly determine the integrity of all subsequent data and analysis [7]. This layer includes not only the in-situ soil and weather sensors but also mobile sensing platforms like drones and ground vehicles equipped with advanced imaging systems. A key challenge at this layer is ensuring sensor robustness and longevity in harsh agricultural environments, which are subject to extreme temperatures, moisture, dust, and chemical exposure. Furthermore, ensuring consistent calibration across a distributed network of sensors is a non-trivial task essential for preventing data drift and maintaining the accuracy required for precision management.
2. **Network Layer**: Functioning as the data transmission backbone, this layer provides the crucial connectivity between the perception layer in the field and the central processing platforms. The choice of communication technology is

critical and context-dependent. In areas with strong cellular coverage, 4G/5G networks offer high bandwidth, suitable for transmitting large data packets like high-resolution images. However, given the often-remote and expansive nature of farms, such infrastructure is not always available or cost-effective. This has driven the adoption of Low-Power Wide-Area Networks (LPWAN), such as LoRaWAN and NB-IoT. These technologies represent a critical enabler for smart agriculture, as they are specifically designed to transmit small data packets (typical of sensor readings) over very long distances (several kilometers) with minimal power consumption. This efficiency allows battery-powered sensors in the perception layer to operate autonomously for years without maintenance, drastically reducing the operational complexity and cost of a large-scale deployment [8].

3. **Service (Platform) Layer**: Often referred to as the system's central nervous system, this is where raw data is transformed into valuable information. Typically residing on cloud-based infrastructure, this layer is responsible for the heavy lifting of data aggregation, secure storage, and advanced analytics. It is here that the AI and machine learning models are housed, trained, and executed [9]. This layer must be capable of ingesting and processing heterogeneous data streams from thousands of sensors in real-time. Its functions include data cleaning (removing erroneous readings), normalization (ensuring data from different sensors is comparable), and integration (fusing different data types for holistic analysis). The scalability of the cloud is essential here, providing the on-demand computational power required to train complex deep learning models on vast historical datasets or to run thousands of simulations for a digital twin.
4. **Application Layer**: This is the human–machine interface, where the synthesized intelligence is delivered to the end-user—be it a farmer, an agronomist, or an autonomous machine. The design of this layer is critical for technology adoption. Complex data must be presented through intuitive, easy-to-understand visualizations, such as color-coded field maps showing nitrogen deficiency, dashboards tracking livestock health, or simple alerts sent to a smartphone. For a farmer, this layer might provide a mobile app suggesting the optimal irrigation schedule for the next 24 h. For an autonomous tractor, the application layer might deliver a precise variable-rate seeding map. The goal of this layer is to make the system's intelligence accessible and actionable, abstracting away the underlying complexity and empowering users to make better, data-informed decisions with confidence.

2.2 *Essential Sensor Technologies for Sustainable Agriculture*

The efficacy of an IoT system is contingent upon the data it collects. Key sensor categories include:

- **Soil Intelligence Sensors**: Crucial for managing soil health, the cornerstone of sustainable farming.

  - *Volumetric Water Content Sensors*: Enable precision irrigation by providing real-time data on soil moisture, preventing waterlogging and nutrient leaching [10].
  - *Multi-parameter Probes*: Advanced sensors that measure not only moisture but also soil temperature, electrical conductivity (EC)—a proxy for salinity and nutrient levels—and pH, offering a holistic view of the root-zone environment.

- **Microclimate and Environmental Sensors**: On-farm weather stations provide hyperlocal data far more relevant than regional forecasts.
  - *Atmospheric Sensors*: Monitoring air temperature, humidity, and leaf wetness duration is critical for predicting fungal disease outbreaks.
  - *Anemometers and Rain Gauges*: Inform decisions on pesticide spraying (to avoid drift) and irrigation scheduling.

- **Crop Health Monitoring Sensors**:
  - *Spectrometers and Multispectral Imagers (e.g., NDVI)*: Often mounted on drones or satellites, these sensors analyze the spectral signature of plant canopies to detect stress from pests, diseases, or nutrient deficiencies long before they are visible to the human eye [11].

- **Livestock and Asset Monitoring**:
  - *GPS and Geofencing*: Used for tracking farm machinery to optimize logistics and for monitoring livestock movement to implement rotational grazing schemes.
  - *Livestock Biometric Sensors*: Wearable sensors (e.g., smart collars, boluses) that monitor physiological and behavioral indicators like body temperature, rumination, and activity levels. These can provide early alerts for illness, reducing the need for broad-spectrum antibiotic use and improving animal welfare [12].

This integrated sensor network forms the empirical bedrock upon which AI can build its analytical models. However, the true power of this network emerges not from individual data streams, but from the practice of **sensor fusion**. This is the process of combining data from multiple, heterogeneous sensors to generate a more accurate, complete, and dependable assessment of the farm's state than could be achieved with any single sensor alone. For instance, soil moisture data [10] becomes significantly more actionable when fused with hyperlocal evapotranspiration rates derived from on-farm weather stations and canopy stress indicators from NDVI imagery [11]. This integrated approach allows the system to differentiate between a simple lack of water and a plant's inability to uptake water due to other stressors, such as soil compaction or disease. This multi-modal, layered understanding is what truly transforms the farm from a black box, managed by heuristics and historical averages, into a transparent, high-fidelity digital ecosystem. It creates a rich, contextualized dataset that is not just a collection of numbers, but a dynamic narrative of the farm's life, ripe for the nuanced interpretation and optimization offered by advanced AI (Table 1).

Table 1 Advanced IoT sensors in agriculture and their role in sustainability

Sensor category	Measured parameter (s)	Key application	Primary sustainability contribution
Soil health	Moisture, EC, pH, NPK, Organic Matter	Precision fertilization and irrigation	Minimizes chemical runoff and eutrophication; conserves water
Crop phenotyping	NDVI, NDRE, canopy temperature, chlorophyll index	Early stress detection and variable-rate apps	Reduces pesticide/ fungicide use through targeted application
Microclimate	Leaf wetness, humidity, temperature, solar radiation	Disease and pest outbreak forecasting	Shifts from prophylactic to preventative spraying; reduces chemical load
Livestock biometrics	Rumination, temperature, activity, location	Individualized health management	Reduces herd-level antibiotic use; enhances animal welfare and efficiency

3 The Cognitive Engine: AI, ML, and the Emergence of Edge Intelligence

If IoT provides the sensory organs, Artificial Intelligence (AI) provides the cognitive engine, processing sensory input and driving intelligent action. AI and Machine Learning (ML) algorithms are the tools that unlock the value embedded within the terabytes of data generated by a smart farm, enabling a shift from reactive to predictive and prescriptive management [13].

3.1 The Spectrum of AI in Agriculture

AI's role in agriculture is multifaceted, encompassing several key functions:

- **Predictive Analytics**: ML models are trained on vast historical and real-time datasets to forecast future events with high probability.
  - *Yield Forecasting*: By integrating soil data, long-term weather forecasts, and in-season imagery, AI models can predict yields with remarkable accuracy, empowering farmers with critical information for logistics, storage, and market planning [14].
  - *Pest and Disease Outbreak Prediction*: AI can identify complex correlations between microclimate conditions (e.g., specific temperature-humidity thresholds) and the lifecycle of pests and pathogens, issuing early warnings that allow for targeted, preventative action rather than costly, reactive cures [15].

- **Prescriptive Analytics and Digital Twins**: This represents a significant evolution from prediction, moving from forecasting what *will* happen to recommending what *should be done*. Prescriptive systems advise on optimal courses of action to achieve specific goals, such as maximizing yield while minimizing water use. The most powerful enabler of this is the concept of the agricultural **Digital Twin**—a comprehensive, dynamic, and virtual representation of a physical farm and all its interconnected processes [16]. Far more than a simple 3D model, a digital twin is a living simulation, continuously fed and calibrated by the real-time data streams from the IoT sensor network. It models the complex interplay between the soil, crop genetics, microclimate, and management interventions. This creates a risk-free virtual environment where a farmer or an autonomous system can test hypotheses and run "what-if" scenarios. For example, a farmer could pose the question: "What is the projected economic and environmental impact of switching to a no-till system next season, considering my specific soil profile and predicted weather patterns?" The AI-powered digital twin can simulate the multi-year effects on soil carbon sequestration, nutrient cycling, and yield, providing a robust, data-backed foundation for long-term strategic decisions that would be impossible to evaluate otherwise. This capability to explore future possibilities in a virtual space before committing resources in the physical world is what makes the digital twin a cornerstone of truly predictive and sustainable farm management.
- **Computer Vision**: This domain of AI enables machines to "see" and interpret their surroundings. Using high-resolution imagery from drones, satellites, or ground-based cameras, computer vision is revolutionizing in-field operations.
  - *Weed and Pest Identification*: AI algorithms can differentiate crop plants from various weed species or identify specific insect pests with superhuman accuracy, enabling precision interventions [17].
  - *Automated Phenotyping*: Computer vision can automatically measure critical plant traits (phenotypes) such as height, leaf area index, and fruit count, accelerating plant breeding and variety selection.

3.2 Edge AI: Decentralizing Intelligence

A critical emerging trend is the shift from purely cloud-based AI to Edge AI. In the traditional model, all sensor data is sent to the cloud for processing. This can be problematic in agriculture due to unreliable rural connectivity and the need for real-time decisions (e.g., for a robot to decide to spray a weed).

Edge AI involves performing AI computations directly on the device (the "edge")—such as on a smart tractor's onboard computer, within a drone's flight controller, or on a local IoT gateway—rather than relying exclusively on a distant cloud server. This decentralization of intelligence is not merely a technical preference; it is a practical necessity for many agricultural applications. The advantages are manifold. Firstly, it dramatically reduces latency. For an autonomous weeding robot to identify a weed and trigger a mechanical hoe in real-time, the decision must be made in milliseconds, a speed unachievable if imagery has to be sent to the cloud and back. Secondly, it improves reliability and autonomy. In regions with intermittent or non-existent internet connectivity, edge devices can continue to perform their intelligent functions without interruption. A smart irrigation valve can continue to make optimized decisions based on local sensor data even if the farm's main internet connection goes down. Thirdly, it offers enhanced data privacy and security. By processing sensitive operational data locally, farmers can mitigate risks associated with transmitting proprietary information to third-party servers. This principle is further advanced by federated learning, a technique where individual edge devices (e.g., multiple farms' systems) contribute to improving a global AI model by only sharing anonymized model updates, not their underlying private data, thus enabling collaborative intelligence without sacrificing data sovereignty [18].

The combination of centralized cloud intelligence for large-scale modeling and decentralized edge intelligence for real-time action represents the future architecture of agricultural AI.

Figure 1 illustrates the conceptual architecture of a modern smart farm that integrates Cloud, Edge, and Digital Twin components. Sensors and robots at the farm (the Edge) continuously capture real-time data on soil, crops, and equipment. The data is transmitted to an Edge Gateway for immediate processing and action. Simultaneously, the Edge Gateway synchronizes with a central Cloud Platform that hosts the Digital Twin of the farm and advanced AI models. Insights from the Cloud are then relayed back to the farm management application and the Edge devices, creating a closed-loop system for predictive and prescriptive decision-making.

4 Sustainable Intensification in Practice: Critical Case Studies

The theoretical promise of AI and IoT finds its grounding in practical application, where the technologies directly confront the complex realities of agricultural production. It is in this operational context that their true transformative potential for sustainability becomes evident. To provide a holistic and intellectually rigorous analysis, this section critically examines three pivotal applications, moving beyond simple success stories to offer a more nuanced perspective. Each case study is framed within a "Challenge-Solution-Critique" model. This structure serves a distinct purpose: it first outlines a persistent sustainability challenge inherent in conventional agriculture; it then details how a specific AI- and IoT-driven solution addresses this challenge; finally, and most importantly, it offers a critique that acknowledges the practical limitations, potential unintended consequences, or socio-economic hurdles associated with the solution. This balanced approach is essential for a realistic assessment, highlighting that technology is not a panacea but a powerful tool whose effectiveness is shaped by design, context, and implementation strategy.

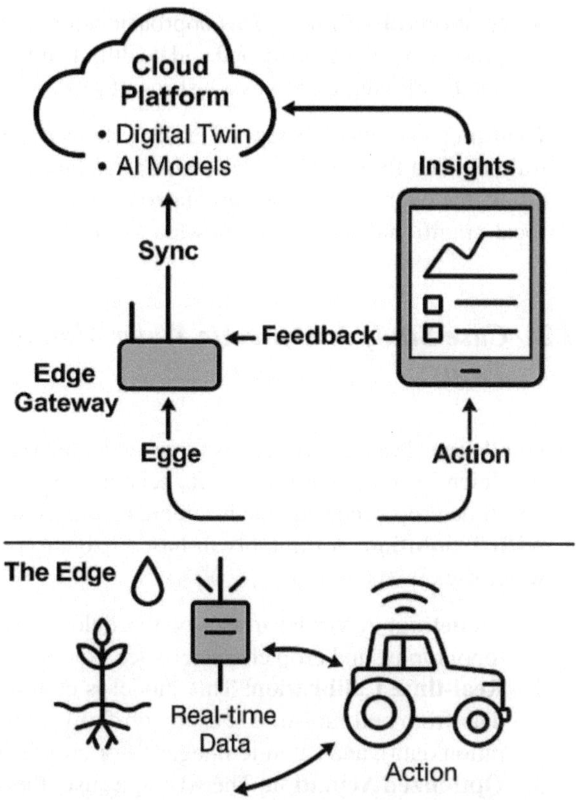

Fig. 1 Conceptual architecture of a modern smart farm integrating Cloud, Edge, and Digital Twin components

4.1 Case Study 1: Integrated Pest Management (IPM) with AI

- **Challenge**: Conventional pest control, often reliant on calendar-based broadcast spraying, is ecologically disruptive and economically inefficient. It contributes to pesticide resistance, harms non-target species (including pollinators), and poses risks to human health [19].
- **AI/IoT Solution**: An AI-driven IPM system shifts the paradigm from eradication to management.

  1. **Sensing and Forecasting**: A network of smart insect traps with embedded cameras and AI-powered identification provides real-time data on pest populations. This is correlated with microclimate data to predict population dynamics.
  2. **Precision Action**: When pest pressure in a specific area crosses an economic threshold, the system recommends a targeted action. This could be the dispatch of a drone to release beneficial predatory insects or, if necessary, to spot-spray a biopesticide on the identified hotspot.

3. **Sustainable Impact**: This approach can reduce pesticide use by up to 90%, preserving biodiversity and soil health. It embodies the IPM principle of using chemical intervention as a last resort [20].

- **Critique**: The effectiveness of these systems depends on high-quality, localized training data for the AI models, which can be expensive to acquire. There is also a risk that over-reliance on automated detection could de-skill farmers and erode local, traditional ecological knowledge about pest management.

4.2 Case Study 2: Dynamic Water Management via Digital Twins

- **Challenge**: Water, a finite resource under increasing strain, is often managed inefficiently in agriculture. Static, schedule-based irrigation fails to account for real-time crop needs and environmental conditions, leading to waste [21].
- **AI/IoT Solution**: A digital twin-based irrigation system enables hyper-efficient water use.

1. **Modeling**: A virtual model of the field is created, incorporating soil type, topography, and crop characteristics.
2. **Real-time Calibration**: This model is continuously updated with real-time data from soil moisture sensors, on-farm weather stations (for evapotranspiration data), and satellite imagery (for crop health).
3. **Optimized Actuation**: The AI engine uses the digital twin to run thousands of micro-simulations, determining the precise water requirement for each zone of the field. It then automatically controls a variable-rate drip irrigation system to deliver exactly that amount of water directly to the root zone at the optimal time.
4. **Sustainable Impact**: Such systems have been shown to reduce agricultural water consumption by 30–60% while often improving yield and quality by eliminating water stress [22]. This conserves water and reduces the significant energy costs associated with pumping.

Critique: Despite its immense potential, the deployment of AI-driven IPM is constrained by significant practical challenges. The primary bottleneck is the development of robust and highly accurate AI models. The effectiveness of these systems is entirely dependent on the quality, quantity, and diversity of the training data [20]. An AI model trained to identify a specific pest in one region under certain lighting conditions may fail dramatically when deployed in a different geography with different crop varieties and environmental conditions. Creating these comprehensive, localized datasets is a labor-intensive and expensive endeavor, often beyond the reach of smaller tech companies or public research institutions.

Furthermore, there is a tangible risk of technological lock-in and a de-skilling of the farming workforce. As farmers come to rely on proprietary "black box" algorithms for pest detection, they may lose their own innate, traditional ecological knowledge of their local pest and predator dynamics. This creates a dependency on technology providers and can erode the farmer's autonomy. If a company ceases to support a product or significantly increases subscription fees, the farmer may be left without a viable alternative, having lost the very skills the technology was meant to augment.

Finally, the economic viability for small-scale, diversified farms remains a significant question. While a 90% reduction in pesticide use is environmentally compelling, the initial capital investment in high-resolution imaging drones and sophisticated AI software may be prohibitive. The return on investment is often clearest for large-scale monocultures where even small efficiency gains translate into substantial savings. For a smallholder farmer managing a complex, multi-crop system, the cost-benefit analysis is far less straightforward, highlighting a critical need for more affordable, scalable, and adaptable IPM technologies tailored to diverse agricultural systems.

4.3 Case Study 3: Precision Nutrient Management and Soil Health

- **Challenge**: The uniform application of synthetic fertilizers is a leading cause of nitrous oxide (N_2O) emissions and water pollution (eutrophication). This practice ignores the inherent spatial variability of soil nutrient content [23].
- **AI/IoT Solution**: AI-powered Variable Rate Technology (VRT) tailors nutrient application to specific needs.
  1. **Mapping Variability**: Fields are mapped using GPS-referenced soil sampling and on-the-go sensors that measure soil electrical conductivity and organic matter. This is overlaid with layers of yield monitor data from previous harvests and in-season NDVI imagery.
  2. **Prescription Generation**: An AI platform analyzes these data layers to create a dynamic nutrient prescription map, dictating the precise blend and amount of N, P, and K needed for every square meter of the field.
  3. **Targeted Application**: This map guides a VRT-enabled fertilizer spreader, which automatically adjusts the application rate as it traverses the field.
  4. **Sustainable Impact**: This approach ensures that nutrients are a productive input, not a pollutant. It enhances fertilizer use efficiency, reduces greenhouse gas emissions, protects water quality, and can improve long-term soil health by preventing nutrient imbalances [24].
- **Critique**: VRT is capital-intensive and its ROI is most easily realized on large, uniform fields. Adapting the technology for the small, fragmented landholdings common in many parts of the world, including India, remains a major challenge that requires novel service models (e.g., "Farming as a Service").

5 The Socio-ethical Frontier: Challenges Beyond Technology

A purely technical analysis of AI and IoT in agriculture is insufficient. The widespread adoption of these technologies presents profound socio-economic and ethical challenges that must be addressed to ensure a just and equitable transition.

5.1 *The New Digital Divide and Data Sovereignty*

- **Economic Disparity**: The promise of digital agriculture is shadowed by the significant risk of deepening the existing divides within the farming community. The high capital outlay for sensors, drones, robotics, and subscription-based software platforms presents a formidable barrier to entry for the world's smallholder farmers, who constitute the majority of global food producers. This risks creating a two-tiered agricultural system: a technologically advanced, hyper-efficient corporate sector and a vast smallholder sector that is unable to access these productivity-enhancing tools. Such a trajectory would not only exacerbate rural economic inequality but could also undermine local food security and the resilience of diverse farming systems that are often better adapted to local ecological niches [25]. Addressing this requires a move beyond simply marketing expensive hardware, towards developing innovative and inclusive service models, such as "Farming-as-a-Service" (FaaS), community-shared equipment, and government-supported digital extension services that are explicitly designed for the smallholder context.
- **Data Colonialism and Governance**: As farms become data-generating assets, the question of data ownership, control, and sovereignty becomes a paramount ethical and economic concern. Farm data—from soil maps to yield histories—is an immensely valuable resource. There is a tangible danger that, in the absence of robust governance frameworks, farmers could unwittingly cede control of this data to large agri-tech corporations in exchange for access to their platforms [26]. This creates a power imbalance where farmers risk becoming "data tenants" on their own land. Aggregated farm data can be used to influence commodity markets, inform the pricing of inputs like seeds and fertilizers, and create derivative financial products, all of which may not benefit the farmer who generated the data. Establishing clear principles for agricultural data governance is therefore critical. These frameworks must prioritize farmer-centric models, ensuring that farmers retain ownership of their data, have full transparency into how it is used, and share equitably in any value created from it. Without such protections, the data revolution could lead to a new form of digital feudalism, undermining the autonomy and economic viability of the very producers it claims to empower.

5.2 Labor, Skills, and Algorithmic Justice

The integration of AI and automation into agriculture will profoundly reconfigure the nature of rural work and introduces critical questions of algorithmic justice. The narrative of robotics alleviating the drudgery of manual labor and solving labor shortages is compelling, yet it concurrently signals a large-scale displacement of the existing agricultural workforce. This transition necessitates a massive, proactive, and well-funded public and private effort in rural education and workforce development. The jobs of the future farm will not be centered on manual labor, but on technical skillsets: data analysts interpreting yield maps, robotics technicians maintaining autonomous fleets, drone pilots certified for agricultural spraying, and farm managers skilled in overseeing complex, interconnected systems. Failure to invest in this reskilling will not only lead to unemployment but will also create an unbridgeable skills gap, rendering even the most advanced technologies useless on the ground. Beyond the question of skills, we must confront the issue of **algorithmic bias and accountability** [27]. An AI model is only as unbiased as the data it is trained on. The vast majority of agricultural data used to train today's leading models originates from the large, monoculture farms of North America and Europe. Consequently, an AI system trained to optimize corn production in Iowa may perform poorly, or provide dangerously flawed recommendations, when applied to the diverse, intercropped millet and sorghum farms of Rajasthan or the terraced rice paddies of Southeast Asia. This creates a form of algorithmic bias that privileges certain agro-ecological zones, crop types, and farming systems over others. It can lead to the marginalization of indigenous crops and farming knowledge, pushing farmers towards a technologically-enforced monoculture. Therefore, achieving algorithmic justice requires a conscious effort to develop AI models that are trained on diverse, localized datasets. Furthermore, it demands transparency and accountability; when an AI recommendation leads to crop failure, who is responsible? The farmer, the software company, the equipment manufacturer? Establishing clear regulatory frameworks for the testing, validation, and liability of agricultural AI is a critical area of governance that is still in its infancy but is essential for building trust and ensuring these systems serve all farmers equitably.

6 The Horizon: Climate-Smart Agriculture and Global Food Resilience

Despite the challenges, the trajectory of AI and IoT integration points toward a more resilient and sustainable agricultural future. These technologies are not merely tools for optimization but are foundational for systemic change.

6.1 Engineering Climate-Smart Agricultural Systems

Climate-smart agriculture aims to achieve a "triple win": sustainably increasing productivity, enhancing resilience to climate change, and reducing agricultural greenhouse gas emissions [28]. AI and IoT are instrumental to all three goals:

- **Productivity**: As demonstrated throughout this chapter, precision management directly enhances yields per unit of input by ensuring that resources like water and nutrients are applied with surgical accuracy, minimizing waste and maximizing their uptake by the crop. This leads to what is known as sustainable intensification—producing more from the same or fewer resources.
- **Resilience**: The ability to withstand and adapt to climate shocks is perhaps an even more critical contribution. AI-powered early warning systems, which analyze long-range weather forecasts and real-time sensor data, can provide farmers with crucial lead time to prepare for events like droughts, heatwaves, or unseasonal frosts. This allows for proactive strategies such as applying anti-transpirants before a heatwave, activating frost protection systems, or performing a timely harvest. On a longer timescale, AI is revolutionizing climate-resilient crop breeding. By rapidly analyzing vast genomic datasets alongside environmental performance data, machine learning can identify the specific genes responsible for traits like drought tolerance or disease resistance, dramatically accelerating the development of new crop varieties tailored to the challenges of a changing climate [28].
- **Mitigation**: The mitigation of greenhouse gas emissions is a direct and powerful outcome of AI- and IoT-driven practices. Agriculture's primary emissions are nitrous oxide (N_2O) from excessive nitrogen fertilizer use and methane (CH_4) from livestock and rice paddies. Precision nutrient management, by ensuring fertilizer is applied only where and when it is needed, directly cuts N_2O emissions, which are nearly 300 times more potent than CO_2 as a greenhouse gas [24]. Similarly, optimizing livestock feed and health through biometric monitoring can reduce methane emissions. Furthermore, these technologies are essential for enabling robust carbon farming initiatives. By using IoT sensors and satellite imagery to accurately measure and verify increases in soil organic carbon resulting from regenerative practices like no-till and cover cropping, farmers can generate high-quality carbon credits. This not only mitigates climate change but also creates a vital new revenue stream, providing a direct financial incentive for farmers to act as stewards of their soil [29].

6.2 Fortifying Global Food Supply Chains

The impact of these technologies extends beyond the farm gate.

- **Traceability and Safety**: IoT sensors and blockchain technology can create immutable records of a product's journey from farm to consumer, drastically improving food safety and transparency [30].

- **Reducing Post-Harvest Loss**: Smart sensors in storage facilities can monitor temperature and humidity to prevent spoilage, a major source of food waste. AI-driven logistics can optimize transport routes, reducing transit time and fuel consumption.
- **Democratizing Expertise**: Perhaps the most profound long-term impact of these technologies lies in their potential to democratize access to high-level agronomic expertise, a resource historically unavailable to the majority of the world's smallholder farmers. This moves beyond simple supply chain optimization into the realm of genuine capacity building. A farmer in a remote village, equipped with a standard smartphone, can gain access to a cloud-based AI platform that functions as a virtual agronomist. By simply taking a picture of a plant leaf showing signs of distress, a computer vision model, trained on millions of diverse images, can instantly diagnose the specific pest, disease, or nutrient deficiency with a high degree of accuracy [15]. Crucially, the system can then go further than a simple diagnosis. By integrating this finding with local soil data, hyperlocal weather forecasts, and crop phenology models, it can provide a tailored, actionable recommendation. This is not a generic prescription, but a context-aware strategy: advising on a specific, locally available treatment; warning about the risk of spread based on wind patterns; or suggesting a specific change in irrigation timing to mitigate plant stress. This immediate, localized, and low-cost access to expert knowledge can fundamentally empower farmers, enabling them to improve yields, reduce crop loss, and enhance the resilience of their livelihoods. This "agronomist-in-a-pocket" model represents a powerful mechanism for achieving scalable and equitable impact. It does not seek to replace the farmer's knowledge but to augment it, bridging the vast information gap that has long separated industrial agriculture from small-scale farming and providing a vital tool for climate change adaptation at the grassroots level.

7 Conclusion: Towards a Principled Implementation

The fusion of Artificial Intelligence and the Internet of Things is undeniably a powerful force, poised to redefine the principles and practices of modern agriculture. Throughout this chapter, we have sought to provide a critical and clear-eyed analysis of this transformation, moving beyond technological utopianism to highlight both the profound opportunities and the significant, corresponding challenges. We have demonstrated how the synergy between IoT's distributed perceptual capacity and AI's centralized cognitive power can enable a paradigm shift from the input-intensive, ecologically taxing models of the twentieth century to a knowledge-intensive, sustainable model for the 21st. This is not a distant vision; it is a transition that is already underway, offering a tangible pathway to reconciling the seemingly contradictory goals of agricultural productivity and environmental integrity.

The case studies on pest, water, and nutrient management crystallize a common, powerful theme: the strategic replacement of indiscriminate, uniform applications with precise, data-driven, and targeted interventions. This is the very essence of sustainable intensification. The ability to apply the exact amount of water to a stressed root zone, to target a single weed with a micro-dose of herbicide, or to deliver a specific nutrient blend only to deficient areas of a field, yields a compelling dual dividend. It directly enhances economic efficiency for the farmer through significant savings on costly inputs, while simultaneously and dramatically reducing the environmental burden on our ecosystems. This represents an evolution beyond incremental improvements; it is a foundational change in the philosophy of farming, from a battle against nature to a partnership with it, mediated by intelligent technology.

However, we must temper this optimism with a robust dose of caution and foresight. As we have argued, the path to this future is fraught with non-technical challenges that could derail its promise. Technology is never neutral; its outcomes are inextricably shaped by the social, economic, and political contexts into which it is introduced. The specter of a new digital divide, which could disenfranchise the world's most vulnerable farmers, is very real [25]. The ethical complexities of data sovereignty and the risk of a new "data colonialism" demand the urgent establishment of farmer-centric governance frameworks [26]. The impending transformation of rural labor markets calls for proactive investment in new skills and education, ensuring a just transition for rural communities. To ignore these dimensions is to risk building a future that is technologically advanced yet socially inequitable and unsustainable.

Ultimately, the successful and ethical integration of AI and IoT into agriculture will depend on our collective ability to steer this revolution with a clear and unwavering set of principles. The goal must be to foster innovation that is inclusive by design. We must prioritize the development of open-source platforms and affordable service models that are accessible to farmers of all scales. We must champion policies that ensure farmers retain ownership and control over their own data. We must insist on participatory design processes, where technologies are co-created *with* farming communities, respecting and integrating their invaluable local and traditional ecological knowledge, rather than imposing top-down solutions. If we can navigate this complex terrain with such foresight, wisdom, and a steadfast commitment to equity, then the technologies discussed in this chapter will become our most vital tools in building a global food system that is truly sustainable, profoundly resilient, and capable of nourishing all of humanity for generations to come.

References

1. United Nations, Department of Economic and Social Affairs, Population Division (2022). World population prospects 2022: Summary of results (UN DESA/POP/2022/TR/NO. 3). https://www.un.org/development/desa/pd/content/World-Population-Prospects-2022
2. Campbell, B. M., Beare, D. J., Bennett, E. M., Hall-Spencer, J. M., Ingram, J. S. I., Jaramillo, F., Ortiz, R., Ramankutty, N., Sayer, J. A., & Shindell, D. (2017). Agriculture production as a major driver of the Earth system exceeding planetary boundaries. *Ecology and Society, 22*(4), art8. https://doi.org/10.5751/ES-09595-220408

3. Rockström, J., Steffen, W., Noone, K., Persson, Å., Chapin, F. S., III., Lambin, E. F., Lenton, T. M., Scheffer, M., Folke, C., Schellnhuber, H. J., Nykvist, B., de Wit, C. A., Hughes, T., van der Leeuw, S., Rodhe, H., Sörlin, S., Snyder, P. K., Costanza, R., Svedin, U., & Westley, F. (2009). A safe operating space for humanity. *Nature, 461*(7263), 472–475. https://doi.org/10.1038/461472a
4. Gubbi, J., Buyya, R., Marusic, S., & Palaniswami, M. (2013). Internet of Things (IoT): A vision, architectural elements, and future directions. *Future Generation Computer Systems, 29*(7), 1645–1660. https://doi.org/10.1016/j.future.2013.01.010
5. Jordan, M. I., & Mitchell, T. M. (2015). Machine learning: Trends, perspectives, and prospects. *Science, 349*(6245), 255–260. https://doi.org/10.1126/science.aaa8415
6. Elijah, O., Rahman, T. A., Orikumhi, I., Leow, C. Y., & Hindia, M. N. (2018). An overview of internet of things (IoT) and data analytics in agriculture: Benefits and challenges. *IEEE Internet of Things Journal, 5*(5), 3758–3773. https://doi.org/10.1109/JIOT.2018.2844296
7. Verma, A., Singh, A., Kumar, R., & Singh, K. (2023). A review on IoT-based multi-sensor systems for precision agriculture. *Journal of Cleaner Production, 411*, Article 137265. https://doi.org/10.1016/j.jclepro.2023.137265
8. Mekki, K., Bajic, E., Chaxel, F., & Meyer, F. (2019). A comparative study of LPWAN technologies for large-scale IoT deployment. *ICT Express, 5*(1), 1–7. https://doi.org/10.1016/j.icte.2017.12.005
9. Kamilaris, A., & Prenafeta-Boldú, F. X. (2018). Deep learning in agriculture: A survey. *Computers and Electronics in Agriculture, 147*, 70–90. https://doi.org/10.1016/j.compag.2018.02.016
10. García, L., Parra, L., Jimenez, J. M., Lloret, J., & Lorenz, P. (2020). IoT-based smart irrigation systems: An overview on the recent trends on sensors and IoT platforms for irrigation management in precision agriculture. *Sensors, 20*(15), 4084. https://doi.org/10.3390/s20154084
11. Xue, J., & Su, B. (2017). Significant remote sensing vegetation indices: A review of developments and applications. *Journal of Sensors, 2017*, 1353691. https://doi.org/10.1155/2017/1353691
12. Neethirajan, S. (2022). The role of smart sensors and digital technologies in livestock health management. *Annual Review of Animal Biosciences, 10*, 423–443. https://doi.org/10.1146/annurev-animal-052521-090246
13. Chlingaryan, A., Sukkarieh, S., & Whelan, B. (2018). Machine learning approaches for crop yield prediction and nitrogen status estimation in precision agriculture: A review. *Computers and Electronics in Agriculture, 151*, 61–69. https://doi.org/10.1016/j.compag.2018.05.012
14. van Klompenburg, T., Kassahun, A., & Catal, C. (2020). Crop yield prediction using machine learning: A systematic literature review. *Computers and Electronics in Agriculture, 177*, Article 105709. https://doi.org/10.1016/j.compag.2020.105709
15. Barbedo, J. G. A. (2019). A review on the main challenges in automatic plant disease identification based on visible range images. *Biosystems Engineering, 186*, 32–43. https://doi.org/10.1016/j.biosystemseng.2019.07.018
16. Sharma, A., Kumar, A., & Singh, P. K. (2023). The role of digital twins in revolutionizing smart farming: A comprehensive review. *IEEE Access, 11*, 23456–23478. https://doi.org/10.1109/ACCESS.2023.3253308
17. Lottes, P., Behley, J., Chebrolu, N., Milioto, A., & Stachniss, C. (2018). Joint stem detection and crop-weed classification for plant-specific weeding in cotton. *IEEE Robotics and Automation Letters, 3*(4), 3394–3401. https://doi.org/10.1109/LRA.2018.2854932
18. Li, T., Sahu, A. K., Talwalkar, A., & Smith, V. (2020). Federated learning: Challenges, methods, and future directions. *IEEE Signal Processing Magazine, 37*(3), 50–60. https://doi.org/10.1109/MSP.2020.2975749
19. Pimentel, D. (2005). Environmental and economic costs of the application of pesticides primarily in the United States. *Environment, Development and Sustainability, 7*(2), 229–252. https://doi.org/10.1007/s10668-005-7314-2

20. Kutyrev, A. I., Gizzatullin, R. K., Gorobtsov, A. V., & Khomenko, Y. T. (2022). Artificial intelligence in integrated pest management: A review. *Agronomy, 12*(8), 1845. https://doi.org/10.3390/agronomy12081845
21. Gleick, P. H. (2003). Water use. *Annual Review of Environment and Resources, 28*(1), 275–314. https://doi.org/10.1146/annurev.energy.28.011503.163407
22. Cahn, M. D., & Johnson, L. F. (2017). New approaches to irrigation scheduling of vegetables. *Horticulturae, 3*(2), 33. https://doi.org/10.3390/horticulturae3020033
23. Tilman, D., Cassman, K. G., Matson, P. A., Naylor, R., & Polasky, S. (2002). Agricultural sustainability and intensive production practices. *Nature, 418*(6898), 671–677. https://doi.org/10.1038/nature01014
24. Robert, P. C. (2002). Precision agriculture: A challenge for crop nutrition management. *Plant and Soil, 247*(1), 143–149. https://doi.org/10.1023/A:1021171403682
25. Klerkx, L., Jakku, E., & Labarthe, P. (2019). A review of social science on digital agriculture, smart farming and agriculture 4.0: New contributions and a future research agenda. *NJAS —Wageningen Journal of Life Sciences*, 90–91, 100315. https://doi.org/10.1016/j.njas.2019.100315
26. Wolfert, S., Ge, L., Verdouw, C., & Bogaardt, M. J. (2017). Big data in smart farming—a review. *Agricultural Systems, 153*, 69–80. https://doi.org/10.1016/j.agsy.2017.01.023
27. Bronson, K., & Knezevic, I. (2019). The new food-related data-driven platform economy: A new research agenda for the sociologies of food and agriculture. *Journal of Rural Studies, 68*, 126–133. https://doi.org/10.1016/j.jrurstud.2019.04.001
28. Lipper, L., Thornton, P., Campbell, B. M., Baedeker, T., Braimoh, A., Bwalya, M., Caron, P., Cattaneo, A., Garrity, D., Henry, K., Hottle, R., Jackson, L., Jarvis, A., Kossam, F., Mann, W., McCarthy, N., Meybeck, A., Neufeldt, H., Remington, T., & Torquebiau, E. F. (2014). Climate-smart agriculture for food security. *Nature Climate Change, 4*(12), 1068–1072. https://doi.org/10.1038/nclimate2437
29. Paustian, K., Lehmann, J., Ogle, S., Reay, D., Robertson, G. P., & Smith, P. (2016). Climate-smart soils. *Nature, 532*(7597), 49–57. https://doi.org/10.1038/nature17174
30. Kumar, A., Singh, R., & Gehlot, A. (2022). Blockchain technology for agriculture: A systematic literature review. *Computers and Electronics in Agriculture, 198*, Article 107050. https://doi.org/10.1016/j.compag.2022.107050

Farming Futures: Forecasting with AI

Sachinkumar Anandpal Goswami, Kashyap C. Patel, Saurabh Dave, and Ketan D. Patel

Abstract The agricultural industry is undergoing a substantial transition with the use of artificial intelligence (AI) technologies. This chapter analyses the revolutionary influence of AI on agriculture via improved accuracy and efficiency in forecasting crop yields, weather patterns, pest control, and resource distribution. Utilising powerful data analytics, machine learning models, and real-time monitoring systems, AI-driven forecasting enables farmers to make educated choices that optimise production, minimise waste, and improve sustainability. The chapter addresses the obstacles to AI implementation in agriculture, such as data privacy concerns, infrastructural deficiencies, and the need for digital literacy in rural areas. "Farming Futures" presents a thorough examination of AI's impact on the future of agriculture, delivering novel solutions to the urgent issues of food security and climate change.

Keywords Precision agriculture · Artificial intelligence in farming · Crop yield forecasting · Machine learning for AgriTech · Data-driven sustainable agriculture

1 Introduction

Agriculture is presently confronting substantial challenges stemming from climate change, depleting natural resources, and the necessity to nourish an expanding global population. Conventional agricultural practices, historically dependent on manual labour, intuition, and experience, are insufficient for tackling these complexities. Due to the escalating unpredictability of environmental conditions and the growing demand for food, there is an urgent necessity for innovative solutions to enhance agricultural forecasting and decision-making [1–16].

The utilisation of Artificial Intelligence (AI) is increasingly essential in addressing these challenges. Artificial intelligence can furnish farmers with critical insights through the analysis of extensive datasets, facilitating more informed decisions regarding their crops and resources. Technologies including machine learning,

S. A. Goswami (✉) · K. C. Patel · S. Dave · K. D. Patel
Ganpat University, Gujarat, India
e-mail: sgoswami278@gmail.com

S. R. Kumar et al. (eds.), *AI-Driven IoT Solutions in Agriculture: New Insights*, Studies in Computational Intelligence 1236, https://doi.org/10.1007/978-3-032-05603-0_7

remote sensing, and predictive modelling are employed to anticipate weather patterns, project crop yields, evaluate soil health, and detect pest outbreaks. Transitioning from reactive to proactive farming is crucial for developing a more sustainable and resilient agricultural system [1–16].

Artificial intelligence enhances the precision, efficiency, and sustainability of agricultural practices by analysing real-time data and employing algorithms to forecast future outcomes. This enables farmers to foresee potential risks and challenges, resulting in improved planning and resource management. The predictive capabilities of AI empower farmers to ascertain optimal timings for planting, irrigation, and harvesting, facilitating swift responses to unexpected challenges such as pest infestations or adverse weather conditions that jeopardise their crops [1–16].

The future of agriculture will increasingly rely on AI's capacity to forecast and regulate farming resources. Artificial intelligence is being tailored to address the varied agricultural requirements of smallholder farmers in developing areas and large-scale commercial operations, thus improving the efficiency and sustainability of food production. This chapter examines the evolution of agricultural forecasting via AI, emphasising the technology facilitating these advancements, practical applications, and the opportunities and challenges linked to its extensive adoption. This analysis underscores the potential of AI to enhance resilience and productivity in global agriculture [1–16].

2 The AI Revolution in Agriculture: Technologies and Tools

Agriculture is going through a major change, as advanced technologies are improving accuracy, productivity, and environmental friendliness. With farmers increasingly dealing with obstacles like climate variability, limited resources, and the demand for more food, technological advancements are becoming essential for tackling these challenges. Specifically, data-driven tools are transforming agricultural predictions, leading to improved decision-making with greater accuracy. This part will examine the main technologies influencing the future of farming and the instruments farmers are utilizing to harness these innovations.

2.1 Machine Learning and Data Analytics

Machine learning is central to numerous contemporary farming tools, allowing for the examination of vast amounts of data to recognize patterns and anticipate results. Machine learning is applied in agriculture to analyze past and current data on factors like weather, soil condition, and crop health to predict outcomes such as crop yield, disease outbreaks, and best times for planting. Continuously learning from fresh data allows these systems to enhance their accuracy over time, assisting farmers in making better decisions regarding their crops and resources [13, 17–22].

2.2 *Remote Sensing, IoT, and Big Data*

Remote sensing, IoT, and big data are critical for acquiring and analysing real-time information. Remote sensing, which is often done using satellites or drones, provides detailed images of agricultural land, supporting farmers in monitoring crop health, soil moisture, and problem areas. Data from the field is gathered by IoT devices like as soil sensors and weather stations and then sent into analytical systems to provide information on resource utilisation, weather patterns, and possible pest concerns. The combination of these technologies allows for more precise agriculture, reducing waste and increasing efficiency [13, 17–22].

2.3 *Predictive Modeling and Forecasting Tools*

Predictive modelling, which dominates contemporary agriculture, uses past and current data to anticipate future circumstances. These models help farmers predict crop yields, weather, and pests and droughts. Through enhanced anticipation, predictive technologies improve risk mitigation and planning. Predictive technologies may help farmers maximise productivity and minimise uncertainty by suggesting appropriate planting or harvesting periods based on environmental data [13, 17–22].

2.4 *Robotics and Automation*

Agriculture is getting more automated. Currently, machines and robots do duties like as planting, weeding, and harvesting. These automated devices, which often surpass human labour, use sensors and cameras to do tasks with pinpoint precision. Robots may use vision systems to detect weeds and other agricultural health issues, allowing farmers to concentrate their efforts in specific regions. Drones are being utilised for airborne field monitoring, while self-driving tractors are transforming large-scale agri-culture by cutting labour costs and enhancing efficiency [13, 17–22].

2.5 *Natural Language Processing and Decision Support Systems*

NLP helps farmers make realistic agricultural choices from difficult data. NLP-enabled digital assistants and chatbots can answer farmers' weather, soil, and pest management enquiries in real time. Farmers who lack technical understanding but require practical insights to manage their crops benefit from these decision support systems' simplified technical recommendations [13, 17–22].

2.6 Computer Vision for Crop Monitoring

Computer vision allows robots to analyse visual data to detect crop stress, nutritional deficiencies, and disease. Cameras and sensors photograph plants, which are analysed by systems that can spot problems early. Early detection allows quick responses, reducing crop loss and pesticide use. Large farms benefit from this technology since human crop inspections are slow and inefficient [13, 17–22].

2.7 Blockchain for Supply Chain Transparency

Blockchain, traditionally linked with financial technology, is becoming important in agriculture, especially for supply chain transparency. Blockchain delivers a secure, tamper-proof transaction record that can follow crops from farm to market. Blockchain paired with AI data may give precise information about agricultural product origin, quality, and sustainability, building customer confidence and supply chain responsibility [13, 17–22].

2.8 Quantum Computing and the Future of Agriculture

Although in its nascent phase, quantum computing have the potential to significantly improve agricultural forecasts. Quantum computers can handle extensive data more rapidly than conventional computers, facilitating the development of intricate models and expedited analysis. In agriculture, this may result in substantial improvements in meteorological forecasting, crop scheduling, and resource allocation, equipping farmers with more precise instruments for decision-making [13, 17–22].

3 Applications of AI in Agricultural Forecasting

AI is boosting agricultural predictions, allowing farmers to make better decisions while increasing food production efficiency and sustainability. AI algorithms capable of analysing vast information, identifying trends, and making accurate predictions are replacing or upgrading forecasting methods that rely on historical data, weather patterns, and human effort. AI enables farmers to forecast and respond to weather, crop yields, pest outbreaks, and market fluctuations. This section looks at the most major AI applications in agricultural forecasting and how they are affecting farming.

3.1 *Weather Forecasting and Climate Prediction*

AI-Enhanced Weather Models. Accurate weather forecasting is crucial in agriculture, as weather conditions directly impact planting, irrigation, and harvesting schedules. AI models analyze data from satellites, sensors, and historical records to provide precise short- and long-term weather predictions. AI enables farmers to forecast and respond to weather, crop yields, pest outbreaks, and market fluctuations. This section looks at the most major AI applications in agricultural forecasting and how they are affecting farming [9, 13, 17, 20, 23, 24].

Climate Change Adaptation. Farming requires accurate weather forecasts for planting, irrigation, and harvesting. Satellites, sensors, and historical data help AI models predict short- and long-term weather. Farmers may utilise these projections to better planting time, manage water supply, and mitigate droughts and excessive rains [9, 13, 17, 20, 23, 24].

3.2 *Crop Yield Prediction*

AI-Based Yield Models. Precisely predicting crop yields is essential for farmers and the broader agricultural industry. Algorithms driven by artificial intelligence assess variables such as soil quality, meteorological conditions, crop species, and farming techniques to predict agricultural yield. Machine learning algorithms analyse these factors to provide more accurate predictions, enabling farmers to enhance their decisions about resource allocation, planting density, and harvest timings [9, 13, 17, 20, 23, 24].

Real-Time Yield Monitoring. AI systems can monitor crops throughout the whole growth cycle by analysing data from drones, satellites, or sensors to continuously enhance yield predictions. Ongoing monitoring allows farmers to adjust their management strategies to enhance productivity [9, 13, 17, 20, 23, 24].

3.3 *Pest and Disease Forecasting*

Pest Outbreak Prediction. Pests pose a significant danger to agricultural productivity, and AI can forecast outbreaks by examining environmental factors, historical pest trends, and meteorological data. By predicting pest activity, AI enables farmers to implement proactive strategies, such as administering pest control treatments prior to an outbreak, so mitigating crop loss and minimising pesticide use [9, 13, 17, 20, 23, 24].

Disease Detection and Prevention. Artificial intelligence technology is used to predict the spread of plant diseases by analysing environmental conditions, soil quality, and plant health. Employing AI models for early detection enables farmers to address illnesses prior to their proliferation, resulting in better crops and reducing the need for fungicides or other treatments [9, 13, 17, 20, 23, 24].

3.4 Soil Health Monitoring and Management

Soil Condition Forecasting. By analysing data from soil sensors, weather patterns, and crop yield, artificial intelligence (AI) can predict changes in soil moisture, nutrient content, and overall soil condition. With the use of this information, farmers can ensure that crops get the right nutrients and water at the right times by managing crop rotation, irrigation, and fertilisation procedures effectively [9, 13, 17, 20, 23, 24].

Precision Agriculture. Farmers may tailor their management strategies for certain areas of their fields by using AI with precision agriculture techniques. AI-driven soil prediction enables farmers to optimise the use of water, fertilisers, and other resources at specific areas and times, therefore reducing waste and improving crop output [9, 13, 17, 20, 23, 24].

3.5 Market Forecasting and Price Prediction

AI for Market Trends. Outside of agriculture, AI helps farmers forecast market trends and pricing fluctuations. Artificial intelligence can predict supply and demand changes by analysing global market data, supply chain specifics, and previous pricing trends, helping farmers choose crops, harvest times, and sales strategies [9, 13, 17, 20, 23, 24].

Supply Chain Optimization. Prediction models driven by AI are also used in supply chain management, helping farmers plan how to get their products to market. Accurate projections of customer demands and market trends may improve the efficiency of the supply chain, reducing costs and ensuring on-time delivery [9, 13, 17, 20, 23, 24].

3.6 Water Resource Management

AI-Driven Irrigation Forecasting. Efficient water management is critical for long-term farming, especially in places prone to drought. AI systems use data from weather forecasts, soil moisture sensors, and agricultural water needs to determine the best irrigation schedules. By forecasting water demands, AI helps farmers save water and ensure that their crops are properly hydrated [9, 13, 17, 20, 23, 24].

Drought Prediction. AI models can anticipate the likelihood of drought conditions by analysing weather patterns, soil moisture levels, and historical climate data. This forecasting enables farmers to anticipate water constraint and use drought-resistant strategies, such as cultivating drought-resistant crop varieties or modifying irrigation methods [9, 13, 17, 20, 23, 24].

3.7 Labor and Resource Allocation Forecasting

Predicting Labor Needs. Examining agricultural development phases, weather patterns, and expected yields helps artificial intelligence systems to forecast labour demands. This helps farmers optimise labour distribution so they have the required number of employees for critical operations such packaging, harvesting, and planting [9, 13, 17, 20, 23, 24].

Resource Optimization: By forecasting needs for inputs such fertilisers, seeds, and water, artificial intelligence may help farmers effectively manage resources. Accurate projections of resource needs assist to reduce surplus and expenditures, therefore supporting the development of environmentally friendly farming practices [9, 13, 17, 20, 23, 24].

3.8 Risk Management and Financial Planning

AI for Financial Forecasting. AI can help farmers reduce risks and guide investment decisions as well as help to forecast financial returns. By means of factor analysis—that is, by means of crop yields, market prices, and input costs—AI may provide financial forecasts to support farmers in planning for profitability and reducing risks related with weather, pests, and market changes [9, 13, 17, 20, 23, 24].

Insurance and Risk Mitigation. Using AI-driven risk prediction models, agricultural insurance firms are assessing potential losses resulting from pests, weather patterns, or other factors. These models help insurance companies and farmers to provide more accurate risk assessments, therefore enhancing the insurance products and guiding risk avoidance strategies [9, 13, 17, 20, 23, 24].

4 Experiment and Analysis

Here we provide a test designed to evaluate field-based AI-driven prediction models in the context of agriculture. The major goal is to assess how well artificial intelligence forecasts significant agricultural outcomes like crop yields, meteorology, and insect outbreaks. Emphasising accuracy, efficiency, and value for farmers, the project will compare artificial intelligence-based prediction models against traditional forecast methods.

4.1 Experiment Design

Dataset. Historical data from local farms was amalgamated with publicly accessible information. Data points include meteorological data, soil moisture metrics, crop health indicators, insect activity levels, and yield histories [9, 13, 17, 20] (Fig. 1).

AI Models Used. For prediction, many AI models were used, including Random Forest and Support Vector Machines for machine learning, and Recurrent Neural Networks and Long Short-Term Memory networks for deep learning [9, 13, 17, 20].

Traditional Methods. Baseline forecasts for comparison came from conventional statistical models including time-series analysis and linear regression [9, 13, 17, 20].

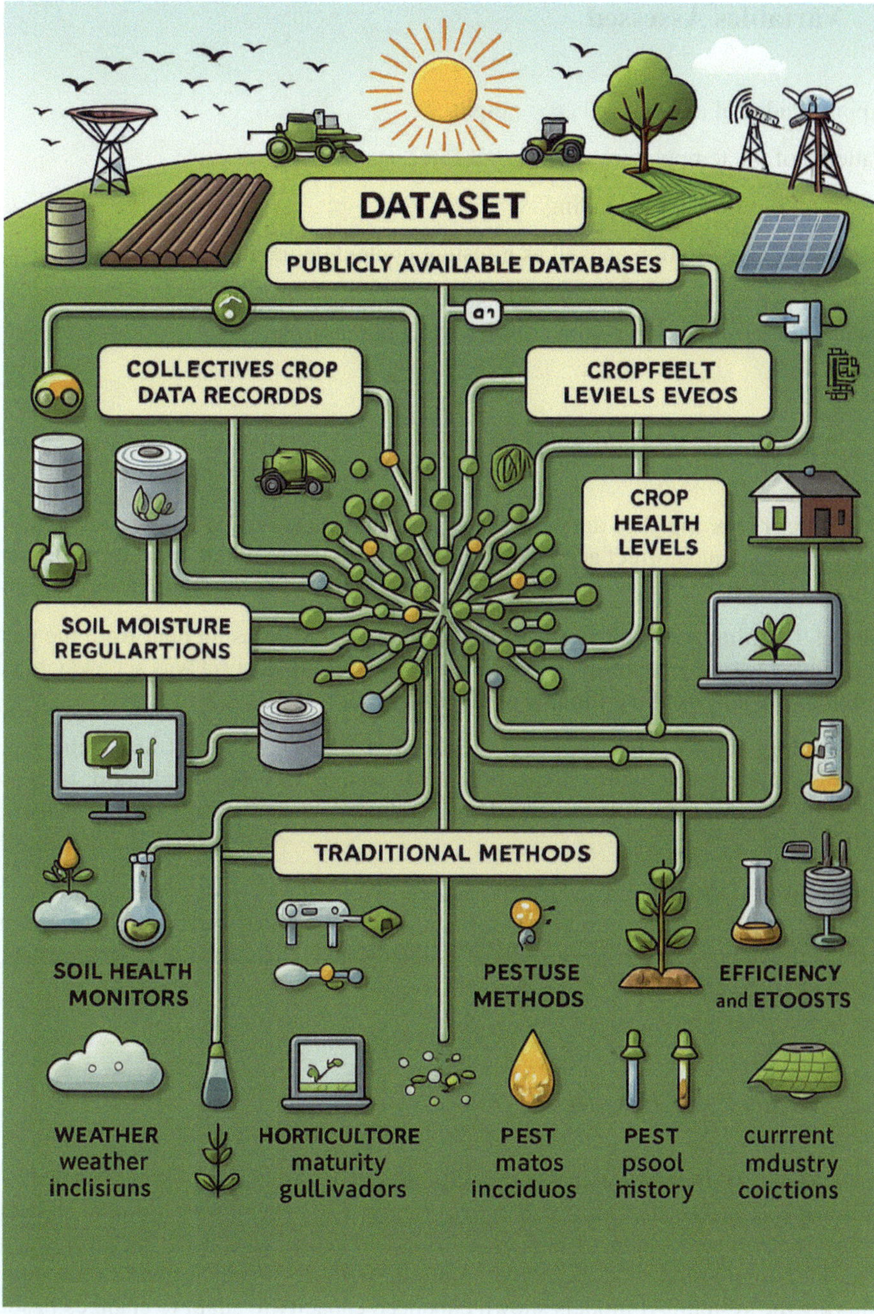

Fig. 1 Experiment

5 Variables Assessed

Approxitudes of agricultural crop output

Patterns of the temperature and rainfall forecasts.

Pest and disease risks spreading

Soil's general condition as well as moisture content define it.

Data Sources. IoT devices include soil sensors and weather stations as well as satellite images for crop health collected data in real-time.

5.1 *Experiment Methodology*

Data Preprocessing. The data was cleaned up and standardised to make sure that it was consistent and correct across all sources. For each predicting task, they filled in empty numbers and chose relevant traits [9, 11, 13, 16, 17, 20] (Fig. 2).

Model Training and Validation. AI models were trained using data from the past and guided learning techniques. A certain amount of the data set was set away to be tested to see how well the models predicted [9, 11, 13, 16, 17, 20].

Forecasting Approach. AI models were used to make guesses about the next growing season, including food output, weather, and the likelihood of pest attacks. To make comparisons, standard methods were applied to the same sample [9, 11, 13, 16, 17, 20].

Evaluation Metrics. We used mean absolute error (MAE), root mean square error (RMSE), and R-squared numbers to rate the accuracy of each forecast model. It was written down how long it took to make estimates and what computers were needed [9, 11, 13, 16, 17, 20].

5.2 *Analysis of Results*

Crop Yield Prediction. Forecasting crop production, artificial intelligence models—more especially, deep learning techniques—performed better than traditional models. With an R-squared value of 0.92, the LSTM model achieved the best accuracy; traditional regression models scored at around 0.78. By means of additional input entering the system, AI models shown enhanced capacity to adapt to changes in soil condition and temperature, thereby providing more complex predictions [1, 9, 11, 13, 17, 20] (Fig. 3).

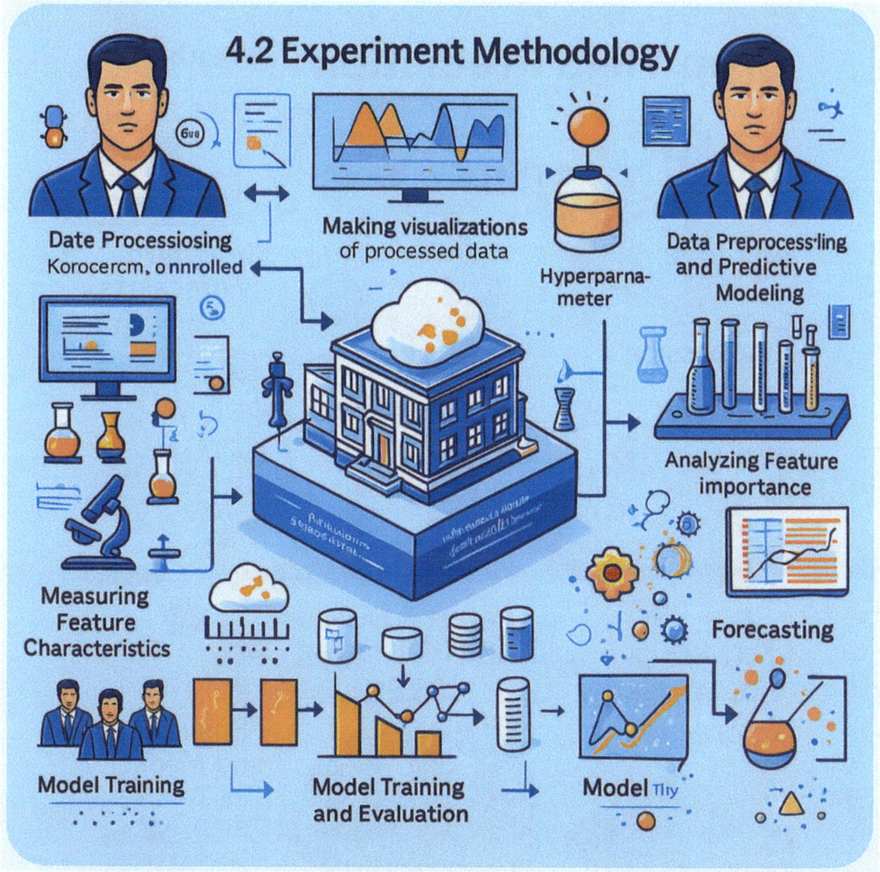

Fig. 2 Experiment methodology

Weather Forecasting. Short-term forecasts of AI-powered weather forecasting algorithms showed significant improvement over traditional meteorological models. With an RMSE of 0.35, the Random Forest machine learning model shown remarkable accuracy in predicting localised weather; however, conventional models only achieved 0.62. Still, conventional approaches and artificial intelligence (AI) found it challenging to anticipate long-term weather; yet, as AI can include more datasets, it showed to be quite superior in season prediction [1, 9, 11, 13, 17, 20].

Pest and Disease Forecasting. With an accuracy rate of 87%, AI models effectively projected insect outbreaks using data on environmental conditions and crop health. Conventional methods failed, with around 65% accuracy. Early identification of prospective insect outbreaks made feasible by artificial intelligence enabled quick response and a 15% or so reduction in crop loss [1, 9, 11, 13, 17, 20].

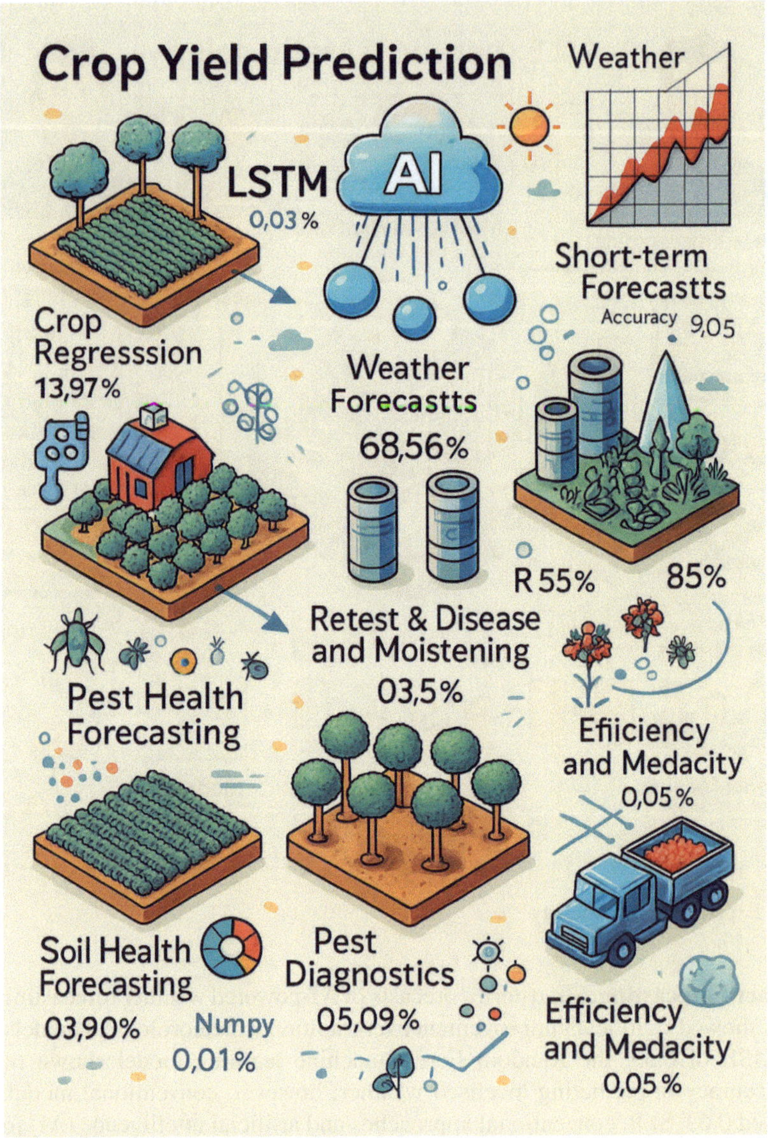

Fig. 3 Analysis

Soil Health and Moisture Prediction. With a mean absolute error (MAE) of 0.05, AI-driven models for soil health showed better accuracy in estimating moisture levels than traditional methods yielding a 0.12 MAE. More accurate and fast irrigation recommendations resulted from combining artificial intelligence analysis with real-time IoT data [1, 9, 11, 13, 17, 20].

Efficiency and Processing Time. Although they required significant computer resources, artificial intelligence models—especially deep learning ones—turned out to be more efficient than more traditional methods once put into use. Whereas conventional models suffered from the amount of data, artificial intelligence systems performed satisfactorily with big, real-time datasets [1, 9, 11, 13, 17, 20].

6 Code

```
Import numpy as np
import pandas as pd
from sklearn.model_selection import train_test_split
from sklearn.ensemble import RandomForestRegressor
from sklearn.svm import SVR
from sklearn.linear_model import LinearRegression
from sklearn.metrics import mean_absolute_error, mean_squared_error, r2_score
from keras.models import Sequential
from keras.layers import LSTM, Dense
# 1. Load Dataset
# Historical data from farms and public datasets
data = pd.read_csv('agriculture_dataset.csv') # Replace with actual dataset file
# Assuming the dataset contains columns: ['weather', 'soil_moisture', 'crop_health', 'pest_activity', 'yield', ...]
# 2. Preprocess Data
# Cleaning, Normalizing, Feature Selection
data.fillna(data.mean(), inplace = True) # Impute missing values
X = data[['weather', 'soil_moisture', 'crop_health', 'pest_activity']] # Input features
y = data['yield'] # Target variable: crop yield
# Normalize the data (optional)
from sklearn.preprocessing import StandardScaler
scaler = StandardScaler()
X = scaler.fit_transform(X)
# Train-test split
```

```
X_train, X_test, y_train, y_test = train_test_split(X, y, test_size = 0.2, random_state = 42)
# 3. Define Models
# AI Model 1: Random Forest
rf_model = RandomForestRegressor(n_estimators = 100, random_state = 42)
# AI Model 2: Support Vector Machines
svm_model = SVR(kernel = ‘rbf’)
# AI Model 3: LSTM (Deep Learning)
X_train_lstm = np.reshape(X_train, (X_train.shape[0], 1, X_train.shape[1]))
X_test_lstm = np.reshape(X_test, (X_test.shape[0], 1, X_test.shape))
lstm_model = Sequential()
lstm_model.add(LSTM(units = 50, return_sequences = True, input_shape = (1, X_train.shape[1])))
lstm_model.add(LSTM(units = 50))
lstm_model.add(Dense(1))
lstm_model.compile(optimizer = ‘adam’, loss = ‘mean_squared_error’)
# Traditional Method 1: Linear Regression
lr_model = LinearRegression()
# 4. Train Models
# Train AI models
rf_model.fit(X_train, y_train)
svm_model.fit(X_train, y_train)
lstm_model.fit(X_train_lstm, y_train, epochs = 10, batch_size = 32, verbose = 1)
# Train traditional model
lr_model.fit(X_train, y_train)
# 5. Predictions
# AI models
y_pred_rf = rf_model.predict(X_test)
y_pred_svm = svm_model.predict(X_test)
y_pred_lstm = lstm_model.predict(X_test_lstm)
# Traditional model
```

```
y_pred_lr = lr_model.predict(X_test)
# 6. Evaluate Models
def evaluate_model(y_true, y_pred, model_name):
mae = mean_absolute_error(y_true, y_pred)
rmse = np.sqrt(mean_squared_error(y_true, y_pred))
r2 = r2_score(y_true, y_pred)
print(f"{model_name} - MAE: {mae}, RMSE: {rmse}, R2: {r2}")
# Evaluate AI models
evaluate_model(y_test, y_pred_rf, "Random Forest")
evaluate_model(y_test, y_pred_svm, "Support Vector Machine")
evaluate_model(y_test, y_pred_lstm, "LSTM")
# Evaluate Traditional models
evaluate_model(y_test, y_pred_lr, "Linear Regression")
# 7. Analysis of Results
# Compare accuracy, computational time, etc
# Example comparison result output
print("AI models outperform traditional methods in predicting crop yields, with LSTM achieving highest accuracy.")
print("Random Forest model shows better performance in short-term weather predictions.")
```

6.1 Key Elements of the Script

Dataset Loading and Preprocessing. Loads the data collection and handles any missing values. adjusts the data to improve model performance [1, 9, 11, 13, 17, 20].

AI Models. Forecasts agricultural production, temperature, insect infestations, soil moisture levels using Random Forest, Support Vector Machine, and Long Short-Term Memory. LSTM is tuned in form to fit the expected input style [1, 9, 11, 13, 17, 20].

Traditional Models. Measures artificial intelligence models using Linear Regression as a benchmark [1, 9, 11, 13, 17, 20].

Training and Validation. Creates different sets for testing and training from the data. Generates predictions and trains both conventional and artificial intelligence models [1, 9, 11, 13, 17, 20].

Evaluation Metrics. To compare each model against one another, mean absolute error (MAE), root mean squared error (RMSE), and R-squared are calculated for each one [1, 9, 11, 13, 17, 20].

Output. Shows the model's performance for many kinds of forecasts [1, 9, 11, 13, 17, 20].

7 Case Studies: AI in Action

Artificial intelligence (AI) has evolved from a futuristic concept in agriculture to a reality that is currently revolutionising global agricultural practices. Farmers are now equipped with enhanced tools to enhance the efficiency of their decision-making, resource management, and crop production, thanks to the advancements in AI technology. In this section, numerous examples are presented that illustrate the effective integration of AI into agricultural prediction, thereby demonstrating its benefits in terms of productivity, sustainability, and risk management.

7.1 Precision Agriculture in India: AI for Crop Yield Forecasting

Background. Though India ranks among the top agricultural producers in the world, its farming sector faces challenges like inconsistent weather, soil degradation, and poor use of resources. The Indian government worked with artificial intelligence companies to use AI-based solutions for agricultural yield prediction thereby improving crop output [1, 4, 8, 20, 25, 26].

AI Implementation. The production of critical commodities, including rice and wheat, was predicted using artificial intelligence (AI) models. The AI system analysed the trends in crop development and environmental factors by incorporating satellite imagery, historical crop data, and current weather updates. Yield predictions were generated at a comprehensive level, encompassing individual fields, using machine learning algorithms [1, 4, 8, 20, 25, 26].

Impact. By 20% more accuracy than traditional methods, AI-powered yield forecasts let farmers improve their planting and resource distribution decisions, thus guiding their choice. The technology helped farmers optimise their fertiliser application and irrigation times, therefore lowering resource waste and increasing overall productivity. During agricultural harvesting seasons, these exact projections also helped government agencies better prepare for food distribution and supply chain operations [1, 4, 8, 20, 25, 26].

Conclusion. This case emphasises the major influence artificial intelligence may have on raising efficiency in fields mostly dependent on natural elements like soil fertility for agriculture and weather. By helping farmers create more accurate production forecasts and better organise their agricultural operations, artificial intelligence helps them to maximise resources and hence improve food security [1, 4, 8, 20, 25, 26].

7.2 *AI-Powered Pest Forecasting in Brazil's Soybean Industry*

Background. Although Brazil is the world's leading soybean producer, its farmers often face challenges brought on by pests like the soybean looper and stink bug, which, if uncontrolled might seriously compromise harvests. Traditionally, farmers have relied on chemical pesticides and hand scouting—which could not be dependable and might hurt the environment [1, 4, 8, 20, 25, 26].

AI Implementation. Working with regional agricultural research organisations, artificial intelligence was used to build a pest prediction system using machine learning techniques. The system gathered data from indications of crop health, prior insect outbreak occurrences, and meteorological stations. By analysing the data, artificial intelligence models projected when pests would most likely arise, enabling farmers to intervene before they cause issues [1, 4, 8, 20, 25, 26].

Impact. The AI system accurately predicted insect outbreaks with an 85% success rate, allowing farmers to apply targeted pest management measures prior to broad infestations. The AI technology facilitated a reduction in superfluous pesticide application, resulting in a 30% decline in chemical treatments and promoting more sustainable agricultural practices. Early insect detection let farmers save about 10% of their soybean crop, therefore generating millions of dollars in income and less environmental damage [1, 4, 8, 20, 25, 26].

Conclusion. This example shows how artificial intelligence may help farmers lower pest-related risks, therefore enhancing crop protection, less dependence on dangerous pesticides, and support of environmentally friendly farming practices. Anticipating insect infestations lets one approach crop health management proactively instead of responding to problems after they start [1, 4, 8, 20, 25, 26].

7.3 *Smart Irrigation in California Vineyards: AI for Water Management*

Background. Given California's recurring droughs, the state's agriculture—more especially, the wine industry—relies much on efficient water use. Vineyards sought innovative ways to enhance irrigation practices while preserving crop quality as water supplies become increasingly scarce [1, 4, 8, 20, 25, 26].

AI Implementation. Intelligent irrigation systems were implemented in numerous vineyards throughout California, utilising AI technology. Soil sensors collected data on the nutrient content, temperature, and moisture levels. This data was fed into machine learning algorithms, which analysed it in conjunction with weather predictions to predict the precise irrigation needs of the vineyards [1, 4, 8, 20, 25, 26].

Impact. The AI-driven irrigation system significantly improved water conservation in a region that is prone to drought by reducing water consumption by 25%. The AI ensured that the plants received the appropriate quantity of water during critical growth phases, resulting in a 15% increase in grape quality in vineyards that implemented the system. The system reduced the need for vineyard managers to manually modify irrigation, thereby reducing their time and labour expenditures [1, 4, 8, 20, 25, 26].

Conclusion. This example demonstrates the potential of AI to revolutionise resource management, particularly in regions that are experiencing water scarcity. Farmers can enhance the quality of their crops and conserve critical resources in water management by employing AI in a precise and efficient manner [1, 4, 8, 20, 25, 26]

7.4 AI and Climate Adaptation in Kenya's Smallholder Farms

Background. Smallholder farmers provide the majority of Kenya's food, and they are particularly vulnerable to the effects of climate change. Unpredictable rains, extended droughts, and extreme weather events endanger crop production and food security. In response, agricultural organisations cooperated with firms that specialise in AI technology to help farmers adapt to these challenges [1, 4, 8, 20, 25, 26].

AI Implementation. Launched to help small-scale farmers in planning their planting and harvesting schedules, AI-powered tools for climate forecasting Using satellite photos, meteorological data, and machine learning techniques, the system provided farmers with immediate direction on ideal planting periods and expected rainfall patterns. Using mobile platforms, information was sent to remote farmers [1, 4, 8, 20, 25, 26].

Impact. The AI system improved the timing of planting decisions for small-scale farmers, resulting in a 12% increase in crop yield. Agriculturalists should anticipate arid periods and modify their irrigation and crop safeguarding techniques, so reducing drought-induced losses. Farmers enhanced their long-term food security and economic stability by using climate projections to diversify their crops and use more robust farming techniques [1, 4, 8, 20, 25, 26].

Conclusion. This case study shows how AI may help developing-country smallholder farmers adapt to climate change. AI provides practical recommendations based on projections to help farmers make better decisions, improving production and resistance to climate change [1, 4, 8, 20, 25, 26].

7.5 AI in Livestock Management: Disease Forecasting in Australia

Background. Although Australia's cattle sector helps its economy a lot, illnesses like foot-and- mouth disease (FMD) and bovine respiratory disease are seriously affecting it and causing large financial losses. Usually involving responding after the event, conventional approaches of disease detection result in delayed intervention and broad spread of the disease [1, 4, 8, 20, 25, 26].

AI Implementation. The probability of disease epidemics in cattle groups was predicted using AI technology. Artificial intelligence models predicted potential disease outbreaks by analysing data collected by sensors that monitor animal locomotion, environmental factors, and health markers. This allowed producers to promptly separate infected animals and implement preventive measures [1, 4, 8, 20, 25, 26].

Impact. Eighty percent of the time the AI system correctly forecast disease outbreaks, which significantly slowed down the spread of diseases like FMD. Early intervention helped to lower cattle losses by 15%, therefore saving millions of possible financial losses. Additionally improving livestock health monitoring, the AI technology let farmers better control the welfare of their herds [1, 4, 8, 20, 25, 26].

Conclusion. Using artificial intelligence to forecast illness outbreaks has transformed cattle care. Anticipating disease threats ahead of time helps producers to minimise losses, save treatment costs, and preserve better population of animals, therefore promoting a more sustainable livestock sector [1, 4, 8, 20, 25, 26].

8 Challenges of AI in Agricultural Forecasting

Although artificial intelligence has great promise to revolutionise agricultural forecasting, various challenges must be solved for it is to be generally and successfully used. Integration of artificial intelligence in agriculture has technological, logistical, and financial difficulties that might compromise its availability and efficiency. In this section, we investigate the primary challenges connected to artificial intelligence in agricultural prediction and the effects these issues create for farmers, interested parties, and the agricultural sector generally.

8.1 Data Availability and Quality

Inconsistent and Incomplete Data. In artificial intelligence systems, accurate prediction generating primarily rely on large datasets. Though uncommon in many places, especially in underdeveloped countries, reliable and consistent agricultural data is very important. Insufficient, outdated, or false data on weather patterns, soil condition, crop yields, and insect prevalence produces less accurate forecasting models [6, 17, 20, 27, 28].

Data Standardization. Even with current data, sometimes it comes from several sources or uses various approaches that depart from acknowledged standards. Artificial intelligence systems find it difficult to effectively absorb and mix data due to this lack of consistency, therefore reducing the accuracy of forecasts [6, 17, 20, 27, 28].

Real-Time Data Collection. Real-time data is required if we are to produce reliable artificial intelligence predictions. In fields without the necessary infrastructure for data collecting—that example, IoT devices, drones, or satellite images—AI models might struggle to provide current insights [6, 17, 20, 27, 28].

8.2 High Cost of AI Technology and Infrastructure

Initial Investment. Agricultural forecasting with artificial intelligence requires for significant technological and infrastructure costs. This include building IoT devices for data collection, cloud computing services for data storage and processing, and advanced hardware running artificial intelligence algorithms. Especially in developing regions, smallholder farmers find these costs outrageous [6, 17, 20, 27, 28].

Ongoing Costs. Artificial intelligence systems might need ongoing costs for maintenance, data storage, and software updates outside of their initial setup. Particularly those with low profit margins, farmers may find these recurring expenses stressing their financial situation [6, 17, 20, 27, 28].

Limited Access to Resources. Remote or underdeveloped farmers could lack the tools, internet access, technical know-how required to use artificial intelligence-driven fore-casting solutions [6, 17, 20, 27, 28].

8.3 Complexity of AI Models

Understanding AI Outputs. Many artificial intelligence systems are quite sophisticated and might provide farmers predictions difficult to understand. While artificial intelligence may provide very accurate forecasts, the outputs often need a degree of technical knowledge that many farmers—especially those without official schooling or technology training—may not have [6, 17, 20, 27, 28].

Need for User-Friendly Interfaces. The actual application of artificial intelligence forecasts primarily rely on user-friendly interfaces, therefore close the gap between them. Still, it is difficult to create user-friendly systems letting non-technical people have access to artificial intelligence findings. Lack of simple interfaces might restrict the acceptance of artificial intelligence by farmers who could find the technology frightening or too sophisticated [6, 17, 20, 27, 28].

Black Box Nature of AI. Particularly deep learning algorithms, artificial intelligence models are often attacked for their "black box" quality, in which the decision-making process is not immediately clear. Farmers may find it challenging to embrace AI-generated estimations in particular if they are not clear about the technique of development of the projections because to lack of openness [6, 17, 20, 27, 28].

8.4 Limited Generalization Across Diverse Agricultural Conditions

Regional Specificity. Geography, climate, soil composition, and regional agricultural practices mostly influence agriculture. AI models trained on data from a certain region may exhibit suboptimal performance in other areas due to significant variations in the variables influencing agricultural outcomes [6, 17, 20, 27, 28].

Adaptability of AI Models. While artificial intelligence models may be trained to meet specific criteria, developing models that generalise well across diverse agricultural contexts is challenging. AI systems developed for standardised or well-researched circumstances may be ineffective for farmers in regions with fluctuating temperatures or diverse agricultural practices [6, 17, 20, 27, 28].

8.5 Ethical and Privacy Concerns

Data Ownership. Many times, farmers are expected to disclose their data to outside artificial intelligence firms in order to generate estimates. This raises issues on privacy and data ownership. Farmers not fully understanding how their data is being used or if it may be sold or shared with other businesses causes possible abuse [6, 17, 20, 27, 28].

Control and Access to Technology. Who controls access to AI forecasting tools and whether small-scale farmers will be left out as major technology companies increasingly enter the agricultural sector remain issues of concern. Should artificial intelligence technology remain under the hands of a limited number of strong corporations, it might worsen already existing inequalities in the agricultural in-dustry, therefore benefitting affluent farmers particularly [6, 17, 20, 27, 28].

8.6 Integration with Traditional Farming Practices

Resistance to Change. Many farmers may be reluctant to embrace new AI-driven systems as they have spent decades depending on conventional forecasting techniques. Lack of knowledge about the technology, mistrust of its dependability, or cultural values stressing long-standing agricultural methods may all contribute to this resistance to change [6, 17, 20, 27, 28].

Blending AI with Indigenous Knowledge. Farmers in many different countries depend on indigenous wisdom handed down over the years. Combining artificial intelligence forecasting systems with these conventional methods is difficult as AI may not completely include local knowledge that is crucial for agricultural decisions [6, 17, 20, 27, 28].

8.7 Unpredictability of Extreme Weather Events

Limitations of Weather Forecasting. While long-term climate forecasts—especially in connection to extreme events—remain challenging, artificial intelligence has advanced significantly in short-term weather condition prediction. Droughs, floods, and unexpected temperature fluctuations do enormous damage to agriculture; artificial intelligence models are not yet fully prepared to foresee such events with the degree of precision needed to properly lower risk [6, 17, 20, 27, 28].

Impact of Climate Change. The great unpredictability that climate change is bringing to weather patterns tests the ability of artificial intelligence systems educated on historical data to provide consistent predictions. Artificial intelligence models might be resource-intensive because to their continuous need for updates and retraining resulting from changing environment circumstances [6, 17, 20, 27, 28].

8.8 Scalability and Customization

Scalability of AI Solutions. Although on big farms with the required infrastructure AI systems may be rather successful, scaling these technologies for small-scale farms remains a difficulty. Many times working with limited resources, smallholder farmers find it challenging to use the advanced tools and systems needed for artificial intelligence forecasting [6, 17, 20, 27, 28].

Customization for Different Farming Needs. Every farm runs using their own set of crops, different kinds of soil, and management techniques. Customising AI solutions to fit the particular requirements of individual farms is difficult; off-the-sheld AI solutions could not provide the particular insights farmers need [6, 17, 20, 27, 28].

8.9 Regulatory and Policy Barriers

Lack of Regulatory Frameworks. The fast expansion of artificial intelligence in agriculture has exceeded the development of rules controlling its use. This may cause doubt on the moral use of AI technology, data consumption, and the responsibility of AI projections [6, 17, 20, 27, 28].

Government Support and Incentives. Many areas lack government encouragement or incentives for farmers to use AI-driven forecasting models. Many farmers would not have the financial capacity or drive to use artificial intelligence into their operations without laws encouraging their acceptance of these technologies [6, 17, 20, 27, 28].

9 Future Directions: AI's Role in Shaping Farming Futures

AI will increasingly affect agriculture's future. As population, climate, and resource constraints rise, the agricultural sector must utilise innovative, data-driven techniques to ensure food security and sustainability. This portion examines the future of AI in agriculture, focussing on trends, breakthroughs, and how AI will transform farming in the next years.

9.1 AI-Driven Precision Agriculture

Greater Precision in Crop Management. AI will improve precision farming techniques going forward so that farmers may micro-manage their land. Integrated with IoT sensors, drone technology, and satellite images, AI algorithms will provide real-time analysis of soil conditions, moisture levels, and crop health. By helping to maximise productivity and minimise waste, this detailed data will enable the best use of pesticide, fertiliser, and water [2, 13, 20, 24, 29, 30].

Automated Decision-Making. AI will eventually determine irrigation, planting, and harvesting strategies without human intervention. These technologies will learn and adapt, creating self-optimizing farms that respond to soil, temperature, and plant health in real time [2, 13, 20, 24, 29, 30].

9.2 Climate-Resilient Agriculture

AI for Climate Adaptation. With climate change introducing greater unpredictability into farming, AI will play a pivotal role in helping agriculture adapt to extreme weather conditions. Future AI models will use data from historical climate records, satellite observations, and advanced meteorological systems to predict long-term climate trends and provide farmers with tailored advice on crop selection, planting windows, and resource management [2, 13, 20, 24, 29, 30].

Sustainable Farming Practices. AI will facilitate the development of climate-smart farming practices, such as regenerative agriculture and no-till farming. By monitoring environmental impacts and optimizing resource use, AI can help farmers reduce their carbon footprint and ensure that farming practices remain sustainable in the face of changing environmental conditions [2, 13, 20, 24, 29, 30].

9.3 Advanced Pest and Disease Management

Predictive Pest and Disease Forecasting. Future artificial intelligence systems will anticipate pests and diseases by integrating real-time data from meteorological stations, satellite imagery, and crop health sensors. Enhanced precision in predicting disease transmission and pest infestations facilitated by machine learning methodologies would allow farmers to implement preventive treatments before the commencement of epidemics. This proactive strategy will mitigate crop loss and hence reduce the need for chemical pesticides [2, 13, 20, 24, 29, 30].

AI-Powered Biological Pest Control. Among the eco-friendly pest control options AI will assist in developing are biologi-cal control agents and precision-targeted treatments Artificial intelligence systems will assist to reduce reliance on hazardous pesticides by identifying and suggesting environmentally benign treatments or natural predators, hence maintaining crop health [2, 13, 20, 24, 29, 30].

9.4 AI and Robotics for Automated Farming

Autonomous Machinery. Farmers will become even more automated as AI-driven robots handle tasks like planting, weeding, and harvesting. Using artificial intelligence, these gadgets will assess base real-time decisions on sensor data, navigate fields, and crop health. Autonomous agricultural instruments will reduce labour costs and increase efficiency, therefore enabling farming to be more sustainable and scalable [2, 13, 20, 24, 29, 30].

AI for Harvest Optimization. Robots equipped with artificial intelligence and computer vision will precisely find the perfect timing to harvest crops depending on maturity, size, and quality. This capability will reduce post-harvest losses and improve the general quality of the products, therefore enhancing farmer profitability [2, 13, 20, 24, 29, 30].

9.5 AI for Sustainable Water Management

Smart Irrigation Systems AI-driven smart irrigation systems will enhance efficiency by using real-time data from soil sensors, agricultural health metrics, and meteorological forecasts to optimise water use. These gadgets will enable farmers to save water while ensuring their crops receive essential hydration for maximum development, especially in drought-prone areas [2, 13, 20, 24, 29, 30].

Water Resource Monitoring and Management. Artificial intelligence will be utilised to oversee and administer larger-scale water resources, such as aquifers and irrigation systems. Artificial intelligence algorithms will forecast water availability, regulate water distribution, and identify irrigation system breaches or inefficiencies by analysing data from multiple sources. This will help ensure the sustainable use of water in agriculture as water supplies become more scarce [2, 13, 20, 24, 29, 30].

9.6 AI in Livestock Management

Health and Welfare Monitoring. AI will increase cattle sector ability to track animal welfare and condition. Early illness indicators, animal behaviour monitoring, and treatment recommendations will all come from AI-driven sensors and cameras alerting farmers. This will improve herd health, help to reduce disease spread, and maximise the general output of cattle enterprises [2, 13, 20, 24, 29, 30].

AI-Driven Nutrition and Breeding. Moreover, artificial intelligence technologies aid to optimum cow breeding methods and food. By use of data on animal genetics, feed efficiency, and environmental factors, artificial intelligence may propose optimum diets and breeding strategies that improve the health and produc-tivity of cattle while reducing environmental effect [2, 13, 20, 24, 29, 30].

9.7 AI-Powered Market Forecasting and Supply Chain Management

Market Trend Predictions. Artificial intelligence will be progressively used in predicting agricultural commodities demand and market trends. Machine learning algorithms will evaluate farmers' perspectives on crop selection, optimal selling times, and commodity pricing using global market data, customer preferences, and economic variables. This will allow farmers to make data-driven choices, so enhancing their profitability [2, 13, 20, 24, 29, 30].

Supply Chain Optimization. Improved logistics and reduced food waste will help artificial intelligence change agricultural supply systems. AI systems will optimise travel routes, project demand for new products, and help to assure efficient transportation of commodities from farm to market. Simplified supply chains arising from this will assist to cut costs and boost food availability [2, 13, 20, 24, 29, 30].

9.8 Democratizing AI for Smallholder Farmers

AI Accessibility. Improving accessibility for smallholders farmers is a main future goal for artificial intelligence in agriculture. Affordable artificial intelligence technologies, mobile apps, and cloud-based platforms will be developed to provide resource- strapped farmers real-time forecasting, crop management insights, and market projections. These technologies will link smallholder farmers with large-scale industrial farms, therefore ensuring that artificial intelligence benefits flow throughout the full agricultural sector [2, 13, 20, 24, 29, 30].

Local Adaptation and Customization. Future AI systems will be customised for local agricultural situations, incorporating indigenous knowledge and localised data into predictive models. By integrating region-specific data, like local climatic patterns, soil types, and cultural traditions, AI may provide tailored suggestions that address the requirements of various agricultural communities [2, 13, 20, 24, 29, 30].

9.9 Ethical AI and Data Governance in Agriculture

Data Privacy and Security. Data ownership and privacy problems will arise when AI systems collect massive amounts of data from farms. Moving ahead, it is vital to establish governance structures and ethical norms to protect farmers' data and ensure that AI-powered solutions are transparent and accountable [2, 13, 20, 24, 29, 30].

Fair Access and Equitable Distribution. Ensuring that AI benefits all farmers—regardless of size, location, or financial resources—will be a critical challenge moving forward. Governments and international organizations will need to collaborate to provide training, resources, and financial support to ensure that AI-driven agricultural innovations are distributed equitably [2, 13, 20, 24, 29, 30].

9.10 AI and Sustainable Food Systems

Reducing Environmental Impact. Artificial intelligence will be essential in mitigating the environmental impact of agriculture. Artificial intelligence may enhance resource utilisation, decrease chemical inputs, and promote sustainable agriculture practices, therefore protecting ecosystems, lowering greenhouse gas emissions, and preserving biodiversity. Utilising artificial intelligence as a guide, the future of agriculture will focus on achieving an optimal balance between productivity and environmental conservation [2, 13, 20, 24, 29, 30].

Food Security and Climate Change Resilience. As global populations grow and climate change disrupts traditional farming practices, AI will be essential in ensuring food security. AI-driven innovations will help increase the efficiency of food production, reduce waste, and improve the resilience of agricultural systems to climate-related challenges. This will be particularly important in regions vulnerable to food shortages and climate impacts [2, 13, 20, 24, 29, 30].

10 Conclusion

Artificial intelligence in agricultural forecasting is transforming farming and offers new prospects to improve productivity, sustainability, and resilience. The increasing environmental instability resulting from global population growth and climate change need innovation within the agriculture industry. Artificial intelligence-based methods and technology allow agriculturalists to precisely forecast weather conditions, crop yields, disease outbreaks, and resource distribution. These solutions enhance efficiency and foster climate-resilient agriculture.

The potential of AI in agriculture is significant; nevertheless, challenges such as data accessibility, elevated implementation costs, and the need for user-friendly interfaces persist. The ethical dilemmas around data privacy and equitable access to AI technologies must be confronted to benefit all farmers, especially smallholders.

The future of AI in agriculture will depend on the effective management of these challenges. Ongoing advancements in AI-driven precision agriculture, autonomous farming, climate adaptation, and sustainable food systems will enhance the efficiency, resilience, and environmental sustainability of global agriculture. Artificial intelligence assists farmers in managing modern agriculture, ensuring food security and environmental sustainability for future generations.

References

1. Akkem, Y., Biswas, S. K., & Varanasi, A. (2023). Smart farming using artificial intelligence: A review. *Engineering Applications of Artificial Intelligence, 120*, Article 105899. https://doi.org/10.1016/j.engappai.2023.105899
2. Application of Machine Learning in Agriculture: Recent Trends and Future Research Avenues. (n.d.). Arxiv.org. https://arxiv.org/html/2405.17465v1
3. Bano, N. A., Naqvi, N. Y., Ahmed, N. A., & Syeda, N. (2023). Enhancing agriculture prediction through AI and parallel distributed computing: A comprehensive study on the impact of weather. *International Journal of Emerging Engineering and Technology, 2*(2), 21–28. https://doi.org/10.57041/7d1afq32
4. Chamuah, A., Venugopal, S., & Chamuah, A. (2024). When algorithms till the land: Speculative futures of AI in Indian agriculture—Transitions Research. Transitions Research. https://transitionsresearch.org/part-3-when-algorithms-till-the-land-speculative-futures-of-ai-in-indian-agriculture/
5. Dutt, S., Kulkarni, P., Akilan, S., Mishra, R., Khetan, P., & Bhadora, A. (n.d.). Agricultural price prediction through artificial intelligence. 560062. https://doi.org/10.37118/ijdr.28018.03.2024
6. Elufioye, O. A., Ike, C. U., Odeyemi, O., Usman, F. O., & Mhlongo, N. Z. (2024). AI-driven predictive analytics in agricultural supply chains: A review: assessing the benefits and challenges of AI in forecasting demand and optimizing supply in agriculture. *Computer Science & IT Research Journal, 5*(2), 473–497. https://doi.org/10.51594/csitrj.v5i2.817
7. Patel, B. R. (2021). Realtime object's size measurement from distance using OpenCV and LiDAR. *Turkish Journal of Computer and Mathematics Education (TURCOMAT), 12*(4), 1044–1047. https://doi.org/10.17762/turcomat.v12i4.599
8. Future Farming: Advancing Agriculture with Artificial Intelligence. (2023). http://www.eurekaselect.com. https://www.eurekaselect.com/ebook_volume/3599
9. Kisliuk, B., Jan Christoph Krause, Meemken, H., Carlos, J., Henning Müller, & Hertzberg, J. (2024). *AI in current and future agriculture: An introductory overview*. KI. Künstliche Intelligenz/KI - Künstliche Intelligenz. https://doi.org/10.1007/s13218-023-00826-5
10. Mohamed-Amine, N., Abdellatif, M., & Belaid, B. (2024). Artificial intelligence for forecasting sales of agricultural products: A case study of a Moroccan agricultural company. *Journal of Open Innovation: Technology, Market, and Complexity, 10*(1), Article 100189. https://doi.org/10.1016/j.joitmc.2023.100189
11. Patel, S. J., & Goswami. (2024). A comparative study of supervised and unsupervised classification techniques for sentiment analysis in IMDB. Ssrn.com. https://papers.ssrn.com/sol3/papers.cfm?abstract_id=4804105
12. Kushalatha, M. R., None Sangamesh, C., Shashank, B., Kerudi, S. M., & Mamatha, A. S. (2024). *Farming futures: Machine learning models for crop yield prediction* (pp. 1–6). https://doi.org/10.1109/icdcece60827.2024.10548990
13. Ryan, M., Isakhanyan, G., & Tekinerdogan, B. (2023). An interdisciplinary approach to artificial intelligence in agriculture. *NJAS: Impact in Agricultural and Life Sciences, 95*(1). https://doi.org/10.1080/27685241.2023.2168568
14. Baswaraju, S., Maheswari, V. U., Chennam, K. K., Thirumalraj, A., Prasad, V., & Aluvalu, R. (2023). Future food production prediction using AROA based hybrid deep learning model in agri-sector. *Human-Centric Intelligent Systems, 3*(4), 521–536. https://doi.org/10.1007/s44230-023-00046-y
15. Thaker, A., Chan, L. H., & Sonner, D. (2024). Forecasting agriculture commodity futures prices with convolutional neural networks with application to wheat futures. *Journal of Risk and Financial Management, 17*(4), 143. https://doi.org/10.3390/jrfm17040143
16. View of The Comparative study of Python Libraries for Natural Language Processing (NLP). (2024). Ijsrcseit.com. http://ijsrcseit.com/index.php/home/article/view/CSEIT2410242/CSEIT2410242

17. Talaviya, T., Shah, D., Patel, N., Yagnik, H., & Shah, M. (2020). Implementation of artificial intelligence in agriculture for optimisation of irrigation and application of pesticides and herbicides. *Artificial Intelligence in Agriculture, 4*(2589–7217). https://doi.org/10.1016/j.aiia.2020.04.002
18. Jose, A., S Nandagopalan, & Venkata, C. M. (2021). Artificial intelligence techniques for agriculture revolution: A survey. *Annals of the Romanian Society for Cell Biology, 25*(4), 2580–2597. https://www.researchgate.net/publication/355521703_Artificial_Intelligence_Techniques_for_Agriculture_Revolution_A_Survey
19. Ranga, M. (n.d.). An assessment of impacts of AI revolution on agriculture in India. https://alochana.org/wp-content/uploads/6-AJ2098.pdf
20. Cavalcante, R., & Silva. (2023). Artificial intelligence in agriculture: Benefits, challenges, and trends. *Applied Sciences, 13*(13), 7405–7405. https://doi.org/10.3390/app13137405
21. Mana, A. A., Allouhi, A., Hamrani, A., Rahman, S., el Jamaoui, I., & Jayachandran, K. (2024). Sustainable AI-based production agriculture: Exploring AI applications and implications in agricultural practices. *Smart Agricultural Technology, 7*, 100416–100416. https://doi.org/10.1016/j.atech.2024.100416
22. Spanaki, K., Sivarajah, U., Fakhimi, M., Despoudi, S., & Irani, Z. (2021). Disruptive technologies in agricultural operations: A systematic review of AI-driven AgriTech research. *Annals of Operations Research*. https://doi.org/10.1007/s10479-020-03922-z
23. Rattan, P., Sharma, G., & Singh, P. P. (2024). Application of artificial intelligence (AI) in the agriculture sector. *Advances in Environmental Engineering and Green Technologies*. https://doi.org/10.4018/979-8-3693-2069-3.ch003
24. Lenniy, D. (2021). *AI in agriculture: Challenges, benefits, and use cases*. Intellias. https://intellias.com/artificial-intelligence-in-agriculture/
25. Karunathilake, E. M. B. M., Le, A. T., Heo, S., Chung, Y. S., & Mansoor, S. (2023). The path to smart farming: Innovations and opportunities in precision agriculture. *Agriculture, 13*(8), 1593. https://doi.org/10.3390/agriculture13081593
26. https://www.mdpi.com/journal/agronomy/special_issues/cropprediction_precisionagriculture
27. Gardezi, M., Joshi, B., Rizzo, D. M., Ryan, M., Prutzer, E., Brugler, S., & Dadkhah, A. (2023). Artificial intelligence in farming: Challenges and opportunities for building trust. *Agronomy Journal*. https://doi.org/10.1002/agj2.21353
28. Sparrow, R., Howard, M., & Degeling, C. (2021). Managing the risks of artificial intelligence in agriculture. *NJAS: Impact in Agricultural and Life Sciences, 93*(1), 172–196. https://doi.org/10.1080/27685241.2021.2008777
29. Dwivedi, Y. K., Sharma, A., Rana, N. P., Giannakis, M., Goel, P., & Dutot, V. (2023). Evolution of artificial intelligence research in technological forecasting and social change: Research topics, trends, and future directions. *Technological Forecasting and Social Change*. https://doi.org/10.1016/j.techfore.2023.122579
30. MacPherson, J., Voglhuber-Slavinsky, A., Olbrisch, M., Schöbel, P., Dönitz, E., Mouratiadou, I., & Helming, K. (2022). Future agricultural systems and the role of digitalization for achieving sustainability goals. A review. *Agronomy for Sustainable Development, 42*(4). https://doi.org/10.1007/s13593-022-00792-6

Ethical Harvest: Ensuring Fairness and Privacy

Subrata Gorain, Suman Dutta, and Sandeep Gawdiya

Abstract The convergence of Artificial Intelligence (AI) and Internet of Things (IoT) in agriculture presents transformative possibilities for global food production, resource optimization, and supply chain management. However, these advancements introduce significant ethical concerns that must be addressed to ensure fairness, privacy, and inclusivity. This chapter explores the concept of ethical harvest, underscoring the global importance of applying AI-driven IoT solutions in an equitable and transparent manner. The exponential growth in data collection from agricultural practices, combined with the predictive capabilities of AI, has heightened concerns over data privacy and ownership. Smallholder farmers, especially in developing regions, face risks of data exploitation and misuse by corporations and governments, which could perpetuate inequalities in access to markets, technology, and critical information. Algorithmic biases in AI models also pose a global challenge. Unintended biases in decision-making processes could deepen disparities, favouring large-scale agribusinesses while marginalizing smaller farming communities. This chapter advocates for global governance frameworks that prioritize algorithmic transparency, fairness, and accountability, ensuring that AI systems are not only effective but also ethical. Additionally, the global digital divide raises the risk of unequal access to these advanced technologies. Bridging this gap requires international cooperation, investment in infrastructure, and capacity-building initiatives to empower smallholder farmers worldwide. This chapter calls for the development of inclusive policy frameworks that integrate ethical AI practices, ensure data sovereignty, and promote equitable technological adoption. By focusing on these global issues, agriculture can embrace AI and IoT innovations while fostering sustainability, fairness, and privacy in food production systems.

S. Gorain
Department of Agricultural Economics, Palli Siksha Bhavana (Institute of Agriculture), Visva-Bharati (Central University), Sriniketan 731236, India
e-mail: subrata.gorain@visva-bharati.ac.in

S. Dutta (✉)
Faculty Center for Integrated Rural Development and Management, Ramakrishna Mission Vivekananda Educational and Research Institute, Howrah, West Bengal, India
e-mail: duttasumansiliguri@gmail.com

S. Gawdiya
Department of Agronomy, Galgotias University, Uttar Pradesh, Greater Noida, India

S. R. Kumar et al. (eds.), *AI-Driven IoT Solutions in Agriculture: New Insights*, Studies in Computational Intelligence 1236, https://doi.org/10.1007/978-3-032-05603-0_8

Keywords AI · IoT · Global agriculture · Ethics · Fairness · Privacy · Algorithmic bias · Data governance · Inclusivity · Smallholder farmers · Sustainability

1 Introduction

The integration of Artificial Intelligence (AI) and the Internet of Things (IoT) into agriculture has revolutionized farming practices, offering precision tools that enhance productivity, sustainability, and resource management [55]. When integrated with the IoT network of connected devices, AI enables real-time monitoring, data analysis, and automation of operations including crop health monitoring, irrigation, and pest management [22]. AI and IoT have recently taken the lead in the evolution of agricultural technology, which has historically progressed from basic mechanisation to sophisticated digital systems [56]. With the help of these technologies, farmers can now make decisions based on data, which maximises productivity and minimises environmental effects [5]. However, the necessity for ethical considerations, especially about data privacy, ownership, and the effects of AI-driven systems on global agriculture practices, comes along with this rapid technological growth. The application of AI and IoT in agriculture raises urgent ethical issues, particularly for smallholder farmers in poor countries [3]. There is a chance that sensitive data could be misused by governments or big businesses when AI is used for data collecting and analysis [54]. Maintaining data sovereignty may be difficult for small-scale farmers, who run the danger of losing control over important data. Furthermore, by favouring huge agribusinesses over smaller farms, AI algorithms are prone to biases that could disproportionately impact marginalised populations and perpetuate inequality [24]. These issues are made worse by the global digital gap, which disadvantages resource-poor farmers by ensuring that access to AI and IoT technology is unequal across areas. Transparency in AI systems, equity in algorithmic decision-making, and global initiatives to close the digital divide through infrastructural investments and capacity-building programs that guarantee inclusivity and avoid exploitation in the agriculture industry are all necessary to address these concerns [26].

Despite the growing adoption of AI and IoT in agriculture, several research gaps persist, particularly in the ethical implications of these technologies. IoT solutions powered by AI have demonstrated potential in raising sustainability and productivity, but little is known about how these technologies affect smallholder farmers, particularly in developing nations [16]. The topic of data sovereignty is still not well understood, especially about how small-scale farmers may maintain control over their data in the face of significant influence from governments and major businesses [30]. The literature that is now available focusses mostly on the technical aspects of AI and IoT, which leaves a knowledge gap regarding how to create frameworks that give vulnerable communities' data ownership and privacy first priority [17]. Furthermore, there is a dearth of study on algorithmic bias in agricultural AI systems [59]. In contrast to huge agribusinesses, marginalised farmers may be disproportionately impacted by unintentional biases in decision-making processes [24]. Nowadays, there is minimal emphasis on guaranteeing equity and inclusivity in the use of AI models; instead, the focus is frequently on increasing the models' efficiency [49]. There is still insufficient research on the worldwide digital divide. Few studies discuss how existing disparities in agriculture are made worse by developed and developing regions' unequal access to AI and IoT technology. Research is required to examine methods for advancing digital inclusion, including capacity building, international collaboration, and legislative measures that guarantee fair adoption of technology. For ethical AI and IoT-driven agriculture practices to be inclusive and sustainable for all stakeholders, these research gaps must be filled.

Addressing algorithmic bias and ensuring fairness in AI models are critical for the sustainable development of agricultural systems, particularly as AI and IoT technologies become more integrated into farming practices [12, 12, 12]. Algorithmic openness is a crucial component of this, since it cultivates trust among users, particularly smallholder farmers who could be apprehensive about AI-driven decisions that affect their livelihoods [33]. Maintaining accountability in AI systems can boost trust and encourage wider use of these tools. Research and case studies demonstrating effective fair AI applications in agriculture might offer crucial information about mitigating biases in algorithm development, which will ultimately result in more moral decision-making procedures [53]. Furthermore, it is hardly possible to overestimate the importance of global governance frameworks, enforcing accountability and equity in a variety of agricultural situations would need the establishment of worldwide regulatory norms on ethical AI [14]. Simultaneously, closing the global digital divide is crucial to enabling resource-poor areas to utilise AI and IoT technology [44]. It is crucial to fund digital infrastructure and capacity-building programs in underdeveloped countries because unequal adoption of these advances exacerbates already-existing inequities. To increase farmers' productivity and market access, such initiatives should concentrate on teaching them how to use AI and IoT tools efficiently [52]. To increase underserved communities' access to technology, digital inclusion initiatives backed by partnerships between NGOs, governments, and the commercial sector are essential [34]. Inclusive frameworks that safeguard smallholder farmers and guarantee fair adoption of technology must be given top priority in policy interventions [8]. To ensure that farmers retain ownership over data, legal safeguards for data sovereignty are also essential. Stakeholders can persuade governments to enact laws that support justice and inclusivity by creating global standards for moral AI practices in agriculture, which would eventually result in a more equitable agricultural environment.

This study aims to investigate and address algorithmic bias and fairness in AI models within agricultural systems, emphasizing the importance of transparency and accountability. It looks into ways to close the digital gap globally and give smallholder farmers fair access to AI and IoT technologies. In order to create a more sustainable and just agricultural environment, the study also intends to provide inclusive regulatory frameworks that safeguard farmers' data sovereignty and encourage moral AI practices.

2 Ethical Concerns in AI and IoT-Driven Agriculture

The convergence of Artificial Intelligence (AI) and Internet of Things (IoT) technologies in agriculture has offered transformative tools for improving productivity, sustainability, and resource efficiency. However, this rapid technological advancement introduces significant ethical concerns that must be addressed to prevent exacerbating inequalities, ensure data protection, and promote responsible AI usage. This section explores three core areas of ethical concern: data privacy and ownership, algorithmic bias and discrimination, and the global digital divide. Addressing these issues is vital to establishing an equitable technological landscape that benefits all stakeholders in the agricultural sector, particularly smallholder farmers in developing regions.

2.1 *Data Privacy and Ownership*

In an IoT-enabled agricultural setting, data from sensors, drones, and farm management applications is constantly collected to support AI-driven decision-making (Table 1). However, this extensive data collection raises ethical questions about privacy, ownership, and potential misuse, especially for smallholder farmers who may lack the resources to enforce data protection. Corporations and governments often acquire agricultural data with the intent of driving advancements in productivity and market insights. However, without proper regulation, there is a risk of exploitation, as this data can be commercialized without adequate benefit to the farmers who generate it. Mehrabi et al. [41] indicate that data-driven agriculture disproportionately favours large agribusinesses, who possess the means to analyse and leverage data for market advantage, often at the expense of smallholders. Additionally, governments may collect data under the guise of monitoring and support but may later use it for regulatory purposes that disadvantage small farmers.

Data sovereignty remains a pressing concern, as smallholder farmers in developing regions often lack control over the data generated on their farms. This data may be stored on external servers, often outside the jurisdiction of the farmers' country, which complicates legal recourse in cases of misuse. Establishing frameworks that support local data storage and provide clear ownership rights is crucial. For example, initiatives such as farmer-led data cooperatives enable communities to collectively manage data resources, retaining both ownership and economic benefits [17, 30]. Without such measures, smallholder farmers risk losing autonomy over their farming practices, as decisions become increasingly data-driven and influenced by external entities.

Table 1 Ethical concerns in AI and IoT-driven agriculture: key issues and proposed solutions

Sl. no	Ethical concern	Description	Proposed solutions
1	Data privacy	Unauthorized collection and use of sensitive agricultural data	Implement data protection laws specific to agriculture, similar to GDPR
2	Data ownership	Ambiguity over ownership rights of data collected on farmlands	Establish ownership rights that recognize farmers as primary data owners
3	Data sovereignty	Loss of control over data due to external data storage	Support farmer-led data cooperatives and local storage solutions
4	Data exploitation	Commercialization of farmer data without compensation	Develop fair compensation and benefit-sharing models for farmers' data contributions
5	Data transparency	Lack of transparency in data collection and usage practices	Mandate transparency reports for companies using agricultural data

2.2 *Algorithmic Bias and Discrimination*

The development of AI systems often involves training models on data from large-scale, commercially oriented farms. This approach can result in algorithmic biases that fail to account for the unique conditions and practices of smallholder farms, which depend on diversified, low-input approaches rather than the high-input, monoculture systems prevalent in industrial agriculture. AI algorithms trained on data from commercial farms may produce recommendations that are unsuitable or even detrimental for smallholders. For example, AI systems optimized for monoculture may lack the nuanced understanding needed to support intercropping practices common among smallholder farms [24]. These biases can lead to advice that inadvertently prioritizes large-scale operations, leaving smallholders with impractical or irrelevant recommendations. Ensuring diverse and representative data in AI models, as well as ongoing bias audits, is essential to reduce unintended bias and enhance the applicability of AI systems for smallholder contexts.

The disparity in AI recommendations between large agribusinesses and smallholder farmers illustrates the impact of algorithmic bias on economic equity. Large agribusinesses benefit from AI-driven insights tailored to capital-intensive practices, while smallholders struggle to derive value from advice that does not align with their resource constraints. Addressing this imbalance requires the inclusion of diverse agricultural datasets that reflect the realities of small-scale, resource-constrained farming. Dara et al. [12, 12, 12] highlights the importance of inclusive data practices and regulatory oversight to ensure that AI tools do not perpetuate existing inequalities within agricultural systems.

2.3 Global Digital Divide

The digital divide in agriculture underscores the stark disparity in access to digital resources between well-resourced and underserved regions. This divide is particularly pronounced in developing regions, where smallholder farmers often lack the infrastructure, digital literacy, and financial resources necessary to adopt AI and IoT technologies effectively. Farmers in low-resource areas face significant barriers to accessing AI and IoT technologies due to the high costs of devices, limited internet infrastructure, and insufficient training on digital tools. The United Nations reports that approximately 3.7 billion people globally lack internet access, impeding access to digital agricultural innovations [60]. Subsidizing digital infrastructure and providing training programs for digital literacy are critical steps in bridging this divide. Governments and NGOs can play a pivotal role in reducing these access gaps by supporting affordable access to essential digital tools for farming communities.

The lack of access to AI and IoT resources in developing regions perpetuates economic inequalities by limiting market opportunities and access to valuable data insights. Farmers without digital connectivity miss out on real-time weather updates, market information, and precision agriculture tools that enhance productivity and resilience. Expanding digital literacy programs and providing affordable, localized solutions are essential strategies for ensuring that smallholder farmers can fully participate in data-driven agricultural practices. Addressing the digital divide is crucial for creating an inclusive agricultural future where all farmers, regardless of their location or resources, can benefit from technological advancements [44].

3 Risks for Smallholder Farmers in Developing Regions

Smallholder farmers in developing regions face significant challenges in the era of AI and IoT-driven agriculture (Table 2). Despite the promise of technology to enhance productivity and resource efficiency, smallholder farmers are particularly vulnerable to exploitation and exclusion due to limited access to digital resources, insufficient regulatory protections, and market-driven biases. This section examines the primary risks to smallholders in terms of data misuse, market access, and marginalization, underscoring the need for inclusive approaches to digital agricultural innovation.

3.1 Data Misuse and Market Access

The growing use of data-driven tools in agriculture has transformed how market information and decision-making processes are shaped. However, for smallholders, these changes can introduce vulnerabilities related to data privacy, ownership, and equity in market participation. While AI and IoT technologies have enabled precision

Table 2 Key risks and interventions for smallholder farmers in the age of digital agriculture

Sl. no	Identified risks for smallholders	Description and impact	Suggested interventions
1	Data ownership vulnerabilities	Risk of data being controlled by corporations, leading to reduced autonomy for farmers	Introduce local data governance frameworks granting farmers data control rights
2	Market access inequality	Limited access to timely market information impairs competitive decision-making for smallholders	Develop digital platforms with real-time, accessible market data tailored for smallholder needs
3	Digital marginalization	Smallholders often lack access to digital tools, resulting in unequal information flows in agriculture	Promote community-based tech hubs and training programs to enhance digital inclusion
4	Economic disparities in tech adoption	AI solutions favor high-input systems, often incompatible with smallholder intercropping practices	Design AI models sensitive to diverse, low-input farming practices and provide inclusive incentives

agriculture for well-resourced farms, smallholders in developing countries often lack access to these technologies due to high costs and infrastructural limitations [47]. Consequently, smallholders are at a disadvantage in accessing timely market information, which can limit their ability to make informed decisions about crop selection, planting schedules, and market timing. This disparity creates a competitive gap, as larger agribusinesses can capitalize on real-time data to optimize their operations and market access. Expanding access to digital resources through subsidies and digital literacy initiatives is crucial for promoting equitable access.

Data ownership remains a critical issue for smallholder farmers who often contribute valuable data without adequate protections or compensations. Corporations and third-party platforms may gain access to farmers' data, such as soil health or crop yield information, which is then used for market forecasting and product development without fair remuneration for farmers. Kshetri [30] suggests that without robust policies to ensure data ownership rights for smallholders, there is a risk that these farmers will continue to lose control over their data, becoming data providers rather than data beneficiaries. Establishing clear policies for data ownership and implementing consent-based data-sharing mechanisms can help safeguard smallholders' interests in the digital agriculture economy.

3.2 *Marginalization of Smallholders*

Digital agriculture is often geared toward large-scale operations, creating a technology gap that risks excluding smallholder farmers. This marginalization stems from unequal information flow, biased market dynamics, and a lack of digital infrastructure in rural areas. The concentration of digital resources and advanced AI tools in large agribusinesses has led to a competitive imbalance that marginalizes smallholders. AI-driven insights into market trends, climate forecasts, and crop recommendations are often inaccessible to farmers in resource-poor areas. This inequality perpetuates an information asymmetry that allows larger players to dominate markets while smallholders struggle to compete. Nyakudya et al. [46] emphasize the importance of community-based platforms that share market data and climate information to support informed decision-making for smallholders. By democratizing access to information, these platforms can help bridge the technology gap and reduce market disparities.

Smallholders are at risk of being further marginalized economically as AI-driven tools optimize processes for large-scale farms but do not address the unique challenges of diversified, low-input farming systems. This bias toward monoculture farming practices disadvantages smallholders who rely on intercropping and other sustainable methods not easily integrated into conventional AI models. Robinson [49] highlights that a one-size-fits-all approach in AI can reinforce existing economic disparities, favoring well-resourced farms over resource-constrained smallholders. Adopting inclusive AI models that account for diverse farming practices and providing financial incentives for sustainable agriculture can reduce these economic disparities.

4 Addressing Algorithmic Bias and Fairness in AI Models

As Artificial Intelligence (AI) and Internet of Things (IoT) technologies become integral to modern agriculture, the need to ensure fairness in algorithmic decision-making has become paramount. Algorithmic biases in AI models can inadvertently exacerbate inequalities, particularly in agriculture, where smallholder farmers may be at a disadvantage compared to large agribusinesses. Addressing these biases requires careful attention to transparency, inclusivity, and governance to foster equitable, trustworthy AI applications across diverse farming systems.

4.1 Transparency in AI Systems

Transparency in AI systems is essential for building trust, especially among smallholder farmers who may rely on AI-driven recommendations to guide their agricultural practices. Transparent AI models allow users to understand the factors influencing AI-generated decisions, which is particularly important when recommendations impact livelihoods and sustainability. AI systems often function as "black boxes," meaning the processes behind their recommendations are opaque to users. This lack of transparency can create mistrust among farmers who may be skeptical of automated decision-making. Lubag et al. [33] highlights that algorithmic transparency is crucial in contexts where AI-driven insights influence critical choices in crop management, pest control, and market access. Transparent AI systems provide farmers with the necessary context to make informed decisions, thus fostering trust and promoting wider AI adoption in agriculture.

Accountability mechanisms within AI models are critical to ensuring that these systems can be held responsible for their outputs. Implementing such mechanisms requires setting up clear guidelines on AI usage and incorporating feedback loops for continual system evaluation. In agriculture, where unforeseen biases can impact economic stability and food security, accountability is essential. Studies suggest that establishing independent review boards to oversee AI deployment in agriculture can provide the necessary oversight to maintain ethical standards and prevent biases that could disproportionately affect smallholders [53].

4.2 Fairness in Algorithm Development

Ensuring fairness in AI requires addressing the inherent biases in algorithm development, which may arise from unrepresentative training data, inadequate testing, or prioritization of profit-driven objectives. In agriculture, fairness is particularly relevant, as biased algorithms can impact access to resources, crop recommendations, and market opportunities, potentially marginalizing smallholder farmers. Bias in AI models often occurs when data inputs are unrepresentative of diverse farming practices, especially in regions where smallholder farmers rely on mixed-cropping systems and other resource-efficient practices. Addressing bias requires incorporating diverse data sources that reflect varied agricultural contexts. For instance, training AI on data from different crop systems and climate zones can result in models that are more inclusive and better suited to smallholder needs [12, 12, 12]. Additionally, ethical guidelines during model development can provide a framework for balancing technological advancement with social responsibility.

Various projects have demonstrated the potential of fair and inclusive AI in agriculture. For example, community-based AI models that integrate local knowledge have shown effectiveness in providing context-specific recommendations. These initiatives not only reduce the risk of biased outputs but also empower local farmers by making AI tools relevant and accessible. The success of such models suggests that participatory approaches in AI development can help build systems that are both fair and effective, supporting sustainable agricultural growth [2].

4.3 Global Governance Frameworks

The development of global governance frameworks is crucial to enforcing fairness in AI-driven agriculture. These frameworks should address international standards on data privacy, model transparency, and accountability, ensuring that AI applications align with ethical principles across different regions and farming scales. International regulatory standards are necessary to ensure consistency in ethical AI practices worldwide. These standards can provide guidelines on data handling, transparency, and fairness, helping to mitigate the risks of exploitation and inequality in agriculture. Organizations such as the United Nations and the Food and Agriculture Organization (FAO) are well-positioned to lead efforts in establishing global principles for ethical AI, as they bring together diverse stakeholders with the shared goal of promoting fair technological access in agriculture [14].

Effective governance requires mechanisms to enforce accountability and monitor AI practices in agriculture. These mechanisms can include certification systems for ethical AI, similar to fair trade certifications, which indicate adherence to ethical standards. Establishing a certification system for ethical AI applications in agriculture could incentivize companies to adopt fair practices, enhancing the social impact of AI-driven tools. Additionally, cross-border data-sharing agreements with built-in ethical safeguards can help create a cooperative environment where AI technologies serve collective agricultural goals rather than competitive interests [34].

The development of big data-driven AI models is a disruptive force in agriculture by providing significant benefits, particularly for smallholder farmers in developing countries [36, 42]. The main challenges in developing novel edge AI tools for agriculture are their limited distribution and the lack of understanding among farmers. These technologies are often created based on specific needs, but their impact is constrained by poor accessibility and insufficient training for effective use. However, significant gap exists between the development of technology-based models, their integration into mobile applications, and their adaptation to specific user needs. Given these challenges, there is a need to develop a system that consolidates all relevant information and recommendations in one place and accessible in multiple languages to enhance usability. This discussion addresses the importance of a digital inclusion agenda led by governments, communities, the private sector, and policymakers to improve access to data-driven agronomic science for farmers and global stakeholders in a centralized platform.

5 What is Digital Divide in Terms of Agriculture?

The digital divide in agriculture refers to the disparity in access to digital tools and technology, especially among farmers and stakeholders in developing countries or remote areas. Bridging this divide ensures that all farmers have equitable access to resources that can enhance productivity, sustainability, and efficiency in agriculture (Fig. 1).

Data-driven agriculture paves new pathways for decision-making in farming, enhancing yields, profitability, and environmental security, ultimately contributing to a sustainable food system [1]. Data-driven interventions in agriculture provides innovative pathways to strengthen global food systems. Leveraging big data, mobile

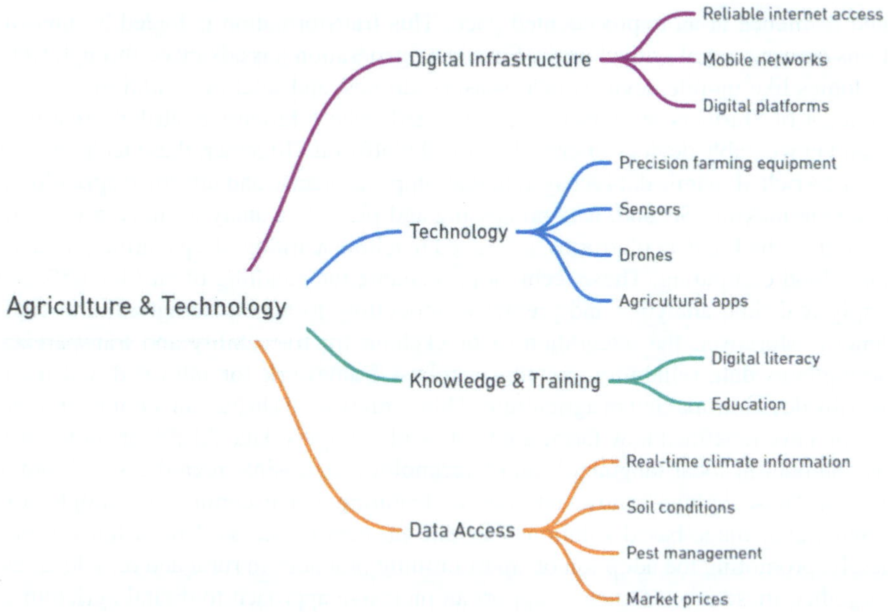

Fig. 1 Bridging the global digital divide in agriculture

technology, and AI models provides effective solutions to diverse agricultural challenges, including input management, low crop yields, disease control, and climate adaptation [36]. These tools enhance decision-making, productivity, and resilience, ultimately supporting a sustainable and equitable food system. However, as internet penetration expands into remote areas, ensuring internet availability, ownership, and affordability becomes essential to addressing key agricultural challenges. Although increased internet access has led to greater smartphone availability and more affordable data usage, the optimal use of digital resources in agriculture remains limited. To fully leverage these advancements, targeted efforts are needed to enhance farmers digital literacy and resource accessibility, enabling data-driven solutions to improve productivity, sustainability, and resilience across global food systems. Nearly 3.7 billion people worldwide lack access to the internet [61]. This growing digital divide raises significant concerns, as it may impede progress toward achieving human rights and the United Nations' Sustainable Development Goals (SDGs), particularly those related to education, equality, and health and well-being. Expanding internet access is therefore essential for strengthening inclusive growth and ensuring that the benefits of digital advancements reach all communities. The without equitable access to digital resources, marginalized communities risk falling further behind, potentially impeding progress in these critical areas. In the era of multimodal, data-driven agriculture, decision-making is increasingly augmented by reliable, multi-source environmental data. This approach holds immense potential to transform global food systems by providing insights that enhance productivity, sustainability, and resilience at an unprecedented pace. This transformation is fueled by innovations across several critical areas. First, data generation has advanced through technologies like mobile devices, field sensors, drones, and satellites. Additionally, the concept of "farmers as sensors" has emerged, where farmers contribute real-time data via wearable devices or crowd-sourced platforms. Together, these technologies create a rich, dynamic data ecosystem that supports precise and informed agricultural decision-making. Second, data processing and predictive analytics have been revolutionized by big data platforms, advanced machine learning, deep learning models, and cloud computing. These technologies enable the handling of vast datasets and apply real-time analytics and predictive modelling to anticipate agricultural challenges. Moreover, the integration of blockchain for traceability and transparency strengthens data reliability, creating a robust framework for informed and trustworthy decision-making in agriculture. Third, innovations in human–computer interaction have redefined how farmers engage with complex data. AI-driven tools, such as chatbots in local languages, make technology accessible regardless of literacy levels. These human-centric interfaces—featuring voice commands, simple text inputs, and image-based solutions—enable farmers to access data insights effortlessly, promoting the adoption of smart farming practices in rural and remote areas. Together, these advancements support an inclusive approach to digital agriculture, ensuring that farmers of all backgrounds can leverage data-driven solutions for improved productivity and sustainability.

Data-driven farming plays a critical role in areas such as weather services, digital extension, data-driven agronomy, market information systems, and agricultural incentives [39]. For example, tools like Liquid Prep—an open-source, low-cost soil moisture sensor—are specifically designed to help smallholder farmers manage irrigation more efficiently. The Liquid Prep app can significantly benefit farmers and communities in water-stressed regions by optimizing irrigation practices and enhancing overall agricultural productivity [6, 29]. These innovations exemplify how targeted digital tools can address pressing challenges, fostering resilience and sustainability in agriculture.

5.1 How to Bridge the Digital Divide in Agriculture?

5.1.1 Region-Specific Language Training and Mentorship

Region-specific language training and mentorship based on farmer's needs, are essential for addressing challenges ranging from food security to climate change. In particular, smallholder farmers often have limited experience with advanced information and communication technology (ICT) tools and may lack the skills to operate new technologies for improved farm management [47]. Bridging this knowledge gap requires providing targeted training and user-friendly tools, enabling farmers to adopt and benefit from digital innovations effectively.

(a) **Long-term Mentorship**: The beyond initial training, continuous mentorship and support are essential for fostering ongoing learning and adaptation among farmers. This support involves creating peer-to-peer networks, where experienced farmers guide others, and establishing expert mentorship programs with digital agronomists [23]. These continuous support systems help farmers troubleshoot issues, stay updated on emerging technologies, and adapt to the rapidly evolving digital landscape, promoting resilience and sustained adoption of digital tools in agriculture.
(b) **Localized and Tailored Approaches**: The training should be tailored to address local challenges, specific crops, climate conditions, and market demands [23]. In areas with low literacy rates, visual or audio-based training methods can be more effective than text-heavy resources, enabling technology to genuinely empower communities by meeting their specific needs. This localized, need-based approach ensures that digital tools are accessible, relevant, and impactful for farmers, enhancing their ability to manage farms effectively and sustainably.

5.1.2 Improving Infrastructure

Infrastructure plays a central role in bridging the digital divide in agriculture [43]. Reliable internet connectivity is essential for farmers to access digital tools, such as weather prediction apps and market price tracking software. Equally important is the development of physical infrastructure, as digital connectivity alone cannot support effective agricultural practices. Without this foundational infrastructure, farmers are unable to fully benefit from digital technologies, limiting the impact of these innovations on productivity and sustainability.

5.1.3 Making Mobile Broadband a Public Good

This approach can empower farmers and bridge the digital divide in rural areas. Broadband services must be universally available, affordable, and accessible, especially in remote agricultural communities where connectivity remains a challenge. Treating connectivity as a public good in these regions would transform rural economies, boost productivity, and equip millions of farmers with the digital tools they need to thrive [20]. This reform has the potential to reshape agriculture, fostering resilience and sustainable growth in the digital age. It should be:

(a) Equitable Access
(b) Improving Productivity
(c) Reducing Costs
(d) Rural Development
(e) Closing the Information Gap.

5.1.4 Developing Customized ICT Solutions

Developing ICT-based solutions that address the specific needs of farmers is essential for overcoming the limitations of one-size-fits-all technologies [28]. The tailoring digital tools and platforms to the unique challenges faced by farming communities, we can ensure these solutions are practical, relevant, and scalable. Such customized tools enhance productivity, sustainability, and income generation, ultimately fostering more efficient and resilient farming practices in the face of diverse challenges. It should be based on:

(a) Localized Solutions
(b) User-friendly Interfaces
(c) Scalable Technology
(d) Data-driven Decision Making
(e) Integration with Financial Services
(f) Building Partnerships.

5.1.5 Providing Cost-Cutting Measures and Subsidies

Cost-cutting methods, such as subsidies, are essential for enabling farmers to access and adopt modern technologies. High costs associated with digital tools often create barriers to adoption, limiting farmers' ability to acquire and utilize these innovations [15]. Subsidy programs and similar initiatives can bridge this divide by making technology more affordable and accessible. Here's how these initiatives can effectively support farmers in embracing digital advancements:

(a) Technology Access Subsidies
(b) Data and Connectivity Subsidies
(c) Support for Energy Costs
(d) Financial Incentives for Digital Adoption
(e) Reducing Input Costs Through Digital Platforms:
(f) Subsidized Training Programs
(g) Public–Private Partnerships.

5.1.6 Disseminating Information in Local Languages

Language barriers often prevent farmers in rural areas from fully utilizing digital tools and platforms. Addressing this challenge through the development of AI-based bilingual chatbots is essential to improve accessibility, encourage the adoption of digital technologies, and foster inclusive participation in modern farming practices [25]. Effective AI driven tool development should be based on:

a. Localized Agricultural Content
b. Voice-Based and Audio Solutions
c. SMS and Mobile Platforms

d. Community-Led Information Dissemination
e. Culturally Relevant Digital Content
f. Custom Apps and Tools
g. Government and NGO Support.

5.1.7 Establishing Service Centers

Establishing service centers is essential to provide on-the-ground support, access to technology, resources, and training for farmers, especially in rural and underserved areas. These centers serve as hubs where farmers can access digital tools, receive technical assistance, and learn to integrate modern, inclusive technologies into their farming practices, regardless of location or economic status [48]. Such centers not only empower farmers but also increase the penetration and adoption of solutions tailored to specific needs, promoting sustainable and accessible agricultural advancements.

5.1.8 Involving All Stakeholders

To ensure that digital technology is accessible to all in rural areas, it is essential to recognize the complexity and diversity of the agricultural sector. The participation and collaboration of multiple stakeholders are crucial in this effort. Each stakeholder plays a vital role in making digital tools and technologies accessible, practical, and sustainable [38]. Governments establish regulatory frameworks and infrastructure, private companies drive innovation and supply products; NGOs and international organizations advocate for inclusivity and provide on-the-ground support; academic institutions develop new technologies; and farmers are both beneficiaries and active participants in the digital revolution. Together, these collaborations can create a robust ecosystem that enhances the adoption of digital solutions in agriculture.

6 Policy Interventions for Ethical AI Practices in Agriculture

Artificial Intelligence (AI) is rapidly evolving and increasingly permeating people's lives. In agriculture, AI is opening new pathways for enhancing productivity, profitability, and sustainability while minimizing costs through informed decision-making and optimized resource utilization [11]. From climate-smart agriculture to predictive analytics for crop management, AI holds immense potential for addressing food security challenges in the context of climate change. However, the rapid integration of AI in agriculture also raises several ethical concerns, including data privacy, fairness, and environmental impacts.

To ensure that advanced technologies are deployed responsibly and inclusively, effective policy interventions are essential. This includes establishing a regulatory framework that promotes innovation while safeguarding farmers and protecting ecosystems. Additionally, the benefits of AI should be equitably accessible throughout the agricultural innovation process. Through this chapter we discuss here policy intervention is necessary to promote ethical AI practices in agriculture with some of examples.

6.1 Ensuring Data Privacy and Ownership

Data privacy, ownership, and the ethical use of AI in agriculture have been significant concerns for decades. AI-driven systems rely heavily on big data from agroecosystems to generate actionable insights [62]. However, without a proper regulatory framework, these systems face challenges, including the potential misuse of data.

(a) **Data ownership rights**: Research trials in both field and controlled environments using sensors, drones, and AI-driven tools generate extensive and reliable datasets. Government policies must clearly define data ownership and usage rights. Additionally, an open-source, submission-based system for datasets is essential to prevent unethical data use and promote collective benefits, such as enhancing regional agricultural productivity.
(b) **Transparent data usage**: AI services must be transparent about data collection, storage, and usage. Companies are mandated to disclose their data handling practices, including how agricultural data will be used and whether it will be shared with third parties.
(c) **Data privacy regulations**: National, regional, and international data privacy regulations, such as the General Data Protection Regulation (GDPR), should be adopted to safeguard agricultural data. These regulations can prevent unauthorized exploitation and ensure that AI companies adhere to ethical standards in handling sensitive information.

6.2 Promoting Fairness and Preventing Bias in AI Models

The AI models often operate as black boxes. This highlighting the need for explainable AI that clearly illustrates how they generate results and patterns relevant to specific regions. In agriculture, biased or incomplete data can result in skewed recommendations that favour certain farmers while disadvantaging others [40]. For instance, AI systems trained on large datasets from commercial farms may not offer accurate guidance for smallholder farmers.

(a) **Inclusivity in AI training data**: The region-specific prescriptive and predictive models based on time series data from diverse environmental conditions are essential. Policy interventions should outline a roadmap for using ethical AI in recommendation systems to optimize farming practices.

(b) **Bias auditing and testing**: Government support for regulations is crucial to ensure bias auditing and testing of AI systems before deployment in agriculture. This involves assessing algorithms for discriminatory outcomes that may marginalize specific farmer groups, such as smallholders, women, and indigenous communities. Policies should promote tools to identify and report biases in AI systems.

(c) **Equitable access to AI tools**: Policies must ensure that advanced AI tools and technologies are accessible at local levels, particularly in less developed and developing countries. Subsidies and incentives for specific practices should focus on infrastructure that promotes digital inclusion and equitable access to technology.

6.3 Safeguarding Labour and Supporting Workforce Transition

The AI-driven agriculture can enhance efficiency through innovative technologies. However, these advancements raise concerns about the displacement of agricultural labour in rural communities that rely on manual work for their livelihoods [50]. The following challenges need to be addressed:

(a) **Job displacement safeguards**: Policies should adopt a balanced approach to AI that prioritizes job security. Technology transfer regulations must require companies to implement gradual transitions to automation, ensuring workers are not abruptly displaced. Additionally, policy interventions should encourage the use of AI to complement human labor, optimizing outcomes rather than replacing workers entirely. For example, optimization tools can assist farm workers with their tasks instead of relying solely on fully automated systems.

(b) **Job Displacement Safeguards**: Policies must balance AI adoption with workforce stability to prevent unemployment crises. Gradual transitions to automation should be regulated, especially in countries with large populations. Additionally, extension agents should facilitate stakeholder meetings to improve literacy in using AI-guided tools.

(c) **Reskilling and Upskilling Programs**: Reskilling and upskilling are essential for helping farmers thrive in the digitally-driven era of AI. Financial incentives for training and career transition programs can support workers in adapting to new roles within the digital agricultural economy.

(d) **Social safety nets**: Policies must establish robust social safety nets for workers affected by future changes in agriculture. Unemployment benefits, income protection schemes, and support for displaced workers are crucial for the future of AI-driven farming.

6.4 Environmental Sustainability and AI-Driven Agriculture

The AI has the potential to make agriculture sustainable by optimizing resource use at low cost; however, without proper regulation, risks may arise in the coming decades [27]. Region-specific prescriptive and predictive models, guided by well-regulated policies, can create a new roadmap toward sustainable farming practices that yield net climate benefits.

(a) **Environmental impact assessments for AI tools**: Governments must introduce laws requiring environmental impact assessments for AI tools in agriculture. This ensures that the potential effects of these technologies on ecosystems and their interconnected biotic factors are evaluated before widespread adoption.
(b) **Promoting Climate-Smart AI guided Agriculture**: Crucial interventions are needed to model ecosystem processes in climate-smart agriculture using AI to address climate shocks effectively. Governments should be incentivized through well-designed, AI-guided frameworks that leverage large-scale, reliable datasets to promote the adoption of sustainable agricultural practices.
(c) **Circular economy in AI-Driven agriculture**: Policies should encourage the integration of AI into circular economy models in agriculture to optimize resource reuse, recycling, and conservation. This approach can serve as a driver for promoting smart management practices and regenerative agriculture aimed at reducing waste. Policy frameworks should incentivize these practices to advance sustainable development goals.

6.5 International Collaboration and Ethical Standards

The AI technologies in agriculture should adhere to international norms for ethical standards to ensure responsible use. Establishing common standards globally can help prevent the exploitation of loopholes in AI technology applications [12, 12, 12].

(a) **International ethical standards for AI in agriculture**: Governments should collaborate with international organizations, such as the Food and Agriculture Organization (FAO), the World Bank, and the United Nations, to establish global ethical standards for AI in agriculture. Key principles like data privacy, fairness, environmental sustainability, and labour protection are essential to ensure AI technologies promote social good across all regions (Table 3).
(b) **Cross-border data sharing agreements**: The AI technologies rely on extensive historical datasets gathered from various countries, making cross-border data-sharing agreements that protect data sovereignty essential. The exchange of reliable datasets, supported by trustworthy and explainable AI, can help address global challenges such as food security and climate change.

Table 3 Principle and constituent ethical issues or guidance as per outlined by Ryan et al. [51]

Principle	Constituent ethical issues or guidance			
Transparency	Transparency, interpretability,	Explainability, communication,	Explicability, disclosure	Showing, understandability
Justice and fairness	Justice, equality, diversity, remedy	Fairness, equity, plurality, redress	Consistency, non-bias, accessibility, challenge	Inclusion, non-discrimination, reversibility, access and distribution
Non-maleficence	Non-maleficence, protection non-subversion	Security, precaution	Safety, prevention	Integrity
Responsibility	Responsibility	Accountability	Liability	Acting with integrity
Privacy	Privacy	Personal or private information		
Beneficence	Benefits social good,	Beneficence common good,	Well-being	Peace
Freedom and autonomy	Freedom	Autonomy	Consent	Choice
Trust	Self-determination trustworthiness	Liberty	Empowerment	
Sustainability dignity	Sustainability dignity	Environment (nature)	Energy	Resources (energy)
Solidarity	Solidarity	Social security	Cohesion	

7 Case Studies on Ethical AI and IoT Implementations

7.1 What is Ethical AI and IoT Implementations?

Ethical AI embodies the principles and practices of responsibly designing, deploying, and governing artificial intelligence (AI) systems and Internet of Things technologies to ensure they are used in a fair, transparent, accountable manner and societal well-being (Table 4; Fig. 2).

8 Future Directions and Recommendations

As the agricultural sector increasingly adopts AI-driven technologies, bridging ethical gaps becomes crucial to ensuring that these innovations benefit all stakeholders equitably [2]. There is a lot of potential in emerging markets for the creation and application of locally relevant ethical AI solutions. While upholding farmers' rights and data sovereignty, these methods can increase productivity.

Table 4 Some of studies related to ethical AI and internet of things implementations in agriculture

Key focus	Methods	Key findings	Ethical concerns addressed	Study by
AI in vertical farms to enhance sustainability	Empirical analysis with interviews	AI supports decision-making, increasing productivity, efficiency, and cost reduction	Promotes ecosystem creation among stakeholders	[9]
Ethical AI in agriculture focusing on privacy, accountability, and bias	Literature review	Guidelines for mitigating ethical issues in AI-based farming tools	Addresses fairness, transparency, privacy, and sustainability	[12]
IoT adoption framework in agriculture	Systematic literature review	IoT has matured to support sustainability and efficiency in agriculture	Promotes sustainable practices without explicit ethical risks	[37]
Integration of IoT and blockchain (BIoT) in agriculture	Experimental design and empirical analysis	BIoT increases transparency, efficiency, and security	Ethical issues with data privacy and stakeholder engagement	[10]
AIoT for improved agricultural sustainability and productivity	Case analysis on AIoT's application	Enhances precision, resource management, and efficiency	Highlights challenges of data quality, privacy, and cost	[32]
Smart agriculture system with IoT automation	System implementation with sensors and microcontrollers	Improves crop productivity through automation	Lacks focus on ethical concerns but suggests sustainability	[31]
Smart agriculture monitoring with machine learning	Algorithm development and field application	Achieves high accuracy in crop monitoring and management	Ethical considerations on data accuracy and privacy suggested	[57]

Technology companies may solve the particular difficulties faced by smallholder farmers in these areas by concentrating on community-driven strategies and developing inclusive, effective systems [35]. Furthermore, underdeveloped nations, where access to cutting-edge agricultural equipment and training is sometimes restricted, urgently need to address their technology gaps. Investing in digital infrastructure and implementing focused capacity-building programs can enable farmers to effectively utilise AI and IoT solutions [22]. Stakeholders may build trust and advance sustainable farming methods that benefit marginalised populations by giving ethical considerations priority when implementing these technologies, guaranteeing that the advantages of AI are equitable and widely distributed.

Technological innovations and emerging trends are poised to revolutionize the agricultural landscape, but they must be guided by ethical principles to maximize

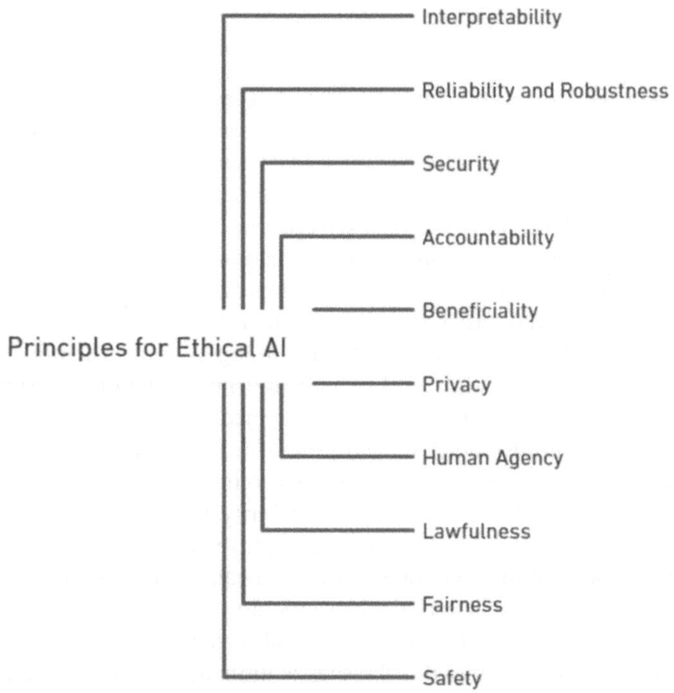

Fig. 2 Principals of ethical artificial intelligence

their impact [19]. AI technologies of the next generation, such as enhanced machine learning algorithms that lessen bias, have enormous potential to improve farming decision-making [4]. These developments may result in improved resource allocation, more precise agricultural yield forecasts, and effective pest control. Furthermore, incorporating blockchain, IoT, and sophisticated sensors into agricultural systems would improve supply chain traceability and transparency [21]. Through these integrations, farmers can keep an eye on up-to-date information about the health of their crops, the state of their land, and the dynamics of the market, enabling them to make morally sound judgements [18, 58]. Strong legal frameworks that uphold privacy and justice are necessary to support these technological advances. To create comprehensive rules that safeguard farmers' rights, maintain data sovereignty, and reduce the dangers of data exploitation, governments and international organisations must work together [7]. Fostering collaborations between the public and commercial sectors, promoting information exchange, and stimulating investment in moral AI practices should be the main goals of policy development recommendations [13, 45]. Agriculture may get closer to a future that is not only productive and efficient but also sustainable and egalitarian for all by integrating ethical considerations with technology advancements.

9 Conclusion

In conclusion, the integration of AI and IoT in agriculture presents both significant opportunities and challenges that require careful navigation to ensure equitable outcomes. It is crucial to close the ethical divides in AI-driven agriculture in order to promote inclusivity and trust among all parties involved, especially smallholder farmers in developing nations. By creating specialised ethical AI solutions and filling up technology gaps in developing nations, we can enable these communities to properly utilise cutting-edge technologies. To create an ecosystem where smallholder farmers may prosper and ensure that the advantages of innovation are equitable and widely distributed, investments in digital infrastructure and capacity-building programs are essential. Furthermore, the transformation of agricultural techniques will be greatly aided by technology advancements and new trends like next-generation AI algorithms and the integration of IoT, blockchain, and sophisticated sensors. But these developments need to be steered by strong legal frameworks that give justice, privacy, and accountability first priority. Governments and international organisations are essential in creating thorough regulations that uphold farmers' rights and encourage moral AI applications. We can promote knowledge exchange and encourage investment in sustainable agricultural technologies by encouraging cooperation between the public and commercial sectors. The ability to balance technological improvements with ethical issues is ultimately what will determine if AI and IoT are successfully implemented in agriculture. We can create a future where the digital transformation of agriculture benefits all stakeholders by emphasizing inclusivity, openness, and fairness. This will result in sustainable practices that boost productivity while upholding the rights and dignity of farmers everywhere. Hence, we must continue to be dedicated to tackling these issues as we go forward in order to guarantee that the agriculture industry develops in a fair, responsible, and advantageous manner for everybody.

References

1. Abiri, R., Rizan, N., Balasundram, S. K., Shahbazi, A. B., & Abdul-Hamid, H. (2023). Application of digital technologies for ensuring agricultural productivity. Heliyon.
2. Aderibigbe, A. O., Ohenhen, P. E., Nwaobia, N. K., Gidiagba, J. O., & Ani, E. C. (2023). Artificial intelligence in developing countries: Bridging the gap between potential and implementation. *Computer Science & IT Research Journal, 4*(3), 185–199.
3. Ali, G., Mijwil, M. M., Buruga, B. A., Abotaleb, M., & Adamopoulos, I. (2024). A survey on artificial intelligence in cybersecurity for smart agriculture: State-of-the-art, cyber threats, artificial intelligence applications, and ethical concerns. *Mesopotamian Journal of Computer Science, 2024*, 53–103.
4. Alibabaei, K., Gaspar, P. D., Lima, T. M., Campos, R. M., Girão, I., Monteiro, J., & Lopes, C. M. (2022). A review of the challenges of using deep learning algorithms to support decision-making in agricultural activities. *Remote Sensing, 14*(3), 638.

5. Araújo, S. O., Peres, R. S., Barata, J., Lidon, F., & Ramalho, J. C. (2021). Characterising the agriculture 4.0 landscape—emerging trends, challenges and opportunities. *Agronomy, 11*(4), 667.
6. Baath, G., & Bawa, A. (2023). Calibrating liquid prep soil moisture sensors for diverse soil types [abstract]. ASA, CSSA, SSSA International Annual Meeting, St. Louis, MO.
7. Banisar, D., & Davies, S. (1999). Global trends in privacy protection: An international survey of privacy, data protection, and surveillance laws and developments. *J. Marshall J. Computer & Info. L., 18*, 1.
8. Birner, R., & Resnick, D. (2010). The political economy of policies for smallholder agriculture. *World Development, 38*(10), 1442–1452.
9. Cavazza, A., Dal Mas, F., Campra, M., & Brescia, V. (2023). Artificial intelligence and new business models in agriculture: The "ZERO" case study. *Management Decision.*
10. Chacko, N. M., Narendra, V. G., Balachandra, M., & Rathinam, S. (2023). Exploring IoT-blockchain integration in agriculture: An experimental study. *IEEE Access, 11*, 130439–130450.
11. Chukwuma, U., Gebremedhin, K. G., & Uyeh, D. D. (2024). Imagining AI-driven decision making for managing farming in developing and emerging economies. *Computers and Electronics in Agriculture, 221*, Article 108946.
12. Dara, R., Hazrati Fard, S. M., & Kaur, J. (2022). Recommendations for ethical and responsible use of artificial intelligence in digital agriculture. *Frontiers in Artificial Intelligence, 5*, 884192.
13. Das, S., Choudhury, M. R., Chatterjee, B., Das, P., Bagri, S., Paul, D., Bera, M., & Dutta, S. (2024). Unraveling the urban climate crisis: Exploring the nexus of urbanization, climate change, and their impacts on the environment and human well-being–a global perspective. *AIMS Public Health, 11*(3), 963.
14. de Pagter, J. (2024). From EU robotics and AI governance to HRI research: Implementing the ethics narrative. *International Journal of Social Robotics, 16*(6), 1281–1295.
15. Deichmann, U., Goyal, A., & Mishra, D. (2016). Will digital technologies transform agriculture in developing countries? *Agricultural Economics, 47*(S1), 21–33.
16. Dhanaraju, M., Chenniappan, P., Ramalingam, K., Pazhanivelan, S., & Kaliaperumal, R. (2022). Smart farming: Internet of Things (IoT)-based sustainable agriculture. *Agriculture, 12*(10), 1745.
17. Dhirani, L. L., Mukhtiar, N., Chowdhry, B. S., & Newe, T. (2023). Ethical dilemmas and privacy issues in emerging technologies: A review. *Sensors, 23*(3), 1151.
18. Dutta, H., Dutta, S., Yadav, H., Prabhu, I. G., Roy, A., Chowdary, N. B., & Das, S. (2024). A promising and forward-looking advancement using drones: Perspectives from Indian sericulture. *Rural and Regional Development, 2*(3), 10014.
19. Eastwood, C., Klerkx, L., Ayre, M., & Dela Rue, B. (2019). Managing socio-ethical challenges in the development of smart farming: From a fragmented to a comprehensive approach for responsible research and innovation. *Journal of Agricultural and Environmental Ethics, 32*(5), 741–768.
20. Fabregas, R., Kremer, M., & Schilbach, F. (2019). Realizing the potential of digital development: The case of agricultural advice. *Science, 366*(6471), eaay3038.
21. Feng, H., Wang, X., Duan, Y., Zhang, J., & Zhang, X. (2020). Applying blockchain technology to improve agri-food traceability: A review of development methods, benefits and challenges. *Journal of cleaner production, 260*, Article 121031.
22. Fuentes-Peñailillo, F., Gutter, K., Vega, R., & Silva, G. C. (2024). Transformative technologies in digital agriculture: Leveraging Internet of Things, remote sensing, and artificial intelligence for smart crop management. *Journal of Sensor and Actuator Networks, 13*(4), 39.
23. George, T., Bagazonzya, H., Ballantyne, P., Belden, C., Birner, R., Castello, R. D., & Treinen, S. (2011). *ICT in agriculture: Connecting smallholders to knowledge, networks, and institutions* (No. 64605, pp. 1–428). The World Bank.
24. Gwelo, F. A. (2025). Algorithmic bias and farmers' autonomy in AI-Driven agricultural marketing and supply chains. In *AI and machine learning applications in supply chains and marketing* (pp. 313–344). IGI Global.

25. Hashmi, E., Yamin, M. M., & Yayilgan, S. Y. (2024). Securing tomorrow: A comprehensive survey on the synergy of Artificial Intelligence and information security. *AI and Ethics*, 1–19.
26. Holzmeyer, C. (2021). Beyond 'AI for Social Good' (AI4SG): Social transformations—not tech-fixes—for health equity. *Interdisciplinary Science Reviews, 46*(1–2), 94–125.
27. Javaid, M., Haleem, A., Khan, I. H., & Suman, R. (2023). Understanding the potential applications of artificial intelligence in agriculture sector. *Advanced Agrochem, 2*(1), 15–30.
28. Ji-Ping, D., Jing-Han, L., Jia-Huan, L., Wei-Feng, Z., & Xiang-Ping, J. (2022). ICT-based agricultural advisory services and nitrogen management practices: A case study of wheat production in China. *Journal of Integrative Agriculture, 21*(6), 1799–1811.
29. Kim et al. (2020). Liquid Prep, a solution that helps farmers optimize water usage during droughts, is now open source. https://developer.ibm.com/callforcode/blogs/cfc-2019-internal-finalist-liquid-prep/
30. Kshetri, N. (2014). The emerging role of Big Data in key development issues: Opportunities, challenges, and concerns. *Big Data & Society, 1*(2), 2053951714564227.
31. Kumari, P., Prajapati, A., Manjhi, S., Saw, A. K., & Karmakar, S. (2020, July). A study on IoT based smart agriculture system. In *International conference on recent trends in artificial intelligence, IOT, smart cities & applications (ICAISC-2020).*
32. Leong, Y. M., Lim, E. H., Subri, N. F. B., & Jalil, N. B. A. (2023). Transforming agriculture: Navigating the challenges and embracing the opportunities of artificial intelligence of things. In *2023 IEEE international conference on agrosystem engineering, technology & applications (AGRETA)* (pp. 142–147). IEEE.
33. Lubag, M., Bonifacio, J., Tan, J. M., Concepcion, R., Mababangloob, G. R., Galang, J. G., & Maniquiz-Redillas, M. (2023). Diversified impacts of enabling a technology-intensified agricultural supply chain on the quality of life in hinterland communities. *Sustainability, 15*(17), 12809.
34. Madon, S., Reinhard, N., Roode, D., & Walsham, G. (2009). Digital inclusion projects in developing countries: Processes of institutionalization. *Information Technology for Development, 15*(2), 95–107.
35. Makate, C. (2019). Effective scaling of climate smart agriculture innovations in African smallholder agriculture: A review of approaches, policy and institutional strategy needs. *Environmental Science & Policy, 96*, 37–51.
36. Mana, A. A., Allouhi, A., Hamrani, A., Rahman, S., el Jamaoui, I., & Jayachandran, K. (2024). Sustainable AI-based production agriculture: Exploring AI applications and implications in agricultural practices. *Smart Agricultural Technology*, 100416.
37. Maroli, A., Narwane, V. S., & Gardas, B. B. (2021). Applications of IoT for achieving sustainability in agricultural sector: A comprehensive review. *Journal of Environmental Management, 298*, Article 113488.
38. Martínez-Peláez, R., Ochoa-Brust, A., Rivera, S., Félix, V. G., Ostos, R., Brito, H., & Mena, L. J. (2023). Role of digital transformation for achieving sustainability: Mediated role of stakeholders, key capabilities, and technology. *Sustainability, 15*(14), 11221.
39. Maru, A., Berne, D., Beer, J. D., Ballantyne, P. G., Pesce, V., Kalyesubula, S., & Chavez, J. (2018). Digital and data-driven agriculture: Harnessing the power of data for smallholders. *Global Forum on Agricultural Research and Innovation.*
40. Mayuravaani, M., Ramanan, A., Perera, M., Senanayake, D. A., & Vidanaarachchi, R. (2024). Insights into artificial intelligence bias: Implications for agriculture. *Digital Society, 3*(3), 48.
41. Mehrabi, Z., Ellis, E. C., Ramankutty, N., & Ricciardi, V. (2021). The global divide in data-driven precision agriculture. *Nature Sustainability, 4*(2), 154–160.
42. Mehrabi, Z., McDowell, M. J., Ricciardi, V., Levers, C., Martinez, J. D., Mehrabi, N., & Jarvis, A. (2021). The global divide in data-driven farming. *Nature Sustainability, 4*(2), 154–160.
43. Meng, X., Wang, X., Nisar, U., Sun, S., & Ding, X. (2023). Mechanisms and heterogeneity in the construction of network infrastructure to help rural households bridge the "digital divide." *Scientific Reports, 13*(1), 19283.
44. Mhlanga, D. (2021). Artificial intelligence in the industry 4.0, and its impact on poverty, innovation, infrastructure development, and the sustainable development goals: Lessons from emerging economies. *Sustainability, 13*(11), 5788.

45. Mikhaylov, S. J., Esteve, M., & Campion, A. (2018). Artificial intelligence for the public sector: Opportunities and challenges of cross-sector collaboration. *Philosophical Transactions of the Royal Society A: Mathematical, Physical and Engineering Sciences, 376*(2128), 20170357.
46. Nyakudya, I., Musademba, D., Chikosha, F., & Nyashanu, A. (2024). Leveraging digital literacy to bridge the information gap in agriculture. *Journal of Agricultural Science and Technology, 26*(4), 345–356.
47. Nyakudya, S., Jambo, N., Madududu, P., & Manyise, T. (2024). Unlocking the potential: Challenges and factors influencing the use of ICTs by smallholder maize farmers in Zimbabwe. *Cogent Economics & Finance, 12*(1), 2330431.
48. Paul, M., Jr., Salama, Y., Abay, K. A., Abdelaziz, F., Zaccari, C., Akramkhanov, A., & Anarbekov, O. (2024). Harnessing digital innovations for climate action and market access: Opportunities and constraints in the CWANA region. *Global Food Security, 41*, Article 100763.
49. Robinson, S. C. (2020). Trust, transparency, and openness: How inclusion of cultural values shapes Nordic national public policy strategies for artificial intelligence (AI). *Technology in Society, 63*, Article 101421.
50. Rotz, S., Gravely, E., Mosby, I., Duncan, E., Finnis, E., Horgan, M., & Fraser, E. (2019). Automated pastures and the digital divide: How agricultural technologies are shaping labour and rural communities. *Journal of Rural Studies, 68*, 112–122.
51. Ryan, M. (2023). The social and ethical impacts of artificial intelligence in agriculture: Mapping the agricultural AI literature. *AI & SOCIETY, 38*(6), 2473–2485.
52. Saiz-Rubio, V., & Rovira-Más, F. (2020). From smart farming towards agriculture 5.0: A review on crop data management. *Agronomy, 10*(2), 207.
53. Sarkar, S., Das, S., Dutta, S., Bera, B., Bera, M., Roy, D., Middya, M. M. H., & Choudhury, M. R. (2024). Economic analysis of the behavior of prices and market arrivals of tomatoes—a case study of West Bengal, India. *Agricultural & Rural Studies, 2*(2), 8–15.
54. Saura, J. R., Ribeiro-Soriano, D., & Palacios-Marqués, D. (2022). Assessing behavioral data science privacy issues in government artificial intelligence deployment. *Government Information Quarterly, 39*(4), Article 101679.
55. Sharma, A., Sharma, A., Tselykh, A., Bozhenyuk, A., Choudhury, T., Alomar, M. A., & Sánchez-Chero, M. (2023). Artificial intelligence and internet of things oriented sustainable precision farming: Towards modern agriculture. *Open Life Sciences, 18*(1), 20220713.
56. Sharma, V., Tripathi, A. K., & Mittal, H. (2022). Technological revolutions in smart farming: Current trends, challenges & future directions. *Computers and Electronics in Agriculture, 201*, Article 107217.
57. Siddiquee, K. N. E. A., Islam, M. S., Singh, N., Gunjan, V. K., Yong, W. H., Huda, M. N., & Naik, D. B. (2022). Development of algorithms for an IoT-based smart agriculture monitoring system. *Wireless Communications and Mobile Computing, 2022*(1), 7372053.
58. Torky, M., & Hassanein, A. E. (2020). Integrating blockchain and the internet of things in precision agriculture: Analysis, opportunities, and challenges. *Computers and Electronics in Agriculture, 178*, Article 105476.
59. Tzachor, A., Devare, M., King, B., Avin, S., & ÓhÉigeartaigh, S. (2022). Responsible artificial intelligence in agriculture requires systemic understanding of risks and externalities. *Nature Machine Intelligence, 4*(2), 104–109.
60. United Nations. (2024). https://www.un.org/techenvoy/content/global-connectivity#:~:text=An%20estimated%203.7%20billion%20people,to%20work%20towards%20these%20outcomes
61. United Nations. (2024). The United Nations sustainable development goals report.
62. Vishnoi, S., & Goel, R. K. (2024). Climate smart agriculture for sustainable productivity and healthy landscapes. *Environmental Science & Policy, 151*, Article 103600.

Beyond the Field: AI and IoT in Food Distribution

Hagar Fathy, Amr Osama Sabour, and Yassmin Sameh Abdellatif

Abstract Artificial intelligence is algorithm development for machines that can process and analyze vast volumes of data, recognize relationships, recognize patterns, and based on that analysis call decisions or predictions based on computer models. Several industries and sectors have seen an increase in the use of artificial intelligence: agriculture finance, manufacturing, transportation, education, retail, and healthcare are just limited Artificial intelligence technology developments that are anticipated to have a big influence in the future on jobs for example, the agri-food industry is increasingly using artificial intelligent to provide more sustainable, efficient and effective products. With concrete examples, it could radically transform the agri-food industry in terms of food safety, crop tracking, precision agriculture, supply chain optimization, predictive analytics, food processing, nutrition quality, and nutrition management. This book chapter reveals that recent advances in artificial Inelegant have increased productivity, enhanced food safety, decreased waste, enhanced quality, and adapted to the food industry which led to summarizing the limitations, and potential applications of AI in the food sectors.

Keywords Artificial intelligent · Sustainable management · Food · Dairy · Internet of things

Abbreviations

AI Artificial intelligence

H. Fathy (✉)
Animal Production Research Institute (APRI), Agricultural Research Center (ARC), Dokki, Giza, Egypt
e-mail: hagarfathy@pg.cu.edu.eg; hagarfathy8@gmail.com

A. O. Sabour
Faculty of Agriculture, Assiut University, Assiut, Egypt

Y. S. Abdellatif
Faculty of Biotechnology, Modern Sciences and Arts University (MSA University), Giza, Egypt

S. R. Kumar et al. (eds.), *AI-Driven IoT Solutions in Agriculture: New Insights*, Studies in Computational Intelligence 1236, https://doi.org/10.1007/978-3-032-05603-0_9

ML	Machine learning
ANN	Artificial neural networks
IoT	Internet of things
COP	Clean-out-of-place
CIP	Cleaning in place
CVS	Concurrent versions system
DT	Digital twin
Ens	Electric noses
ET	Electric tongues
NGS	Next-generation sequencing
BSF	Black soldier fly
TDNN	Time-delayed neural network

1 Introduction

The population of the globe is predicted to increase quickly, reaching approximately 9.7 billion people by 2050 [1, 2]. The question of ways to meet the expanding need for food while maintaining sustainability, and food security is becoming increasingly pressing. This is why the agri-food business might undergo a revolution and become more sustainable through the application of artificial intelligence (AI) in numerous ways. In addition to increasing efficiency and lowering waste, it can assist farmers, food producers, and distributors in making better decisions and enhancing food security and sustainability. In 1965, economist Herbert Simon, who won the Nobel Prize, predicted that "machines will be capable of doing any work a man can do." The incredible advancements made possible by AI applications have realized his futuristic viewpoint [3]. AI is the capacity of programs of computers to do tasks like reasoning, learning, decision-making, and problem-solving that are typically performed by humans. AI consists of several subfields, such as computer vision, robotics, natural language processing, deep learning, cognitive computing, and machine learning (ML). AI technology can make use of a variety of techniques, including Logic Programming, Artificial Neural Networks (ANN), Swarm Intelligence, Fuzzy Logic, Reinforcement Learning [4], and expert systems [5]. The selection of an algorithm is contingent upon the particular task or issue at hand, as each algorithm possesses distinct benefits and drawbacks. AI finds widespread use in fields including video analysis, image analysis [6], speech recognition, financial forecasting, autonomous vehicles [7], and many more [8]. AI may be applied in the agri-food sector, just like it can in any other business, to increase productivity [9], create new, more nutrient-dense crops [10], cut waste [11], and guarantee safety [12]. AI can be utilized to enhance distribution and logistics [13] and maximize crop yields [13, 14]. AI can be used in precision agriculture to evaluate data from satellites, drones, and sensors to maximize agricultural techniques including fertilization, irrigation, and pest control. Higher yields, lower prices, and a smaller environmental effect are possible outcomes [15–17]. AI-powered sensors and cameras for agricultural monitoring can track crops in nutrient shortages, and nutrient shortages. This makes farmers able to avoid crop loss and act swiftly [18–20]. To forecast crop yields and consumer demand, AI systems can examine historical data, soil conditions, and weather trends. This can assist farmers in scheduling their planting and harvesting activities and in making the most of their pricing plans [21]. AI has the potential to improve supply chain efficiency through waste reduction, logistics optimization, and demand prediction. AI algorithms, for instance, can be used to optimize inventory levels, route trucks, and forecast the best times to harvest crops [22, 23].

AI can potentially improve food processing activities, including grading and sorting, and identify flaws or pollutants in food products [24]. Fruits and vegetables can also be identified and sorted using AI according to size, color, and other characteristics. This can lessen waste while enhancing the uniformity and quality of food products. By evaluating data sensors also evaluating cameras to find possible pollutants or additional dangers, Food safety can be monitored with AI. By doing so, foodborne illness can be avoided and public health can be enhanced [25, 26]. Furthermore, AI can be utilized to deliver individualized dietary recommendations by analyzing customer data such as gender, activity level, and age. This can enhance general health and enable customers to make better-informed dietary decisions [27]. Even if these studies offer some insightful information on applications of AI in the agri-food sector, and dairy industry a thorough analysis is required to comprehend the most recent developments in AI technology in this field. Thus, the goal of this analysis is to draw attention to the latest advancements in the food and agriculture industries as well as the use of AI technology, with particular examples drawn from databases spanning the years 2010 to 2023. The assessment also provides an overview of the field's constraints, difficulties, and opportunities going forward [28].

2 AI's Function in the Processing of Food

AI is a branch of computer science that simulates human knowledge bases, learning abilities, and cognitive processes. AI provides a wide range of methods, such as ANNs to address a wide range of issues in the food and agriculture sectors, including the world's food demand. AI has been applied in supply chain management, manufacturing optimization, food quality improvement, and industrial hygiene maintenance. Using a computerized system, the industry can evaluate and deliver the product under ideal conditions at every stage. For instance, it can monitor all aspects of seed selection, crops, agricultural operations (including watering and temperature monitoring), new product development, and adherence to food industry safety standards, all of which can improve the quality of the products. However, there are barriers to AI's adoption in the food sector, including concerns about data privacy, the automation of jobs, errors in the algorithms used to make judgments, and challenges with legal and regulatory compliance. Implementing AI is made considerably more difficult by the ethical considerations of food sustainability and safety. The future of the food business hinges on adopting a comprehensive strategy that maximizes AI's potential while upholding moral and ecologically responsible business practices [29].

The food processing business is progressively embracing AI and ML technology. These technological advancements aid process optimization, effectiveness, and quality control. In addition to the tools and goods produced by different AI technologies, The multitasking capabilities of intelligent systems such as product sorting, intelligent food packaging, development of new food products, foreign object detection, supply chain management, and equipment cleaning are explained [16, 30].

2.1 Smart Sensors in Dairy Sectors

In dairy companies, sensors are extremely promising for managing waste and by-products, particularly when combined with AI technologies or robots. For example, a Fourier-transform infrared spectroscopy sensor in conjunction with chemometric techniques was used to monitor the synthesis of galacto oligosaccharides from cheese whey in real time. Dairy products are susceptible to microbial deterioration; therefore, various heating techniques are frequently applied to prolong their shelf life and minimize dairy waste. A variety of spectroscopy-based sensors can be used to monitor the thermal treatments applied to dairy products [31]. Using smart sensors and MV in precision dairy production might reduce waste, boost breeding efficiency, and save costs [32].

2.2 Recognition of Foreign Objects

Foreign object contamination results in serious problems, such as food recalls, customer rejection, customer injury, and a decline in brand trustworthiness. During food processing, handling, or preparation, foreign objects such as insects, glass, metal, or rubber may inadvertently find their way into the meal or packaging material. Eating food contaminated with foreign objects can cause choking or other issues, even though the dangers connected with them vary depending on their size, kind, clarity, and hardness. Since it can be difficult to identify these pollutants visible to the inexperienced eye. Technologies like AI and ML can be quite helpful in detecting them by examining food product photos to find any possible extraneous objects [33]. Utilizing image recognition systems that have been extensively trained on photo datasets of uncontaminated, and contaminated food goods is one method of employing AI and ML to detect foreign objects. Then, using real-time picture analysis of the food goods, the system can detect any potential foreign objects.

2.3 Food Printing: 3D and 4D Extrusion Technology

When integrated AI with technologies such as 3D and 4D printing, AI can be especially interesting. It is feasible to improve printer performance, lower the possibility of errors, and enable automated production by incorporating AI into the 3D/4D printing process. Combining AI with 3D/4D printing technology can lead to the creation of startup companies along with study initiatives that use AI in 3D/4D printing products and services [34–37]. 3D food printing technology is currently being adopted by the food processing industries globally to enhance operations more methodically and autonomously. By enabling computerized food customization, delivering on-demand food manufacturing, and lowering food waste, this technology has the potential to increase profitability and customer satisfaction in active food chains of value. The main distinction between a standard 3D printer and a 3D/4D food printer is the type of printing media that is utilized. A 3D/4D food printer uses food items as the printing base instead of melted plastic. Customers can use websites or mobile applications to actualize designs from an e-commerce platform, which reduces the cost of warehousing, packaging, and delivery [38]. By improving 3D printing techniques and equipment, food printing becomes more efficient while food processing costs and time are reduced [39]. Another point is that 3D printing technology will enable personalized products to be delivered faster than traditional food processing methods.

Food safety will be enhanced as a result of the products being delivered considerably faster and possibly eliminating the need for chemical preservatives and synthetic polymer-based packing materials. Moreover, food product multiphase processing could be reduced to a single step [40]. Food Packaging with Intelligence.

One of the demanding jobs and labor-intensive processes in the production divisions of the food industry is correctly arranging and packing food goods. Food packaging has four essential purposes: it must protect food, present the product, maintain cleanliness, and facilitate transportation [41]. Throughout the entire food logistics chain, it guards against physiologic, chemical, and biological harm to the food. The package's visual appearance serves as both the initial impression of the product and a tool for consumers to assess the food's quality. Therefore, effective packaging is vital to the processing of food [42]. The food packaging business is using AI and ML more and more to enhance the production, design, and usability of packaging materials. To enhance packaging design, the food packaging sector is utilizing AI and ML technology. Liu used a deep convolution generative adversarial network (DC-GAN) based package design model [42]. Visual communication technology can be used to improve the image of packaging design, leading to increased visual communication capacity, an enhanced impact of packaging design, and an increased degree of integration of visual information. Because fruits and vegetables vary so much in size, color, and form, creating AI-based systems is a laborious effort in this industry. As a result, a vast amount of data is needed to correctly train the system and carry out the task in an organized way.

Three-dimensional printing: Most dairy by-products have diverse components (such as proteins, fats, carbs, and minerals) that are reusable as components in a variety of food production procedures and recipes. Researchers looked into the 3D printing of food pastes with high starch, fiber, and protein, content to create nutrient-dense diets low in fat or sugar [43]. At varying proportions (ranging from 10 to 45%), the pastes contained oat protein, faba bean protein, rye bran, and semi-skimmed milk powder. It was discovered that milk powder's fat and carbohydrate content affected its printability and structural stability, with a higher fat content allowing for smoother material flow through the syringe tip [44]. Liu et al. used extrusion-based 3D printing and used milk-protein concentrate instead of whey protein isolate obtaining similar outcomes [31]. They discovered that the capacity of various protein types to bind water varies. Further, over time, WPI improved printability and deformation resistance, while less WPI produced a paste with reduced viscosity and increased mobility when added to MPC. With accuracy, fermentation culinary waste, like cheese whey, is being utilized more and more as a fermentation medium these days to create a range of culinary ingredients and bioactive substances, including enzymes, natural preservatives, colorants, and antioxidants. Precision fermentation, as opposed to conventional fermentation, entails genetically engineering microorganisms to generate complex organic compounds, such as particular enzymes or protein ingredients [45]. Making sustainable macromolecules, and microbial proteins from dairy waste and other agro-industrial waste and byproducts, this novel biotechnological approach is being used more frequently. These precision-fermented chemicals have better purities than traditionally produced components, meaning that they can be utilized in much smaller amounts to improve the end product's functional and sensory qualities [45].

Drones and other intelligent technologies like robotics can also play a vital part in drastically lowering packing costs [46]. Robotics reduces the demand for work from humans, automates the packing process, and increases the process's speed and efficiency. Food goods can be inspected and sorted by robots to make sure they are packaged correctly and live up to quality requirements. Robotics can be used to control inventory, lowering the possibility of shortages or surplus stock and guaranteeing that packaging materials are available when needed [47].

Degradation reactions, microbiological contamination, and robots in dairy manufacturing and processing could happen during different unit operations, like cutting. For instance, there may be hazards to food safety due to cross-contamination of products or product components. This issue has been noted with Listeria monocytogenes-contaminated Gorgonzola cheese rinds. Innovative robotic systems that control the cutting stage while avoiding food waste, and sanitary emergencies, and improving sustainability through energy savings can be introduced to address this problem. If robots are given the ability to understand the mechanical properties, chemical makeup, and microbiological quality of food products, they ought to be able to arrange cutting operations more efficiently. Big data, robots, and AI can all be combined to enhance the results. An intelligent Tetra Pak® vertical blender was developed by the Tetra Pak® company to reduce cheese loss from complete emptying and increase cottage cheese output. Along the same lines, Fairme®, a French start-up with headquarters in Grenoble, used robotics, the Internet of Things (IoT), and AI technology to build a fully autonomous dairy processing factory that can process many products, including butter, cream, cheese, and yogurt. After a buyer places an online purchase, the production facilities can be put into use. By avoiding overproduction and unsold food products, this robotic farm reduces dairy waste by processing the proper amount of milk and producing the right quantity of finished products [48].

2.4 *Demand–Supply Chain Management*

Policies about food safety are essential for the open and honest execution of all food logistics operations and are currently a source of worry for the food business [49]. Artificial Intelligence is being utilized to monitor all process steps, including resource management and cost control. From the point of cultivation to the point of consumer gathering, it tracks and predicts the flow of goods [50]. Retailers can get comprehensive AI solutions from one company. Many retail operations, such as inventory control, resource management, transportation, and billing, can be made better with their solutions. By enhancing shelf life and optimizing packaging, the technique can help the retail industry [51]. AI algorithms that access data on product features, and ambient conditions can assist in determining the best packing materials and designs to enhance product quality and shelf life. AI can also contribute to better food safety by enhancing visibility and transparency across the supply chain and logistics operations. Businesses can detect and resolve potential problems with food safety before they become serious by tracking products from farm to table. To achieve the aim of a broader agri-food supply chain that includes more parties and spans the whole supply chain lifespan, more contributions using a variety of data sources will be needed [52, 53].

2.5 *Sorting for Products*

To assess product parameters at the input level, a variety of technologies, including X-ray systems, laser technology, Infrared spectroscopy, and high-resolution cameras can be integrated by AI-based systems. [54]. By assisting in the identification of flaws, impurities, and inconsistencies in the products, these technologies can encourage thoughtful choices and raise the general caliber of the output. Sorting techniques that depend on input-level evaluation may suffer greatly from uneven product homogeneity. Products that aren't homogeneous can lead to poor sorting judgments, which might increase waste or degrade product quality [55]. Some sorting systems address this issue by utilizing a variety of technologies and sensors to assess the caliber of the product and spot flaws from various angles. Furthermore, by training ML algorithms to identify trends and variances in product attributes, sorting decisions can become more accurate and consistent. An AI system is a leading global supplier of cutting-edge collecting and sorting devices, that can effectively complete the sorting operation with 90% efficiency. Industries have benefited from using such systems in several ways, including higher yields, lower labor costs, and expanded production. It has been found that the problems associated with potato segregation and arrangement can increase by 5–10% [56].

To classify apples utilizing datasets via recognizing patterns and decision-making, the deep learning industry is a subfield of ML and AI. Similar to this, Deep Convolutional Neural Network uses Concurrent Versions System support to help identify the type of apple. The data from apple detection, ripeness classification, and image processing were processed by the deep learning model. From a given digital image, the classifiers can produce the best outcome, which is the apple's ripeness class [55, 57, 58]. Additionally, standards, categories, flaws, and the type of beverage produced are taken into consideration while classifying coffee beans. Arabic espresso varieties are categorized into eight groups based on their imperfections, ranging from two to eight.

2.6 Development of Novel Food Products

Since the food business entirely depends on public opinions for new product development, the information gathered by the different decision-making processes is helpful when introducing new items. The ML-based module could determine the precise answers to questions like "What exactly the consumers are looking for" and come to intelligent Decisions by examining the data that the system collected. Soft drink delivery is provided by automatic vending machines installed by one of the multinational corporations across the USA. Customers can choose from thousands of available flavors. Cherry Sprite, a new product that the company launched, is an illustration of how the machine could be inspected using the data collected by ML module, and deep learning algorithms. Additionally, it has been proposed that several food sectors will benefit tremendously from the application of ML in the coming decades when it comes to the development of new food products [59]. Using AI technology, a biotech company has unveiled the first-ever bioactive peptide. A special peptide network made from rice protein is used as an element in sports nutrition to enhance immune function and reduce inflammation by modifying cytokine responses. The business has established itself as the pioneer in the world for showcasing AI's ability to enhance human wellness [59]. Another US-based IoT technology startup, Hello Egg, has partnered with an AI-powered "home cooking sidekick," mobile application and web pages that completely automate kitchen tasks. With the use of voice technology, the home assistant may make personalized nutrition recommendations depending on each user's preferences. In addition, it can organize the shopping cart, maintain the pantry, provide recipes in video form, and help with grocery delivery [59]. The purpose of sensory panelists working in the food and beverage sectors is to assess new goods using their senses under customer taste preferences. Unfortunately, it is challenging to forecast how the target group will be seen and what their preferences will be. This prompted the development of a solid approach by the industries for gauging and forecasting customer preferences using an AI-based Gastrograph system that interprets market choice using ML and prediction algorithms [60].

2.7 Cleaning and Maintenance of Equipment

For effective maintenance, equipment, and The food processing industry processing instruments require routine cleaning. AI-driven technologies, such as clean-out-of-place [61], and cleaning-in-place (CIP) systems, support the food industry in maintaining product quality and hygiene [61]. To carry out the tasks, a variety of cameras and sensors are mounted for its implementation. A European business that specializes recently announced SOCIP, a cleaning solutions specialist, a Self-Optimizing-CIP system that uses AI to automatically optimize the cleaning procedure for equipment used in food preparation. To determine how many food particles and microbiological detritus are inside the apparatus, SOCIP uses optical fluorescence and ultrasonic sensing imaging techniques. Using sensors to scan the interior of the apparatus and produce a real-time image of the surface, the SOCIP system operates. The amount of cleaning required to get the appropriate degree of cleanliness is then calculated using this information.

3 AI's Impact on Food Safety and Quality

AI is a technology that industries would like to use for decision-making, cost reduction, process estimates, high profitability, and overall quality enhancement because of its compelling capabilities. When AI and data science are paired, the quality of food and services provided by eateries, online food delivery services, and grocery stores can all be improved, which might boost revenue, profitability, and customer pleasure. AI can enhance supply chain optimization, menu optimization, sales forecasts, and tailored suggestions by evaluating data and using algorithms [62]. Food quality management, food security management, and food waste management are the three main categories AI has significantly contributed to the food industry [63]. AI plays important roles and has potential applications in several food sectors, including dairy, beverage, and bakery.

AI has the potential to optimize production and enhance product quality by analyzing vast volumes of data from dairy production operations and identifying trends and forecasts. Dairy farmers can achieve higher quality and more efficient production by utilizing AI techniques like FL and ANN to generate more accurate forecasts and modify their procedures accordingly. In the beverage sector, AI solutions like e-nose, e-tongue, concurrent versions systems (CVS), and picture analysis can assist producers in streamlining their operations and comprehending customer preferences. Additionally, AI may benefit the bakery sector by boosting efficiency and improving product quality. Bakeries can also enhance manufacturing methods and produce more inventive goods using AI-powered equipment like robots and visualization.

3.1 Quality Management of Food

Because fresh produce is highly perishable, improper storage and monitoring can cause it to spoil quickly. A large amount of food was wasted in the past since many merchants lacked the equipment needed to keep an eye on the condition of fruits that were being stored in real-time. But thanks to technological improvements, sellers may now monitor the condition of fresh fruits and vegetables in real-time with several options [64]. A promising technique for keeping an eye on the state and quality of fresh produce along the cold chain is digital twin (DT) technology. Fruit tissue damage is influenced by several variables, including their physical and biochemical characteristics as well as environmental elements including humidity, temperature, and postharvest handling. It is possible to see any problems early on and take corrective measures to stop more harm by employing DT technology to monitor these parameters [65]. DT technology can be utilized to enhance fresh fruit and vegetable storage and transportation, in addition to monitoring the cold chain [65, 66]. In the fruit and food sector, thermal imaging is a non-contact, developing technology that is gaining popularity for assessing fruit quality. It provides an alternative to extraction, which could result in irreversible damage, for a non-destructive method of inspecting a product. By detecting the temperature difference between healthy and injured tissues as a result of variations in thermal diffusion coefficients. Infrared thermal imaging can identify the existence of damage or flaws in fresh vegetables, and fruits [66]. Similarly, the food business is using CVS technologies more frequently to evaluate different food items. By using several CVS types, such as conventional CVS, hyperspectral CVS, and multispectral CVS, it is possible to identify potential issues early on and take corrective action to prevent additional loss or contamination, to examine food external quality commodities [67]. Among the most promising AI innovations for the food business are electric tongues (ET) and electric noses (Ens) [68]. The EN is an instrument comprising a variety of electronic chemical detectors with low specificity and an appropriate recognition of patterns method, able to detect both simple and complex odors. Ets are multisensory techniques for liquid analysis that utilize chemical sensing arrays and the recognition of patterns [69]. The many flavors and scents that are present in the food or beverage being examined can be detected by the sensors employed in Ens. After that, this data is moved to a data center so that ML algorithms can access it [70]. These algorithms can examine the data and conclude the knowledge they have gained. Furthermore, food products can be made with higher overall quality by utilizing EN technology. Based on minute variations in the scents and aromas of several batches of food products, businesses can modify their production processes to guarantee that their items are reliable and of the greatest caliber [71]. The Specialty Coffee Association of America developed the coffee cupping technique. This is a tried-and-true method for assessing the flavor of well-roasted coffee. It entails stepping grounded beans of coffee in hot water and assessing the flavor, fragrance, and other sensory aspects of the resulting brew. AI tools like ANNs and Ens have been employed to improve coffee cupping in recent years. The volatile organic compounds that give coffee its aroma can be found and examined using e-noses. A great degree of accuracy can be achieved in predicting the flavor and quality of roasted coffee by combining an ANN and EN [72]. Another innovative technology in the food business that shows promise is eTs, which may be used to evaluate the characteristics of different kinds of beverages, such as dairy and alcoholic ones. Ets are capable of identifying a variety of flavor attributes, including sweetness, saltiness, sourness, and bitterness, all of which can play a significant role in assessing a product's overall quality. For instance, different flavor components, such as theaflavin and thearubigin, have different concentrations depending on how old the tea is. These chemicals can be quantified and utilized to determine the type of tea being examined by combining a UV–VIS spectrophotometer-based analysis with a pulse voltammetric ET [73, 74]. The tongue's capacity to identify dissolved solids and volatile compounds—which produce scents and have the potential to evaporate is one of its main advantages. This makes it a helpful tool for examining a sample's scent and overall flavor profile [75]. Many studies have been conducted on the application of Ets for food recognition, including the distinction between liquid and meat foods. An ET was utilized to examine a variety of liquid food, flesh samples, meat products, and juice [74]. Based on taste profiles, the study's findings demonstrated that the ET could effectively distinguish between various liquid and meat food types. The uses of electronic tongues and noses (e-noses and e-tongues, respectively) in identifying attributes linked to food quality. The study discovered that because e-noses and e-tongues can detect and identify a wide range of volatile and non-volatile substances that affect food flavor and scent, they are being employed more and more in the food sector [75].

3.2 Managing Food Safety

Food safety guarantees that it satisfies the mandatory nutritional criteria and is free of any hazardous or poisonous ingredients. To guarantee food safety and proper hygiene throughout food processing, storage, and sale, as well as to reduce the possibility of biotic and abiotic pollutants that might result in food poisoning, a multidisciplinary approach is required. Food product attributes including size, shape, color, and texture may all be examined using image processing techniques. Food items can be quantified in terms of size and shape for quality control purposes by measuring their anticipated area and perimeter [76].

It is important to remember, nevertheless, that image processing techniques by themselves are unable to identify dangerous microbes or other possible threats to food safety. To ascertain a single cultivated isolate's entire genome sequence (such as a virus, and a bacterial colony), next-generation sequencing (NGS) is utilized. This process is also referred to as "whole genome sequencing" or "metagenomics," whereby the sequences of various microorganisms found in a biological sample are produced using NGS. Additionally, the analysis of NGS data can be completed much more quickly with the help of AI and automation, which makes it possible to identify possible risks to food safety earlier. This enables prompt intervention before contaminated items reach consumers, perhaps preventing widespread disease [77].

3.3 Managing Food Waste

Food waste is a serious problem that has an impact on food security, the environment, and the viability of the economy. 1.6 billion tons of food, roughly are wasted every year, with inedible by-products of food production processes accounting for the majority of this waste (81%). Within the food sector, there is a growing realization that wasted food is a cost of conducting business that cannot be avoided, but also a major resource underuse problem and sustainability issue. Utilizing contemporary methods like omics can help address the problems associated with managing and reducing food waste. To comprehend the biological reactions that occur during the breakdown of food waste and to pinpoint possible dangers and food waste contaminants, for instance, disease omics, proteomics, metagenomics, waste omics, and transcript can be employed. To lessen and manage the problems associated with food waste, experts and government agencies have created and tested several various concepts and solutions. For example, Black Soldier Fly (BSF) farming is a promising technique because of its adaptability and multiple uses, including managing food waste and producing sustainable food. When it comes to productive production and effective waste management, BSF farming can greatly benefit from the application of IoT technologies. In terms of effective production and waste management, BSF farming can greatly benefit from the adoption of IoT technologies. Farmers may monitor and manage several aspects of BSF farming remotely by combining sensors and devices with software and mobile applications. Higher yields and higher-quality biomass are produced as a result of the ability to more precisely and individually manipulate the BSF larvae's growth environment [78, 79].

When paired with other Industry 4.0 technologies, IoT has the potential to significantly reduce food waste associated with milk and dairy products. Buildings, machinery, automobiles, and even living things, like cows and dairy farms, can all be linked together to monitor resources more successfully. Businesses may find that tracking systems are particularly helpful in reducing food waste. IoT might be able to keep farmers updated on the health of each animal through wireless sensor devices. Before they influence the milk supply, cow illnesses can be successfully and reliably detected by sensor-based technology. IoT based food-waste monitoring systems can cut food waste by as much as 60.7%, which has favorable effects on the environment and the bottom line [80].

3.4 Shelf Life Prediction

Food shelf life has been frequently predicted using AI and ML approaches, such as ANNs. These methods create predictive models that can determine the anticipated shelf life by utilizing data from the food product's many sensory, and physicochemical properties combined with models of mathematical [81–84]. Using time-delayed neural network (TDNN) models to forecast processed cheese's shelf life. The goal of the project was to create a model that would be able to predict processed cheese's shelf life with accuracy and eliminate the necessity for labor-intensive physical testing. The study's findings demonstrated that the TDNN model could correctly forecast processed cheese's shelf life [85]. based on ripeness levels and respiration rates in different supply chain situations, AI models were constructed to forecast the duration of mangoes' shelf life in different conditions. To classify the maturity degrees of mangoes, a deep-CNN was refined using 1524 picture data [86].

3.5 AI's Function in Personalized Nutrition

A person's diet can be very unique and complicated, and what works for one person might not work for another. A technique known as "personalized nutrition" takes into account each person's particular dietary requirements, preferences, and health objectives. Technology breakthroughs like AI and ML are making it possible to create customized dietary plans. These programs generate tailored nutrition recommendations based on information about a person's food, lifestyle, microbiota, and genetics [87]. A biotechnology business that provides individualized nutrition plans based on a person's microbiome. The business analyzes a person's gut microbiome using AI and ML to generate customized food recommendations depending on the kinds and quantities of gut bacteria. A different digital health organization provides various individualized health coaching services, such as hypertension control, weight management, diabetes management, and tailored nutrition coaching. This startup offers individuals individualized health coaching through voice AI and speech recognition technologies. Through speech recognition technology, the company's mobile app analyzes a user's voice and offers tailored coaching depending on the user's responses. One is a website and smartphone app that lets users keep track of their daily food consumption and nutrition objectives. Using a database of more than 800,000 food items, users of the software may effortlessly record their meals and monitor their calorie, macronutrient, and micronutrient intake. Im2Calories analyzes the visual characteristics of food photos and estimates the caloric content of the food by combining computer vision algorithms with deep learning [88].

4 Blockchain in Dairy Sectors

Information on traceability and transparency is required for all food supply chains, but as milk products are among the most susceptible foods, this information is especially crucial. Pollution, hazardous chemicals, poor industrial practices, and poor animal husbandry can all cause issues for the supply chain for milk. Blockchain technology might be used to track milk products in the dairy supply chain, ensuring the quality of the final product and minimizing byproducts and waste. Blockchain technology makes it possible to swiftly identify the farm's origin and the precise batch that is implicated in cases of dairy food safety problems. Food waste is reduced in this way since only one batch—more specifically, one cow is thrown out rather than the entire herd or output [80].

5 Conclusion

AI holds great promise to transform the food and agriculture industry by enhancing productivity, fostering sustainability, and optimizing efficiency. But there are also some worries about how AI will affect the food and agriculture industries in the future. For instance, there are worries that AI can lead to greater inequality and a loss of jobs in rural areas. The high expense of putting AI systems into practice is a significant barrier. The substantial hardware, software, and training costs associated with AI may be too expensive for businesses of all sizes. Concerns have also been raised concerning the accuracy and dependability of systems of AI, especially for food safety, and agriculture management decision-making. To raise sustainability, efficiency, and production, in the sector of agri-food. AI can completely transform the agri-food industry by making it more inventive, efficient, and sustainable, but it also brings up significant ethical, legal, and societal issues that must be properly explored and resolved. To guarantee long-term advantages, it is crucial to make sure that these technologies are created and applied ethically and sustainably. The creation and application of AI technologies, the laws and rules governing their application, and how society responds to the changes AI brings about will all have an impact on how sustainable AI becomes. AI sustainability takes into account some social, cultural, and environmental aspects. The sustainability and future of AI require careful consideration of many important factors. The skills gap must be closed, and enough talent must be available to create and implement AI systems sustainably and responsibly. This necessitates funding educational and training initiatives that can provide participants with the know-how required to operate in the AI industry. Even though there are still obstacles to be solved, like worries about data privacy, expensive costs, moral dilemmas, and the requirement for specialized training, AI's future in this sector is bright. In the years to come, food production and distribution should see major advances as ever more farmers employ gadgets with AI.

References

1. Godfray, H. C. J., Beddington, J. R., Crute, I. R., Haddad, L., Lawrence, D., Muir, J. F., et al. (2010). Food security: The challenge of feeding 9 billion people. *Science, 327*(5967), 812–818.
2. Albalaty, J., Forsan, F., Awad, S., Elsadek, E., Rahman, A., Badawy, M., et al. (2023). SARS-CoV-2 vaccines from A to Z: A review of the current challenges. *Global Nest Journal* 148–171.
3. Pournader, M., Ghaderi, H., Hassanzadegan, A., & Fahimnia, B. (2021). Artificial intelligence applications in supply chain management. *International Journal of Production Economics, 241*, Article 108250.
4. Pantazi, X.-E., Moshou, D., & Bravo, C. (2016). Active learning system for weed species recognition based on hyperspectral sensing. *Biosystems Engineering, 146*, 193–202.
5. Kumar, K., & Thakur, G. S. M. (2012). Advanced applications of neural networks and artificial intelligence: A review. *International Journal of Information Technology and Computer Science, 4*(6), 57–68.
6. Saeed, W., & Omlin, C. (2023). Explainable AI (XAI): A systematic meta-survey of current challenges and future opportunities. *Knowledge-Based Systems, 263*, Article 110273.
7. Miglani, A., & Kumar, N. (2019). Deep learning models for traffic flow prediction in autonomous vehicles: A review, solutions, and challenges. *Vehicular Communications, 20*, Article 100184.
8. Goodell, J. W., Kumar, S., Lim, W. M., & Pattnaik, D. (2021). Artificial intelligence and machine learning in finance: Identifying foundations, themes, and research clusters from bibliometric analysis. *Journal of Behavioral and Experimental Finance, 32*, Article 100577.
9. Dhanaraju, M., Chenniappan, P., Ramalingam, K., Pazhanivelan, S., & Kaliaperumal, R. (2022). Smart farming: Internet of Things (IoT)-based sustainable agriculture. *Agriculture, 12*(10), 1745.
10. Chukkapalli, S. S. L., Mittal, S., Gupta, M., Abdelsalam, M., Joshi, A., Sandhu, R., et al. (2020). Ontologies and artificial intelligence systems for the cooperative smart farming ecosystem. *IEEE Access, 8*, 164045–164064.
11. Sharma, P., Vimal, A., Vishvakarma, R., Kumar, P., porto de Souza Vandenberghe, L., Gaur, V. K., et al. (2022). Deciphering the blackbox of omics approaches and artificial intelligence in food waste transformation and mitigation. *International Journal of Food Microbiology, 372*, 109691.
12. Jagadeesan, B., Gerner-Smidt, P., Allard, M. W., Leuillet, S., Winkler, A., Xiao, Y., et al. (2019). The use of next generation sequencing for improving food safety: Translation into practice. *Food microbiology, 79*, 96–115.
13. Wang, Y., Jin, L., & Mao, H. (2019). Farmer cooperatives' intention to adopt agricultural information technology—Mediating effects of attitude. *Information Systems Frontiers, 21*, 565–580.
14. Xiong, Y., Ge, Y., Grimstad, L., & From, P. J. (2020). An autonomous strawberry-harvesting robot: Design, development, integration, and field evaluation. *Journal of Field Robotics, 37*(2), 202–224.
15. Zhang, P., Guo, Z., Ullah, S., Melagraki, G., Afantitis, A., & Lynch, I. (2021). Nanotechnology and artificial intelligence to enable sustainable and precision agriculture. *Nature Plants, 7*(7), 864–876.
16. Ben Ayed, R., & Hanana, M. (2021). Artificial intelligence to improve the food and agriculture sector. *Journal of Food Quality, 2021*(1), 5584754.
17. Safwat, M.-A., Forsan, H. F., Abdul-Rahman, A. R., Abduljabar, R., & Riad, O. (2021). COVID-19: Immunology, pathology, severity and immunosuppressants. *Azhar International Journal of Pharmaceutical and Medical Sciences, 1*(1), 1–12.
18. Forsan, H. F. (2023). *Applications of nanomaterials in medicine. Sustainable nanomaterials for biomedical engineering: Impacts, challenges, and future prospects*. Apple Academic Press.
19. Subeesh, A., & Mehta, C. (2021). Automation and digitization of agriculture using artificial intelligence and internet of things. *Artificial Intelligence in Agriculture, 5*, 278–291.

20. Talaviya, T., Shah, D., Patel, N., Yagnik, H., & Shah, M. (2020). Implementation of artificial intelligence in agriculture for optimisation of irrigation and application of pesticides and herbicides. *Artificial Intelligence in Agriculture, 4*, 58–73.
21. Forsan, H. F. (2024). Applications of nanosensors in agriculture and food sectors applications of nanosensors in agriculture and food sectors. In *Handbook of nanosensors: Materials and technological applications*. Springer Nature Switzerland.
22. Eli-Chukwu, N. C. Applications of artificial intelligence in agriculture: A review. *Engineering, Technology & Applied Science Research, 9*(4).
23. Toorajipour, R., Sohrabpour, V., Nazarpour, A., Oghazi, P., & Fischl, M. (2021). Artificial intelligence in supply chain management: A systematic literature review. *Journal of Business Research, 122*, 502–517.
24. Antonucci, F., Figorilli, S., Costa, C., Pallottino, F., Raso, L., & Menesatti, P. (2019). A review on blockchain applications in the agri-food sector. *Journal of the Science of Food and Agriculture, 99*(14), 6129–6138.
25. Kakani, V., Nguyen, V. H., Kumar, B. P., Kim, H., & Pasupuleti, V. R. (2020). A critical review on computer vision and artificial intelligence in food industry. *Journal of Agriculture and Food Research, 2*, Article 100033.
26. Friedlander, A., & Zoellner, C. (2020). Artificial intelligence opportunities to improve food safety at retail. *Food Protection Trends, 40*(4).
27. Qian, C., Murphy, S., Orsi, R., & Wiedmann, M. (2023). How can AI help improve food safety? *Annual Review of Food Science and Technology, 14*(1), 517–538.
28. Sak, J., & Suchodolska, M. (2021). Artificial intelligence in nutrients science research: A review. *Nutrients, 13*(2), 322.
29. Hassoun, A., Garcia-Garcia, G., Trollman, H., Jagtap, S., Parra-López, C., Cropotova, J., et al. (2023). Birth of dairy 4.0: Opportunities and challenges in adoption of fourth industrial revolution technologies in the production of milk and its derivatives. *Current Research in Food Science*, 100535.
30. Kataria, P. (2024). Technology at the table: Incorporating AI into contemporary food industry operations. *Food Science and Nutrition Cases*.
31. Forsan, H. F., & Awad, S. S. (2023). *Cyanidin: Advances on resources, biosynthetic pathway, bioavailability, bioactivity, and pharmacology*. Springer International Publishing.
32. Liu, Y., Liu, D., Wei, G., Ma, Y., Bhandari, B., & Zhou, P. (2018). 3D printed milk protein food simulant: Improving the printing performance of milk protein concentration by incorporating whey protein isolate. *Innovative Food Science & Emerging Technologies, 49*, 116–126.
33. Zhang, Y., Zhang, Q., Zhang, L., Li, J., Li, M., Liu, Y., et al. (2023). Progress of machine vision technologies in intelligent dairy farming. *Applied Sciences, 13*(12), 7052.
34. Sabanci, K., Kayabasi, A., & Toktas, A. (2017). Computer vision-based method for classification of wheat grains using artificial neural network. *Journal of the Science of Food and Agriculture, 97*(8), 2588–2593.
35. Bedoya, M. G., Montoya, D. R., Tabilo-Munizaga, G., Pérez-Won, M., & Lemus-Mondaca, R. (2022). Promising perspectives on novel protein food sources combining artificial intelligence and 3D food printing for food industry. *Trends in Food Science & Technology, 128*, 38–52.
36. Wilms, P., Daffner, K., Kern, C., Gras, S., Schutyser, M., & Kohlus, R. (2021). Formulation engineering of food systems for 3D-printing applications–A review. *Food Research International, 148*, Article 110585.
37. Forsan, H. F., & Hassan, R. S. (2023). *Novel nutraceutical milk compound in Alzheimer's prevention. Handbook of neurodegenerative disorders*. Springer.
38. Forsan, H. F., Abd El-Hak, A. E., Ahmad, E., Alswerky, E. M., Elagezy, F. K., Yassin, M., et al. (2024). Toward better science-based advice on nutrition. In *Nutrition and psychiatric disorders: An evidence-based approach to understanding the diet-brain connection*. Springer Nature Singapore.
39. Li, G., Hu, L., Liu, J., Huang, J., Yuan, C., Takaki, K., et al. (2022). A review on 3D printable food materials: Types and development trends. *International Journal of Food Science & Technology, 57*(1), 164–172.

40. Nachal, N., Moses, J., Karthik, P., & Anandharamakrishnan, C. (2019). Applications of 3D printing in food processing. *Food Engineering Reviews, 11*(3), 123–141.
41. Taneja, A., Sharma, R., Ayush, K., Sharma, A., Mousavi Khaneghah, A., Regenstein, J. M., et al. (2022). Innovations and applications of 3-D printing in food sector. *International Journal of Food Science & Technology, 57*(6), 3326–3332.
42. Firouz, M. S., Mohi-Alden, K., & Omid, M. (2021). A critical review on intelligent and active packaging in the food industry: Research and development. *Food Research International, 141*, Article 110113.
43. Barska, A., & Wyrwa, J. (2017). Innovations in the food packaging market—intelligent packaging—a review. *Czech Journal of Food Sciences, 35*(1).
44. Lille, M., Nurmela, A., Nordlund, E., Metsä-Kortelainen, S., & Sozer, N. (2018). Applicability of protein and fiber-rich food materials in extrusion-based 3D printing. *Journal of Food Engineering, 220*, 20–27.
45. Alami, A. H., Olabi, A. G., Khuri, S., Aljaghoub, H., Alasad, S., Ramadan, M., et al. (2024). 3D printing in the food industry: Recent progress and role in achieving sustainable development goals. *Ain Shams Engineering Journal, 15*(2), Article 102386.
46. Boukid, F., Hassoun, A., Zouari, A., Tülbek, M. Ç., Mefleh, M., Aït-Kaddour, A., et al. (2023). Fermentation for designing innovative plant-based meat and dairy alternatives. *Foods, 12*(5), 1005.
47. Ahmad, U., Alvino, A., & Marino, S. (2022). Solar fertigation: A sustainable and smart IoT-based irrigation and fertilization system for efficient water and nutrient management. *Agronomy, 12*(5), 1012.
48. Tavill, G. (2020). Industry challenges and approaches to food waste. *Physiology & Behavior, 223*, Article 112993.
49. Bernini, V., Dalzini, E., Lazzi, C., Bottari, B., Gatti, M., & Neviani, E. (2016). Cutting procedures might be responsible for Listeria monocytogenes contamination of foods: The case of Gorgonzola cheese. *Food Control, 61*, 54–61.
50. Rejeb, A., Rejeb, K., Zailani, S., Keogh, J. G., & Appolloni, A. (2022). Examining the interplay between artificial intelligence and the agri-food industry. *Artificial intelligence in agriculture, 6*, 111–128.
51. Shankar, V. (2018). How artificial intelligence (AI) is reshaping retailing. *Journal of Retailing, 94*(4).
52. Yoo, H., & Park, D. (2021). *AI-based 3D food printing using standard composite materials. Data science and digital transformation in the fourth industrial revolution* (pp. 123–135).
53. Monteiro, J., & Barata, J. (2021). Artificial intelligence in extended agri-food supply chain: A short review based on bibliometric analysis. *Procedia Computer Science, 192*, 3020–3029.
54. Forsan, H. F., & Hasan, R. S. (2023). *Biodegradable polysaccharides nanocomposites*. Springer International Publishing.
55. Cai, W., Wang, J., Jiang, P., Cao, L., Mi, G., & Zhou, Q. (2020). Application of sensing techniques and artificial intelligence-based methods to laser welding real-time monitoring: A critical review of recent literature. *Journal of Manufacturing systems, 57*, 1–18.
56. Dewi, T., Risma, P., & Oktarina, Y. (2020). Fruit sorting robot based on color and size for an agricultural product packaging system. *Bulletin of Electrical Engineering and Informatics, 9*(4), 1438–1445.
57. Kumar, I., Rawat, J., Mohd, N., & Husain, S. (2021). Opportunities of artificial intelligence and machine learning in the food industry. *Journal of Food Quality, 2021*(1), 4535567.
58. Forsan, H. F. (2024). Diet and sleep disorders. In *Nutrition and psychiatric disorders: An evidence-based approach to understanding the diet-brain connection*. Springer Nature Singapore.
59. Forsan, H. F. (2024). *Zebrafish as a model for a novel neuroprotective natural product*. CRC Press.
60. Taneja, A., Nair, G., Joshi, M., Sharma, S., Sharma, S., Jambrak, A. R., et al. (2023). Artificial intelligence: Implications for the agri-food sector. *Agronomy, 13*(5), 1397.

61. Trencher, G. (2019). Towards the smart city 2.0: Empirical evidence of using smartness as a tool for tackling social challenges. *Technological Forecasting and Social Change, 142*, 117–128.
62. Smithers, G. W., Ballard, F. J., Copeland, A. D., de Silva, K. J., Dionysius, D. A., Francis, G. L., et al. (1996). New opportunities from the isolation and utilization of whey proteins. *Journal of Dairy Science, 79*(8), 1454–1459.
63. Mavani, N. R., Ali, J. M., Othman, S., Hussain, M., Hashim, H., & Rahman, N. A. (2022). Application of artificial intelligence in food industry a guideline. *Food Engineering Reviews, 14*(1), 134–175.
64. Pérez-Gomariz, M., López-Gómez, A., & Cerdán-Cartagena, F. (2023). Artificial neural networks as artificial intelligence technique for energy saving in refrigeration systems A review. *Clean Technologies, 5*(1), 116–136.
65. Hussein, Z., Fawole, O. A., & Opara, U. L. (2020). Harvest and postharvest factors affecting bruise damage of fresh fruits. *Horticultural Plant Journal, 6*(1), 1–13.
66. Melesse, T. Y., Bollo, M., Di Pasquale, V., Centro, F., & Riemma, S. (2022). Machine learning-based digital twin for monitoring fruit quality evolution. *Procedia Computer Science, 200*, 13–20.
67. Gowen, A., Tiwari, B., Cullen, P., McDonnell, K., & O'Donnell, C. (2010). Applications of thermal imaging in food quality and safety assessment. *Trends in Food Science & Technology, 21*(4), 190–200.
68. Addanki, M., Patra, P., & Kandra, P. (2022). Recent advances and applications of artificial intelligence and related technologies in the food industry. *Applied Food Research, 2*(2), Article 100126.
69. Jiang, H., Zhang, M., Bhandari, B., & Adhikari, B. (2018). Application of electronic tongue for fresh foods quality evaluation: A review. *Food Reviews International, 34*(8), 746–769.
70. Calvini, R., & Pigani, L. (2022). Toward the development of combined artificial sensing systems for food quality evaluation: A review on the application of data fusion of electronic noses, electronic tongues and electronic eyes. *Sensors, 22*(2), 577.
71. Jambrak, A. R., Šimunek, M., Petrović, M., Bedić, H., Herceg, Z., & Juretić, H. (2017). Aromatic profile and sensory characterisation of ultrasound treated cranberry juice and nectar. *Ultrasonics Sonochemistry, 38*, 783–793.
72. Ullah, A., Liu, Y., Wang, Y., Gao, H., Wang, H., Zhang, J., et al. (2022). E-Taste: Taste sensations and flavors based on tongue's electrical and thermal stimulation. *Sensors, 22*(13), 4976.
73. Deisingh, A. K., Stone, D. C., & Thompson, M. (2004). Applications of electronic noses and tongues in food analysis. *International Journal of Food Science & Technology, 39*(6), 587–604.
74. Rudnitskaya, A. L. A., Seleznev, B., & Vlasov, Y. (2002). Recognition of liquid and flesh food using an electronic tongue. *International Journal of Food Science & Technology, 37*(4), 375–385.
75. Tan, J., & Xu, J. (2020). Applications of electronic nose (e-nose) and electronic tongue (e-tongue) in food quality-related properties determination: A review. *Artificial Intelligence in Agriculture, 4*, 104–115.
76. Zhu, X., Yuan, X., Zhang, Y., Liu, H., Wang, J., & Sun, B. (2022). The global concern of food security during the COVID-19 pandemic: Impacts and perspectives on food security. *Food Chemistry, 370*, Article 130830.
77. Kondakci, T., & Zhou, W. (2017). Recent applications of advanced control techniques in food industry. *Food and Bioprocess Technology, 10*, 522–542.
78. Van Wynsberghe, A. (2021). Sustainable AI: AI for sustainability and the sustainability of AI. *AI and Ethics, 1*(3), 213–218.
79. Hassoun, A., Prieto, M. A., Carpena, M., Bouzembrak, Y., Marvin, H. J., Pallarés, N., et al. (2022). Exploring the role of green and Industry 4.0 technologies in achieving sustainable development goals in food sectors. *Food Research International, 162*, 112068.
80. Varavallo, G., Caragnano, G., Bertone, F., Vernetti-Prot, L., & Terzo, O. (2022). Traceability platform based on green blockchain: An application case study in dairy supply chain. *Sustainability, 14*(6), 3321.

81. Sobhy, H. M., Abd, M. E., Elsabie, W., & Fathy Forsan, H. (2021) Study of high nutritive value of almond milk beverage. *Plant Archives, 21.*
82. Hagar, F., Forsan, M. M. E. A., Elsabie, W. B., & Sobhy, H. M. (2024). Apricot addition for enrichment yogurt with amygdalin. *Al-Kitab Journal for Pure Sciences, 8*(1), 63–70.
83. Forsan, H. F., El Abd, M., Elsabie, W. B., & Sobhy, H. M. (2024). Cassava addition for amygdalin enrichment in Dairy Beverages. *Anbar Journal of Agricultural Sciences, 22*(2).
84. Ferreira, C., & Gonçalves, G. (2022). Remaining useful life prediction and challenges: A literature review on the use of machine learning methods. *Journal of Manufacturing Systems, 63*, 550–562.
85. Goyal, S., & Goyal, G. K. (2012). Time-delay simulated artificial neural network models for predicting shelf life of processed cheese. *International Journal of Intelligent Systems and Applications, 4*(5), 30.
86. Dutta, J., Deshpande, P., & Rai, B. (2021). AI-based soft-sensor for shelf life prediction of 'Kesar' mango. *SN Applied Sciences, 3*(6), 657.
87. Verma, M., Hontecillas, R., Tubau-Juni, N., Abedi, V., & Bassaganya-Riera, J. (2018). Challenges in personalized nutrition and health. *Frontiers Media SA.*
88. Ma, T., Wang, H., Wei, M., Lan, T., Wang, J., Bao, S., et al. (2022). Application of smart-phone use in rapid food detection, food traceability systems, and personalized diet guidance, making our diet more health. *Food Research International, 152*, Article 110918.

AI-Driven IoT Solutions in Agriculture: Legal Pathways to Safeguard Children from Labour Exploitation

Navonita Mallick and Shashwata Sahu

Abstract Even globally, agriculture is still one of the biggest employment sectors, especially in developing countries. At the same time, it is a sector where child labour prevails, with nearly 70% of children engaged in agriculture, according to ILO data. Even though laws against child labour exist or are in signing, following and implementing them specifically in rural interior places is a daunting task. Advancements in IoT technology driven by Artificial Intelligence offer highly hopeful possibilities to tackle such issues. Moreover, these technologies are not confined to increasing agricultural produce but will also give new dimensions for monitoring, preventing and eliminating child labour. This paper discusses the intersection of AI, IoT and legal frameworks where these technologies are integrated in such a way that they can help to monitor real-time data for child labour law implementation. The research examines current legal frameworks, where they are weak and how technology can enhance enforcement. It uses a doctrinal research approach and deals with statutes, international conventions, etc., in child labour, along with case studies on technological innovations in agriculture. This discussion showcases that AI and IoT solutions have been used for agricultural efficiency. However, the potential of these technologies can be utilized more effectively in this domain to counteract child labour. This will be perfect, although the detection of child labour can improve if technologies like these are incorporated within legal frameworks, assuming they do so ethically and consider privacy concerns. The paper proposes that the use of AI-driven IoT technologies in agriculture, using strong legal frameworks would provide an accountable system to ensure transparency and strengthen enforcement for child labour.

Keywords AI-driven IoT · Child labour · Agriculture · Legal frameworks · Labour exploitation

N. Mallick · S. Sahu (✉)
School of Law, KIIT-DU, Bhubaneswar, India
e-mail: contactshashwata@gmail.com

S. R. Kumar et al. (eds.), *AI-Driven IoT Solutions in Agriculture: New Insights*, Studies in Computational Intelligence 1236, https://doi.org/10.1007/978-3-032-05603-0_10

1 Introduction

The best way to predict the future is to create it.
— Abraham Lincoln and Peter Drucker

Agriculture is a vital sector for most developing countries, and it refers to the production of plants and animal commodities that help provide food products. Agriculture provides immense employment opportunities, and many of the population depends on this system [1]. Besides accounting for a major portion of the national GDP, agriculture provides the bulk of livelihood in rural areas, where most people depend on farming and related occupations to earn a daily living. However, this industry has a widespread issue in the form of child labour. Agriculture is said to absorb nearly 70% of the world's child labour force, as highlighted in this recent report by The International Labour Organization (ILO) [2]. Reports have suggested that in India, where agriculture continues to be a mainstay of the economy, millions of children are involved with agricultural practices and, as such, miss out on education, health and well-being. These children are usually from marginalized communities, and generational poverty ensnares youth into working at a very young age.

New technologies can help overcome these systemic barriers, and the potential for new technology to alleviate some of these pains is timely; particularly with how quickly Artificial Intelligence (AI) has started to pair together with The Internet of Things (IoT), we now have a promising opportunity in agriculture [3]. AI-led solutions like automated devices, systems and predictive analytics increase agricultural lands' productivity and assist farmers in monitoring their farms in real time. Agricultural sites are equipped with IoT networks, including smart sensors, drones, and other connected devices; these collect continuous data from the fields [4]. Basically, we know that using these technologies can help make farming more efficient, save various resources and enable better decision-making in agricultural applications. Most importantly, they are opportunities to fight child labour and control precisely, even on the farms, noting physical labour done by machines [5]. However, the potential of AI and IoT technologies in fighting against child labour has not really been explored except for some edge cases. While contemporary research appears to emphasize the labour-saving effects that these technologies bring, there is limited application of current technology with respect to safeguarding children's rights while enforcing compliance for child labour and adherence to the law in agriculture [6]. Challenges like financial constraints, limited infrastructure and a lack of digital literacy in rural parts raise the adoption level even further. Meanwhile, policymakers have yet to quicken the pace of technology integration in child labour enforcement frameworks, leading to a wedge between legal intent and on-the-ground execution.

This article addresses this gap by investigating how AI-powered IoT solutions can help prevent child labour use in agriculture. The study outlines challenges to enforcing child labour standards in agriculture and examines relevant legal frameworks. It seeks to underscore the potential use of technology for closing these enforcement lacunae, paying particular attention to real-time data monitoring support to ensure compliance in implementation. Furthermore, the study will also provide how these technologies can be used to testify lawful actions against culprits, thus resulting in increased efficiency of regulatory frameworks.

2 Research Objectives

This study aims to understand how AI-powered IoT solutions could help in the fight against child labour in agriculture. Moreover, it plans to analyse the extant legal frameworks on child labour and how they could be harnessed vis-à-vis AI and IoT. Through an analysis of modern technology trends and legal frameworks, it aims to propose a strong legal mechanism that can be implemented in conjunction with emerging technologies to enhance compliance with child labour legislation and thus alleviate the problem children face from being forced into agriculture.

3 Research Methodology

This paper employs the doctrinal research method based on reviewing existing statutes, international conventions and cases concerning child labour and preaching technological innovations in agriculture. In the final chapters, legal texts, policy documents and secondary literature from scientific monographs will be subjected to critical analysis to establish whether these frameworks are fit for purpose or capable of change as technology advances. The research investigates different AI and IoT implementations in agriculture, especially as they could possibly be used to track child labour laws.

4 Conceptualization

Artificial Intelligence and the IoT are revolutionizing farming through advanced technology tools to make real-time decisions for better agricultural productivity. AI uses algorithms and machine learning to analyse data, allowing computers to analyse data, predict outcomes and automate tasks without human intervention [7]. AI in agriculture uses predictive analysis to determine crop yield as well as artificial pest detection and weather forecasting models. Whereas IoT is referred to as the Internet of Things, which means a network in which things are connected through the Internet and collect real-time data. It involves the formation of a digital ecosystem fuelled by sensors, which communicate with each other and pass on information to help in auto-controlling and remote monitoring [8]. In addition, AI and IoT-driven technologies are making large changes in agricultural activities. Sensors deployed in fields measure soil moisture, temperature and crop health while providing farmers an opportunity for data-driven insights. Large farms started using drones with cameras and sensors, which fly over the farm, doing aerial reconnaissance to determine if something is wrong, e.g. pest plagues or irrigation problems [9]. Robotic harvesters, seeding machines and other automation tools cut down on much manual labour associated with farming, increasing efficiency. IoT-based monitoring systems monitor all farm activities in real time and alert the farmers about any abnormalities observed well before they stray out of control [10]. These technologies liberate the way of doing agriculture and have immense scope to keep checking and monitoring labour activities, including child labour.

Child labour, child labour refers to the exploitation of children through any form of work that deprives children of their childhood and interferes with their ability to attend regular school. Child labour in agriculture occurs mainly in an informal setting on family farms where children work with parents planting, harvesting or else weeding and taking care of livestock [11]. This work can involve demanding tasks and put children at risk of pesticide exposure, long working hours, and hazardous machinery. Child labour in agriculture arises for several reasons, such as poverty, lack of quality education and the necessity for additional household income [12]. Even if we can eliminate it in the formal economy, many rural families have kids minding their siblings and tending livestock while parents work the fields or engage elsewhere.

Both at the International and National levels, efforts have been made to eliminate Child Labour in Agriculture. The ILO establishes such standards through, among others, Convention No. 138 on the Minimum Age for Employment and Convention No. 182 on the Worst Forms of Child Labour that call for prohibition as good elimination of child labour [13]. There are also related targets on eradicating child labour by 2025, within the United Nations Sustainable Development Goal 8. In many countries, at the national level, too, laws such as the Child Labour (Prohibition and Regulation) Act 1986 in India have been enforced long back, followed by amendments to prevent child exploitation [14]. However, despite their efforts, enforcement has proved hugely problematic in the sprawling countryside and fields of farming villages (Fig. 1).

The above graphic helps to combine a few of the causes influencing child work, illustrating how residents' dynamics affect child labour since it simultaneously features education, health and nutrition, social protection and beginning childhood development. All of these factors also affect the vulnerability to labour exploitation that children have, and they create cycles of poverty and lack of access to education. The root causes of child labour lie in poor health and weak education. Social security is as important to curb this evil demand. This highlights the need to combine social development efforts with legal frameworks to break away from this cycle of child labour.

Fig. 1 Key factors influencing child labour

5 Theoretical Framework

The innovations are applicable in rural agriculture is explained by other theories of technological adoption, such as Rogers' Diffusion of Innovations and the Unified Theory of Acceptance and Use of Technology (UTAUT) model [15]. The proposed theories emphasize barriers such as the infrastructure, digital literacy and influence in adoption among other factors that apply to child labour monitoring. In addition, Human Rights Theory and Labour Law Theory are concerned with the need for children to be protected from exploitation and mirror global regulatory frameworks, including UNCRC [16]. This model combines AI, IoT and legal offerings to track child labour practices while enforcing compliance and policy-based assistance that will bring a transparency dimension that aims for holistic protection of children in agriculture.

5.1 Technological Adoption Theories

Understanding technological adoption plays a significant role in answering how we can leverage innovations like AI and IoT directly into farming touchpoints, especially those from a rural background. Rogers' Diffusion of Innovations Theory. The first theoretical model is to guide our understanding of how new ideas, technologies, or practices become adopted by a population [17]. Roger classifies the process of technology adoption into 5 stages: awareness, interest, evaluation, trial and finally acceptance. These are the stages an individual or community passes through before they conclude to adopt new Software. Agricultural diffusion, especially in rural areas, is slow due to the limited availability of information and high financial constraints with a lack of infrastructure. As with many new technologies, the trend is set by large commercial farms and, more specifically, early adopters, while smaller farms and rural communities fall to slow in change [18]. This idea matters because if we want AI and IoT tools to be used for child labour monitoring and enforcement in a way that is not only theoretical but also productive on the ground at scale, it needs buy-in from different quarters. Farmers all the way down to their lives practising conditions themselves; next possible minority communities interest groups operating as intermediaries between farmers working usurious hours as identified propagating intricate web of intertemporal havoc and supply-chain pegged corporate executives with investments targeting sustainability virtue-signalling yet production lines-maintaining [19].

Further insights on the issues faced in adopting technology come from UTAUT. UTAUT mainly includes four factors: expected performance, effort expectance, social impact, and facilitating conditions determining technology acceptance. Performance expectancy in AI and IoT for agriculture was the perception that these technologies will increase productivity and ease of monitoring [20]. Effort expectancy refers to the ease of use of these technologies, which can be a major barrier in rural contexts as digital literacy is low. Social influence, including community leaders and peer recommendations, shape attitudes towards adoption. Finally, facilitators like easier access to capital and governmental backing need attention if the innovation is likely to be adopted successfully [21]. This theory also illustrates that benefits are not enough to ensure adoption. It points out how some barriers limit uptake despite the many potential advantages of AI and IoT for rural communities.

5.2 *Labour Rights and Child Protection Theories*

Human Rights Theory values all persons' intrinsic dignity and basic rights, including children. It insists upon their right to education, protection and freedom from exploitation as children. In addition, child labour in agricultural settings is not only preventing children from going to school but also causing them to suffer physical dangers and work themselves out physically on the land [22]. This theory also fits with global frameworks such as the UNCRC, advocating for children's welfare and protection. Rights: At the very least, AI and IoT technologies can guarantee these rights by enabling real-time monitoring of child labour practices, thus preventing exploitation.

Labour Law Theory addresses both the creation and implementation of legal systems designed to protect workers, including child labour, from exploitation. Labour law goals include creating safe conditions of work, regulating employment practices and prohibiting the use of child labour in hazardous employment [23]. The Child Labour (Prohibition and Regulation) Act of 1986 in India bans the employment of children only in certain industries like agriculture and a few others where work is done under hazardous conditions. However, enforcement of these laws is more difficult, particularly in isolated rural areas. According to the Labour Law Theory, effective monitoring and implementation of legal provisions are important factors in checking child labour legislation [24]. Combining AI and IoT technologies with labour law enforcement is a strategic weapon against possible offenders who, based on historical data, could violate the law. Integration of those means can assist in real-time violation detection, allowing us to react quickly and support legal actions towards implementation.

5.3 Conceptual Model

This research employs AI, IoT, and a legal framework for monitoring child labour laws in agriculture, which is this study's overall idea or conceptual model. The model works across three levels: data collection and monitoring, enforcement of compliance and legal interventions.

IOT-based data collection and monitoring: IOT devices that include sensors and drones collect real-time data from farms. These devices track labour activity, growing cycles and environmental factors. These data are then analysed with the help of AI algorithms to spot patterns that look indicative of child labour [25]. It could, for instance, identify working days that were especially long or types of work typically assigned to children.

Compliance outrage: When a deviation is identified, it is easily reported to the regulators and law enforcement departments. This helps in early intervention to tackle child labour practices. In addition, AI-powered systems can produce reports and evidence supporting legal actions to improve enforcement processes [26].

Notification to Farmers and Farm Workers: The notifications will alert farmers, farm workers and others who consume the products about compliance with child labour laws.

Policy and Legal Intervention: AI and IoT systems data collected through AI and IoT systems are valuable to law hackers. The model sheds light on the extent of child labour and identifies where interventions are needed most. This information can then be utilized by policymakers to both strengthen existing laws, introduce new regulations and better allocate resources [27]. There are also opportunities for governments, NGOs and technology providers to work together to ensure AI and IoT solutions can be implemented effectively.

Conceptual model of the combination and management style between technology and legal frameworks to eliminate child labour in agriculture. This demonstrates the power of AI and IoT in better supervision and regulation to safeguard children from labour exploitation [28]. The integrated approach, again, not only promotes transparency and accountability but also fits in with broader human rights goals and labour standards.

6 AI and IoT Solutions in Agriculture

Agriculture is also being impacted through AI and the IoT. It has uses as diverse as improving efficiency to addressing issues around ethical labour practices. Smart sensors, drones and predictive analytics tools will help monitor crops, fields and labour activities in real time [29]. Irregularities are caught via aerial imagery from drones, and worker activity is logged by wearable IoT devices to show that labour laws have been followed. These technologies are used to reduce or eliminate child labour by automating activities such as monitoring, identifying offences and issuing timely alerts to take immediate action [30]. Examples from countries such as India, the Netherlands and Kenya show how AI and IoT technology can boost agricultural production in a local area without exploiting labour [31].

6.1 Overview of AI and IoT Technologies

Innovative methods of smart sensors, drones, automated pest control systems and predictive analytics are contributing to a major change in agricultural practices with the advent of AI and IoT. Smart sensors, which contribute real-time information regarding soil moisture, temperature and humidity health and crop status [32]. The data enables farmers to make informed decisions regarding irrigation, fertilizer and pest management, ensuring perfect crop growth. IoT networks link these sensors and transfer the data accumulated in one place to be processed and analysed through AI Algorithms [33]. That means farmers can constantly access field performance data and act ahead of time against new trends (Fig. 2).

Aerial imagery of huge fields from drones with high-resolution cameras and sensors alerts to areas such as pest infestation, water stress or plant diseases. These insights enable farmers to take preventive measures, helping them increase resource utilization and curb crop loss [34]. To monitor them, a drone could observe the labour and detect different patterns or irregularities, e.g., children's work, when monitoring their crop areas. The biggest advantage of this aerial perspective is that it covers a vast land and makes monitoring by humans at such large farms difficult [35]. Devices like robotic sprayers used by automated pest control systems travel to specific areas and spray minimal pesticides. These tools not only decrease manual work but also environmental concerns. First, AI tools controlled by algorithms have been embedded in IoT systems to forecast pest outbreaks depending on the weather conditions, which affect crop cycles using historical records [36]. This provides an opportunity for farmers to prepare in advance before they are affected, hence a preventative measure. Machine learning also drives predictive analytics, increasing the precision of farming

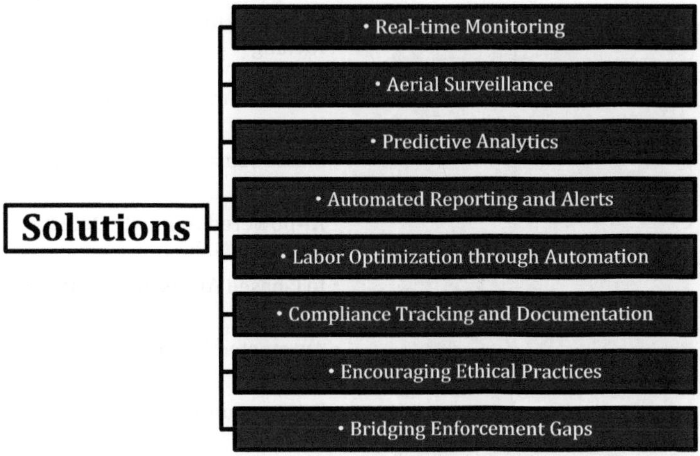

Fig. 2 How AI and IoT offer promising solutions for addressing child labour in agriculture

decision-making as it can predict crop yield, weather patterns, and market trends. These forecasts help farmers plan for better crop stress management and revenue generation [37]. Together, they increase agricultural efficiency and are the building blocks of adequate monitoring for labour practices.

6.2 Labour Monitoring Using AI and IoT

It is a solution that increases productivity in AI and IoT and provides an effective solution for battling child labourers across agriculture. These technologies can capture labour patterns, earn real-time data monitoring, and catch sight of possible child labour happening on the field, leading to compliance with relevant laws. The IoT-connected sensors placed in fields can track movement and working hours, ensuring that appropriate labour standards are followed. Wearable IoT devices even monitor how hard workers work, for example, a Fitbit on children, warning when tasks appear too demanding to comply with minimum age or maximum hour laws [38] (Fig. 3).

Using this data, AI algorithms detect the anomaly or pattern discovered while processing, with which they can recognize that children have been working in agriculture. It can also trigger alerts if it finds people of a certain age doing the same repetitive tasks or working unusually long. This helps raise an alarm for the farm owner or authorities as a timely intervention can occur [39]. This monitoring ability can be further enhanced when drones offer a perspective on labour activities from the sky, enabling and providing authorities with a bird's eye view for overseeing labour activities in large areas that are too vast to monitor manually.

Fig. 3 Key aspects of labor monitoring using AI and IoT

Additionally, they help in documenting the violations. AI-enabled tools, for example, can create comprehensive reports of suspected incidents related to child labour, which could be brought as evidence by regulatory bodies during their inspections. Plus, predictive models can predict when child labour will likely happen e.g. during peak harvest seasons, helping governments and NGOs target their interventions accordingly [40]. This results in early child labour detection and more effective enforcement of legal frameworks for ethical labour practices within agriculture (Table 1).

As the table at the top shows, adding AI and IoT technology can also greatly improve agricultural functionality. In short, there are always opportunities appropriate for children. It came so that, with the large or small operation efficiency of technology in Zambia, child labour cannot solve the issue. Using smart sensors, drones, and predictive analytics, farmers can have optimal operations and monitor labour practices. Capitalizing on these technologies, early detection of child labour violations is possible through real-time tracking and automated alerts with interventions [41]. In addition to fighting child labour, these solutions tackle the root of evils with automation tools like robotic harvesters. These innovations support integration into agricultural systems, engagement in ethical farming practices, and corresponding productivity with social responsibility.

Table 1 AI and IoT technologies for agricultural efficiency and child labor prevention

Technology	Application in agriculture	Role in child labor prevention
Smart sensors	Monitor soil moisture, temperature, and crop health	Detect long working hours and unsafe tasks for children
Drones	Aerial monitoring of fields and crops	Identify child labour on large farms through aerial imagery
Wearable IoT devices	Track movement and working hours of labourers	Alert supervisors to child involvement in prohibited activities
Predictive analytics	Forecast weather and crop yield	Predict high-risk periods for child labour (e.g., harvest season)
Robotic harvesters	Automate harvesting tasks	Reduce dependency on manual labour, minimizing child labour risks
AI-powered cameras	Monitor farm operations in real-time	Capture evidence for compliance with labour laws
IoT-based alerts	Notify farmers of operational issues remotely	Immediate alerts to regulators for suspected child labour

7 Case Studies

The agricultural counterparts are not left behind, as AI and IoT have been implemented in the agrarian sectors around the globe. This is a testament that when used well, it can achieve high productivity while supporting sustainable practices [42]. For example, in the Netherlands, greenhouses are equipped with IoT sensors connected to a global network that monitors temperature, humidity and CO_2 emission in real time, ensuring the perfect conditions for optimal plant growth from anywhere while saving on labour costs. Use of drones to monitor their farms and labour working on fields, overseeing operations, thus becoming legally compliant with local rules [43]. In the US, once again, big farms utilize highly targeted AI-powered robotic systems to harvest fruits and vegetables. Automatic systems like these minimise human labour and prevent the abuse of vulnerable workers. IoT and mobile data put Kenyan weather stations in farmers' pockets [44]. Others integrate labour monitoring features, allowing farmers to ensure their work complies with the local labour laws. In these cases, we have AI and the IoT to thank for increasing agricultural efficiency and ensuring ethical labour standards.

Connectivity in large mechanically farmed farms and government-led projects has grown, given that India is adopting AI and IoT technologies in agriculture at a growing pace. AgriStack, an initiative by the Indian government, aims to build a digital ecosystem integrating IoT devices like drones and AI-powered analytics for better farm practices [45]. In this vein, one category of initiatives involves state-level experiments with IoT-based monitoring systems for controlling various farm activities like labour practices. One exciting example is pilot projects in Andhra Pradesh, where drones supervise farm activities so they do not violate labour laws. The e-Choupal initiative by ITC Limited provides farmers' market information and advisory services for improving farm productivity using digital tools [46]. While the central concern here is to enhance agricultural productivity, it also introduces ethical labour practices in areas such as non-children working in its field embryonic primers for eventual compliance with labour laws. AI models are used in Maharashtra to predict crop yields and optimize water resources [47]. Part of each system is a labour tracking module that helps keep the farm operation compliant with state child labour laws.

Rather than an academic piece, consider these as short case studies illustrating what is happening on the ground regarding using AI and IoT technologies for agriculture while showing how both principle 4, i.e. transformation due to new technology and PUWP regarding labour standards, are tackled [48]. These initiatives are nascent, to be sure, but early signs suggest they have great potential. Farmers reported higher yields, decreased costs and greater ease in labour law compliance. However, these examples also bring to light the obstacles to the implementation of new technology, especially in rural regions [49]. Problems including low digital literacy, infrastructure underdevelopment and resistance to change are essential factors for expanding the applications of AI and IoT in agriculture.

8 Legal Framework Governing Child Labour

International frameworks like the ILO Conventions No. 138 and 182, UNCRC, emphasize preventing children's exploitation and ensuring their education at any cost. These frameworks and the SDGs themselves encourage cooperation against child labour [50]. India has enacted laws such as the Child Labour (Prohibition and Regulation) Act, 1986 and the Right to Education (RTE) Act explicitly intended to counter child labour, but they struggle with significant gaps in enforcement concerning agriculture that are compounded by social-economic pressures [51]. Despite global legal frameworks, enforcement is poor in rural areas. Advanced technology like AI and IoT can potentially be used as a remedy to monitor the gaps in enforcement better.

8.1 International Legal Frameworks

Several intercontinental agencies have made the struggle against child labour a major focus, with the ILO in front. ILO Conventions establish the minimum age for work to ensure that children are not subjected to any form of employment which would interfere with their development and education. It demands the minimum working age to be identical to when compulsory education usually ends, which is somewhere between 14 and sixteen years of old [52]. Convention 182 is about the worst forms of child labour, including hazardous work that can lead to harm to the health and moral well-being of children. These conventions hold signatory countries like India to implement and enforce such laws [53]. Another important international framework is the "United Nations Convention on the Rights of the Child (UNCRC)" adopted in 1989. It acknowledges children as people with individual rights, including the right to live without exploitation. Among other provisions, Article 32 of the UNCRC states that member states must protect children from economic exploitation and work likely to be hazardous or interfere with their education [54]. States should also develop means for the monitoring and protection of these rights.

The global imperative to eliminate child labour has been re-emphasized in recent years by the Sustainable Development Goals (SDGs), with Goal 8 i.e. Decent Work and Economic Growth at its core. One target under this goal, i.e. Target 8.7, aims to eliminate child labour in all forms by 2025 [55]. The Agenda 2030, with its SDGs, plays a central role in cooperation between governments, civil society and the private sector in finding sustainable solutions that address the root causes of child labour [56]. Although these frameworks provide a strong background for overcoming child labour, implementation in sectors like agriculture, which are frequently hidden, can still become an issue.

8.2 Indian Legal Frameworks

India has taken a leadership role in enacting legislation to combat child labour, including the Child Labour (Prohibition and Regulation) Act 1986. This Act prohibits the employment of child labourers below 14 years of age in hazardous industries and regulates the working conditions for children employed under permissible conditions [57]. Unfortunately, family-based enterprises and agriculture can use this loophole to justify the exploitation of children. To fill these gaps, the act was amended in 2016, and the minimum age for work increased to 14 years, while adolescents between 14 and 18 were prohibited from carrying out hazardous work [58].

The "Right of Children to Free and Compulsory Education Act, 2009 (RTE Act)" is also important in this regard as it mandates free education for all children from the age group between 6 and 14 years [59]. The law is intended to reduce the economic pressures leading children towards work rather than schooling. To not only prevent these practices but also provide victims of labour exploitation with rehabilitation and care for their treatment as children in need of protection, new legislation was enacted called Juvenile Justice (Care & Protection) Act 2015, which makes special provisions that are beneficial for recovery purposes. Children in agriculture are directly affected by the Bonded Labour System (Abolition) Act, 1976, due to bonded labour systems and hence indirectly covering provisions relating to child labour. The other salient feature is a multi-pronged strategy to combat child labour listed in the National Child Labour Policy, 1987 [60]. It includes legislative measures, rehabilitation of children withdrawn from work, and linking them to services such as education, health care and vocational training even though child labour continues despite said legal frameworks and is largely unenforced, especially in agriculture where enforcement is limited, and monitoring mechanisms are flawed (Table 2).

Table 2 Comparative analysis of child labor regulations in agriculture across jurisdictions

Country/ jurisdiction	Key legislation	Minimum age for employment	Prohibition on hazardous work	Enforcement mechanisms	Challenges in implementation
India	Child Labour (Prohibition and Regulation) Act, 1986	14 years	Banned under 18 years	Labour inspectors and NGOs	Limited rural monitoring, economic dependence
United States	Fair Labor Standards Act (FLSA)	16 years (non-hazardous jobs)	18 years	Department of Labor inspections	Family farms exempt from many labour laws
European Union	EU Directive on Young Workers	15 years	Banned under 18 years	National labour inspectorates	Lack of uniform enforcement across member states
Brazil	Estatuto da Criança e do Adolescente (ECA)	16 years (part-time from 14)	Banned under 18 years	Federal inspections, social programs	Informal labour practices in rural areas
South Africa	Basic Conditions of Employment Act, 1997	15 years	Banned under 18 years	Department of Employment and Labour	Lack of resources for rural enforcement
Kenya	Employment Act, 2007	13 years (light work)	Banned under 18 years	Child protection officers, labour inspectors	High poverty rates leading to non-compliance
Australia	Fair Work Act, 2009	No national minimum (state laws vary)	18 years	State-level labour authorities	Variations in child labour laws across states

This comparative evaluation, as shown above, reveals that despite several jurisdictions in this sector having created a legal infrastructure for regulating child labour, huge differences exist amongst these practices due to enforcement and challenges. The challenge in countries such as India and Kenya are that the thread of child labour weaves deeply into other social fibres like lack of rural scrutiny, economic dependency, etc., so it may not be easy to make laws on paper executed effectively at all levels [61]. While developed areas such as the United States and the European Union have more stringent regulations, there remain problems with family farm exemptions or uneven enforcement among member states [62]. This review underscores the necessity for novel approaches to complement existing programs by deploying AI and IoT technologies, which can provide support activities such as on-field monitoring of traditional methods while aiding compliance management that has been lacking from routine applications across multiple agricultural practices.

9 Challenges in Legal Enforcement

Although the fact of having comprehensive legal frameworks discloses a standing commitment to eradicating child labour, enforcement is by far one of the greatest challenges that needs addressing, not least in rural and agricultural zones. In remote places, there may be issues of their own that make it difficult to identify acts that violate child labour laws. Much of the agricultural work is done on family-owned farms or small plots spread over vast distances, which inspects by authorities more challenging [63]. Therefore, child labour practices remain invisible, and those found to be exploiting children almost entirely escape prosecution (Fig. 4).

The illustration above shows that child labour is a vicious cycle which affects a nation's growth and development. Child labour served as a way of exploiting children that deprived education and skill development and, along with this, got substantially poor wages and labour standards. That creates adult unemployment, poverty in the home children would provide for and less power to unions, which allows exploitation of kids. Combating this vicious cycle requires complementary efforts between education, social welfare, and labour policies to secure sustainable economic growth while observing children's rights.

Another important area is the lack of institutional support for child labour laws. Labour inspectors often focus on urban areas, constrained by resources and capacity

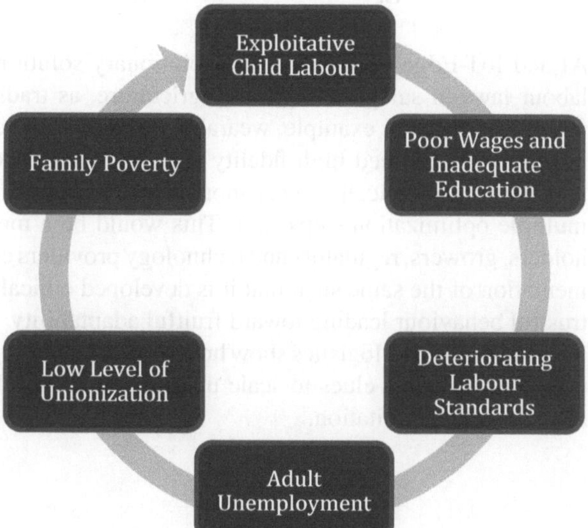

Fig. 4 The vicious cycle of child labour and its impact on national development

to effectively monitor labour standards being met on rural agricultural properties. This gets more complicated as the informal nature of agricultural work hampers enforcement. Those working children could be on family farms that would not necessarily fall under legal definitions as labourers [64]. The loophole makes it difficult to detect and respond to cases of child labour in agriculture. Economic and social factors further facilitate weak enforcement. Poverty from a lack of education and limited adult employment opportunities leads people to involve their kids in agricultural work. This harsh socio-economic situation helps to frustrate far-reaching child labour laws, which are often difficult to implement given that children's economic contribution provides a necessary reliance for homes [65]. Additionally, rural communities tend to have low awareness of child labour laws, and cultural acceptance also perpetuates the tradition of children working.

The adoption of technology provides an opportunity for enforcement, but it is also a challenge. Although AI and IoT tech facilitate farm activity monitoring in real-time, adherence to child labour laws demands less infrastructure support and quite a lot of investments in training. However, in rural areas where smart monitoring is most cost-effective, there may be limited access to the digital infrastructure required or a backlash against surveillance-based monitoring from farmers [66]. Second, for enforcement to become effective, it needs to be complemented by an array of stakeholders, including government agencies, NGOs and community groups working together, but this often does not happen. Without concerted action, initiatives to combat child labour in agriculture are fragmented and less effective [67]. In addition, building stakeholder relationships is the key to addressing these challenges and creating long-term solutions.

10 Technology as a Tool for Legal Enforcement

AI and IoT-Edge can provide a revolutionary solution to monitor or enforce child labour laws in safeguarding rural agriculture, as traditional inspections have been found wanting. For example, wearable trackers and drones monitor labour practices in real-time and feed high-fidelity data back to AI-powered analytics algorithms, which can automatically detect anomalies or predict child-labour-prone periods with multiple optimization steps [68]. This would be a medium in which all the stakeholders, growers, regulators and technology providers come together to enable implementation of the same such that it is developed ethically for an ethical use, creating trustful behaviour leading toward fruitful adaptability. Lessons from sectors such as manufacturing and logistics show how similar technologies can improve compliance, providing valuable clues to scale up good agricultural labour practices in the fight against child exploitation.

10.1 AI and IoT for Compliance Monitoring

The same holds when it comes to regulating labour laws for children, and AI, as well as IoT, can make this extremity a reality. Moreover, the traditional methods of enforcement, such as manual inspections of farms and land parcels in rural agricultural settings, are impractical because agriculture is extensive and has limited resources available to authorities [69]. Meanwhile, IoT-capable appliances, including sensors or drones, can access real-world data on farm works that map the application of labour practices as live information. For instance, wearable devices that come with IoT can be used to monitor the activities of workers in terms of their movement and hours worked to ensure no child is involved. Immediate intervention can be provided if the devices send notifications of young workers carrying out prohibited activities [70]. They could also provide aerial surveillance over vast agricultural fields, spotting child labour in operations that can slip under the radar. The steady data flow from IoT devices guarantees compliance with legal farming regulations; offending in these regards is less painful.

This monitoring is more vigorously accomplished with the use of AI-powered systems that analyse patterns in the data collected. Machine learning algorithms can help spot anomalies in work hours or tasks that could signal child labour. Such systems can also predict the periods when child labour is at greatest risk, such as during planting and harvesting seasons, thereby allowing authorities to carry out focused inspections [71]. IoT data can be combined with AI analytics to provide enforcement agencies with detailed reports that stand up in court, making the compliance audit process more efficient. With the above technologies, we move to a more proactive mode of child labour enforcement as opposed to waiting for something bad to happen and then doing an inspection.

10.2 Collaboration Between Stakeholders

Collaboration between multiple players is essential for the effective use of AI and IoT in law enforcement. Farmers, regulators, and technology companies need to cooperate to apply these advancements ethically and responsibly. Farmers have a prime contribution to adopting these technologies on their farms [72]. However, their willingness to embrace monitoring tools varies depending on trust, convenience, and potential benefits. First, however, there are many farmers in rural areas such as this one who will be categorized into suspicion or threat groups initially because they believe their privacy could somehow change or quite possibly because of a penalty. Thus, there is a need to educate farmers about AI and IoT tools and encourage them to use them for regulatory compliance and farm productivity [73]. Another way to increase participation in regulated markets is by providing financial incentives, such as technology adoption subsidies.

The responsibility of regulators is to design pathways for how these technologies can be used ethically under current law. In the meantime, they will need to cooperate with farmers very closely and prove that these tools are meant for enforcing child labour laws and not used arbitrarily as grounds for penalizing a farmer [74]. Regulatory bodies could also invest in training labour inspectors to properly utilize AI or IoT systems for more effective monitoring, especially in remote areas which are hard to monitor by means of the old-fashioned on-site inspections. Acknowledging the challenge of monitoring farm operations, technology providers become an important partner in this process by developing tools to make it easier to look at how it works in the field. They need to develop user-centric solutions that meet the environments typical of agriculture [75]. Furthermore, by working with the government and NGOs on ways to monitor child labour in agriculture, you can develop customised solutions that target the challenges of this type of work. Farmers and regulators need to meet with technologists regularly so that problems can be hammered out quickly, ensuring the broader adoption of these tools.

11 Lessons from Other Sectors

Building these rules from experience is one way in which lessons can be drawn into decision-making, borrowing insights on compliance monitoring and ethical practices as currently being worked out AI and IoT technologies applied to non-agricultural industries. The case studies show how technology supports compliance across varied sectors and, by extension, agriculture [76]. Manufacturing facilities are taking advantage of IoT-based systems to monitor the movements and behaviour of workers on factory floors. Track where employees have been and how many hours they worked to ensure everyone is paid fairly, legally, within the limits of labour law, and for safety compliance. AI-powered analytics help agencies identify possible risks, such as excessive working hours or unsafe conditions [77]. That way, management can address potential issues before they become real problems. These technologies have reduced workplace accidents, and they enforce labour standards. The same model can be applied in the case of agriculture, wherein AI and IoT tools are deployed to help keep track of health conditions while also tracing acts that violate child labour laws.

Retail and logistics companies might use IoT sensors to track their supply chain data, from ethically sourced products to meeting industry regulations. From origin to destination, these systems will follow where the goods pass through and ensure that suppliers abide by fair labour regulations. High-risk suppliers, or regions where a breach is more likely to come, are identified when AI algorithms analyse supply chain data [78]. These technologies generate insights that make it easy for companies to choose better suppliers more objectively and without violating ethical guidelines. A similar model can be used in agriculture for monitoring farm operations & also to verify child labour laws being complied with right from production chains. The construction industry similarly offers lessons in tech-based compliance oversight.

Worker safety and compliance with labour laws at construction sites are always challenges. Wearable sensors and cameras are two IoT devices that follow work in real-time to see if they conform with set safety standards [79]. AI systems analyse the collected data to identify potential risks and predict future accidents. By taking this proactive stance, the number of workplace accidents has decreased, and safety compliance rates have improved. Regarding agriculture, this could be implemented through AI and IoT technologies that help enforce compliance with labour laws at farm operations, like ensuring no child is used for grossly underpaid work, thus protecting children from exploitation.

12 Ethical and Privacy Considerations

The same applies to AI and IoT technologies for labour monitoring, in which different countries have raised ethical and privacy issues by balancing necessary oversight with intrusive surveillance. These tools raise serious questions regarding the privacy and trust of workers. Clear guidelines must constrain monitoring to compliance and openness requirements. In addition, using encryption and anonymization methods in all measures is important to prevent misuse because such acts are also imperative for compliance with privacy standards set around the world, like GDPR [80]. Similarly, community engagement, training and policy frameworks are needed to promote responsible adoption and ensure equitable and ethical usage.

12.1 Balancing Surveillance with Privacy

With AI and IoT-powered monitoring of labour activities, questions arise about the definition of necessary oversight versus intrusive surveillance. While these technologies can catch child labour abuse, ongoing surveillance could also violate the rights of farm workers to privacy. Workers are frequently very uncomfortable with being monitored, particularly in informal labour settings common to rural agricultural areas [81]. This often results in opposition or criticism of any technology uptake. Moreover, we also run a risk that those behind these farms or authorities might abuse this technology by using it as surveillance overwatch for child labour law compliance, creating unnecessary oversight.

Finding this balance between surveillance and privacy lies in putting well-defined lines around when the monitoring can take place. There is minimal scope for using this data for any purpose other than child labour detection and compliance with the Minimum Wages (Jharkhand) Act, 2016 and Rules made thereunder [82]. Regulations need to prescribe what AI and IoT systems can only be used for that respect the dignity of transparency so workers are not in the dark about their use if they affect them. Trust is paramount, and trust will not be easily won unless there is transparency about how those technologies are utilised among workers and communities [83]. Employers should speak candidly about the purposes of surveillance and emphasise that it benefits society, protects children from exploitation, safeguards good labour standards, etc.

12.2 Data Privacy and Protection Issues

IoT devices used for tracking agricultural activities will inevitably gather personal data, which might include the movements and working hours of workers in the field and what kind of work they do. The privacy and security of this data must be protected to avoid harm to workers. Unauthorised access and breaches to IoT systems can result in disclosing personal or sensitive information, which could harm workers. Moreover, the data compiled by these systems could easily be misinterpreted or utilized in different ways, e.g. hiring practices and unjust reprisal towards workers [84]. These risks can be mitigated by following stringent data protection guidelines and best practices. Employers and regulators can also take measures to encrypt the data obtained from IoT devices, a notable technique for placing appropriate segregation of duties. The data should be shared only with those responsible for the enforcement of child labour laws and must not go beyond that end. Finally, the data can be anonymised, so they will not have to be linked to personal surveillance [85]. Workers should also be informed about what data is being collected, how it will be used and who can access the information to ensure transparency and trust. At the same time, governments and other regulatory bodies need to create legislation that governs how data gathered by AI or IoT services in agriculture can be used. These frameworks should be coherent with international data privacy regulations such as the "General Data Protection Regulation (GDPR)" and guarantee workers' rights to privacy [86]. Strong Data Governance practices would reduce the risks of data misuse and ethical technology adoption.

12.3 Ensuring Fair Use of Technology

Responsible adoption of AI and IoT technologies will not only come through technical solutions but also have a lot to do with community engagement and training programs at their disposal. Farm-folk or rural, green-acre workers may be unfamiliar with these technologies and, therefore, are likely to resist a sense of monitoring [87]. An example is training programs for farmers and workers to understand how these technologies work, what benefits they can provide, and which aspects comply with the law. By delivering Compliance Monitoring solutions, stakeholders make their way around resistance to change and pave the path for hassle-free technology adoption [88].

Engagement with the community is crucial to fostering trust and good digital manners. Engaging with workers and communities in planning for, deploying, and using these technologies is one way to ensure their fears are addressed. There needs to be collaboration between the government, NGOs, and technology providers to create solutions in rural communities that align with their cultural and social nuances for mass adoption [89]. Community leaders and farmer organizations can forge alliances for the responsible use of AI and IoT tools, serving as intermediaries between farmers, workers and regulators.

Policies around the ethical use of technology are just as important. Governments should regulate to make sure AI and IoT can be used in fair ways so they do not inadvertently perpetuate societal problems such as discrimination or job insecurity [90]. Periodic audits and supervision of technology use can be performed to detect any misuse and infringe on ethical codes. Additional incentives to farmers who responsibly adopt these technologies, e.g., through subsidies or financial support, will also encourage appropriate use and fight resistance.

13 Findings and Discussion

The report identifies AI and IoT as game-changing technologies in combatting child labour on farms by moving enforcement from punitive inspections to real-time reporting through data infusion. Drones, sensors and AI-based analytics give your ongoing surveillance to identify when these high-risk periods will occur so that if anything does happen, there is a clear evidential record for digital failed evidence. However, there are barriers, including missing laws, lacking basic digital access, and being unaffordable [91]. The researchers conclude that this suggests the importance of institutional support and legal clarity. Integrating AI and IoT in legal frameworks makes it easier for enforcers to make their job and compliance feasible to be followed up on by warning violators with financial consequences of systematic audits as well as prioritising stakeholder engagement, forging a step ahead better approach against eliminating child labour.

13.1 Key Findings from Conceptual, Technological, and Legal Analysis

The paper points out AI and IoT as game changers that can help solve agricultural child labour problems. They also provide a mechanism for proactive monitoring through the real-time collection of data as opposed to enforcement based on traditional inspections. IoT sensors and drones can monitor labour activities, which means you have constant checks over what is happening on your farm. This enables them to detect violations and allows authorities to intervene before the exploitation gets out of hand [92]. AI can further examine this data to detect patterns and any security threats. It can predict when child labour is most likely to happen, for example, during the peak harvesting season, and hence, it directs regulators' attention to these days. What's more, these reports from AI are proof of objective facts on violations that can be utilised in court proceedings [93]. The solutions are then plugged into these technologies to render operations on the farm more transparent, such that children cannot be taken advantage of and labour standards meet requirements.

In this respect, legal analysis shows that the current frameworks are mainly reactive rather than preventive. While India may have the Child Labour (Prohibition and Regulation) Act, more importantly, international conventions like ILO Conventions No. 138 & 182 act as guides; they are poorly enforced, especially in agricultural contexts [94]. The findings provide a case for complementing legal frameworks with technology to reinforce enforcement and make laws more effective.

13.2 Challenges Identified

Despite the promise of AI and IoT technologies, several challenges hinder their widespread adoption and effectiveness in agricultural settings. One of the primary barriers is the gap in legal frameworks. While laws exist to regulate child labour, they do not explicitly address using digital tools to monitor compliance [95]. This lack of legal clarity discourages regulators and farmers from adopting technology-driven solutions. Furthermore, there are loopholes in the law, such as permitting children to work in family-based agricultural activities, which makes it difficult to distinguish between legitimate work and exploitative practices.

Deploying AI and IoT technologies in rural areas also faces several practical challenges. Many rural regions lack the digital infrastructure to support IoT networks, such as reliable internet connectivity and power supply [96]. Additionally, farmers and local communities often lack the technical expertise required to operate these systems, resulting in resistance to adoption. There are also concerns about the costs associated with acquiring and maintaining IoT devices and AI tools, which can be prohibitive for small-scale farmers. Another challenge is the lack of institutional support for technology adoption [97]. Regulatory bodies may not have the resources or capacity to implement AI and IoT systems on a large scale. Labour inspectors are often concentrated in urban areas, leaving rural regions under-monitored. Without adequate institutional support, the benefits of AI and IoT technologies may remain unrealized.

13.3 Synergies Between Technology and Legal Frameworks

AI and IoT-based technologies have the advantage of being able to fill gaps in law enforcement. These will provide regulators with the tools they need to be enabled enough to monitor compliance effectively. Conventional enforcement measures like on-site inspections are costly and infrequent, restricting the ability to enforce child labour practices in remote agricultural areas [98]. On the other hand, with IoT sensors and drones in place augmented by AI-driven analytics, continuous monitoring is provided rather than relying on periodic physical inspections.

The reconciliation of technology with legal frameworks relates to AI and IoT's capacity for evidence-based enforcement. This evidence can then be employed in a court of law to enforce compliance with relevant regulatory measures, streamlining the enforcement process and offering greater transparency [99]. This can also enable regulators to localise problem areas and target their interventions better, thereby improving the efficiency with which resources are deployed. Integrating AI and IoT in their operations allows both enforcement bodies to transform from reactive compliance into a preventative regime that stops child labour well before it starts. Technology can send reminders to farmers about labour laws for compliance and red flags in case of any suspected flouting. This helps instil accountability and transparency and makes inadvertent violations less likely [100]. Incorporating AI and IoT-based solutions into legal frameworks could further ease collaboration among stakeholders, including government agencies, NGOs, and technology providers, to enforce child labour laws more strongly.

14 Recommendations

An AI and IoT-led approach towards addressing the issue of child labour in agriculture, therefore, assumes significance only if it is combined with a multi-pronged strategy that simultaneously tackles operational challenges as well as policy needs [101]. This part also offers some recommendations for government, policymakers, and other stakeholders in integrating technology and capacity-building mechanisms and collaborating with best sustainable practices.

14.1 Strengthening Infrastructure and Promoting Technology Adoption

The major hurdle to setting up AI and IoT deployments in rural agriculture is digital infrastructure. Governments and private sector partners should invest in affordable digital infrastructure, reliable internet connectivity and power supply. To cater to the needs of small-scale farmers, public–private partnerships can work wonders in developing cost-effective solutions. Incentivizing farmers with financial support and subsidies, especially small-farm holders, to embrace AI and IoT can drive its adoption widely [102]. The next phase will demand technical support from their respective governments with training programs to ensure that farmers and agricultural workers know how to utilize these technologies well. Integrating digital literacy training sessions, supplemented by AI and IoT tools that are easy to use in the field, will reduce pushback towards widespread adoption. Engage with community leaders and farmer organizations to build trust and acceptance towards adopting technology-centric monitoring.

14.2 Developing Policy Frameworks and Ethical Guidelines

Governments and policymakers should have comprehensive guidance to define the role of AI and IoT technologies in child labour monitoring. They have kept ethical considerations around privacy preservation, fair use, and global labour standards at their core. These technologies should be in the training of labour inspectors, and data governance frameworks may need to be implemented for this personal information not to fall improperly. Evaluating the technology used regularly also ensures you would be compliant with labour laws and ethical standards.

14.3 Collaborative Efforts with NGOs and International Organizations

NGOs and multilateral agencies, such as the ILO and UNICEF, help secure social licenses to operate scattered transformations in tangible practice. The public and nonprofit sectors can collaborate with governments and tech companies to spread the message about what AI and IoT can do to keep children from becoming child labourers. Expedited R&D and implementation: NGOs can provide technical expertise directly, facilitating the application of these technologies in underserved regions. As privacy and surveillance concerns grow, we will likely need all the parties involved in this process to collaborate. By involving the community in an initiative to apply AI and IoT while adopting a participatory planning approach, stakeholders can ensure that the local context is taken into account and agrees with the values of the communities.

14.4 Building Sustainable Public–Private Partnerships

Sustainable adoption demands public–private partnerships that enlist the help of governments, tech corporations and research bodies to design sector-specific solutions. Continuous improvement of monitoring systems to keep them effective and relevant over time, with the help of these partnerships [103]. These include regular audits and evaluations of technology adoption best practices, challenges, compliance with labour laws, etc. These kinds of partnerships can help governments build systems for sharing knowledge and coordinating action among farmers, regulators, NGOs and technology developers. This visibility will help build confidence among stakeholders and provide a pathway to an agri-sector free of child labour, which also meets the Labour standards.

15 Conclusion

> The test of our progress is not whether we add more to the abundance of those who have much; it is whether we provide enough for those who have too little.
>
> — Franklin D. Roosevelt

The paper also illustrates that AI and IoT technologies have the transformative power to intervene in the long-standing problem of child labour, specifically within agriculture. Directing the DOL to investigate labour laws is a thing of the past, and this real-time monitoring tool would take that out by merely proactively supervising by local labour laws, even in rural areas. AI-supported analytics pick up on patterns in the constantly streaming data from sensors and drones that could indicate child labour on farms. Used together, these tools provide a means of early intervention to protect our kids [104]. Research demonstrates how AI and IoT hardware could be used to enforce labour laws, which can contribute important arguments in legal and technology policy discussions. The principle clarifies that these tools should be integrated rather than used as an alternative to existing legal frameworks, stressing cooperation between governments, NGOs, and technology providers [105]. These findings also inform the development of policies, compelling stakeholders to use innovative strategies that can be embedded within global labour standards and ensure children's rights.

However, it should be acknowledged that this research is not without limitations. Infrastructure and lack of digital literacy are still hurdles to implementing AI and IoT solutions, specifically in rural locations. Inconsistencies in monitoring and privacy or surveillance concerns may also create data gaps. These challenges underscore the importance of more insight into scalable and accessible solutions that can be implemented in diverse agricultural settings. Future research could examine how AI and IoT technologies can be applied to other sites of child labour, such as construction, domestic work or mining. It is, therefore, key to shift this approach beyond the scope of agriculture, as child exploitation extends across various industries, and technology can prevent that irrespective of where it occurs, thus achieving our larger goal of ending all forms of child labour.

References

1. Gamage, A., Gangahagedara, R., Gamage, J., Jayasinghe, N., Kodikara, N., Suraweera, P., & Merah, O. (2023). Role of organic farming for achieving sustainability in agriculture. *Farming System, 1*, Article 100005. https://doi.org/10.1016/j.farsys.2023.100005
2. (2024). Child labour I international labour organization. https://www.ilo.org/projects-and-partnerships/projects/child-labour. Accessed 30 Oct 2024
3. Haleem, A. (2022). Artificial intelligence (AI) applications for marketing: A literature-based study. *International Journal of Intelligent Networks, 3*, 119–132. https://doi.org/10.1016/j.ijin.2022.08.005
4. Subeesh, A., & Mehta, C. R. (2021). Automation and digitization of agriculture using artificial intelligence and internet of things. *Artificial Intelligence in Agriculture, 5*, 278–291. https://doi.org/10.1016/j.aiia.2021.11.004
5. Rakholia, R., Tailor, J., Prajapati, M., Shah, M., & Saini, J. R. (2024). Emerging technology adoption for sustainable agriculture in India—a pilot study. *Journal of Agriculture and Food Research, 17*, Article 101238. https://doi.org/10.1016/j.jafr.2024.101238
6. Păvăloaia, V.-D., & Necula, S.-C. (2023). Artificial intelligence as a disruptive technology—a systematic literature review. *Electronics, 12*, 1102. https://doi.org/10.3390/electronics12051102
7. Elbasi, E., Mostafa, N., Zaki, C., AlArnaout, Z., Topcu, A. E., & Saker, L. (2024). Optimizing agricultural data analysis techniques through AI-powered decision-making processes. *Applied Sciences, 14*, 8018. https://doi.org/10.3390/app14178018
8. Sharma, K., & Shivandu, S. K. (2024). Integrating artificial intelligence and Internet of Things (IoT) for enhanced crop monitoring and management in precision agriculture. *Sensors International, 5*, Article 100292. https://doi.org/10.1016/j.sintl.2024.100292
9. Benti, N. E., Chaka, M. D., Semie, A. G., Warkineh, B., & Soromessa, T. (2024). Transforming agriculture with Machine learning, deep learning, and IoT: Perspectives from Ethiopia—challenges and opportunities. *Discov Agric, 2*, 63. https://doi.org/10.1007/s44279-024-00066-7
10. Elbasi, E., Mostafa, N., AlArnaout, Z., Zreikat, A. I., Cina, E., Varghese, G., Shdefat, A., Topcu, A. E., Abdelbaki, W., Mathew, S., & Zaki, C. (2023). Artificial intelligence technology in the agricultural sector: A systematic literature review. *IEEE Access, 11*, 171–202. https://doi.org/10.1109/ACCESS.2022.3232485
11. Admassie, A. (2003). Child labour and schooling in the context of a subsistence rural economy: Can they be compatible? *International Journal of Educational Development, 23*, 167–185. https://doi.org/10.1016/S0738-0593(02)00012-3
12. Labor NRC (US) and I of M (US) C on the H and SI of C. (1998). Agriculture. In: Protecting youth at work: Health, safety, and development of working children and adolescents in the United States. National Academies Press (US)
13. Hoque, M. M. (2024). A critical review of Bangladesh's child labor regulations and policies. *World Development Sustainability, 5*, Article 100177. https://doi.org/10.1016/j.wds.2024.100177
14. Nations U World Day Against Child Labour. In: United Nations. https://www.un.org/en/observances/world-day-against-child-labour. Accessed 30 Oct 2024.
15. Li, L., Min, X., Guo, J., & Wu, F. (2024). The influence mechanism analysis on the farmers' intention to adopt Internet of Things based on UTAUT-TOE model. *Science and Reports, 14*, 15016. https://doi.org/10.1038/s41598-024-65415-4
16. McGarry, D. A framing theory and discourse analytic study of the child labour policy debate.
17. Sahin, I. (2006). detailed review of rogers' diffusion of innovations theory and educational technology-related studies based on rogers' theory. *The Turkish Online Journal of Educational Technology, 5*.
18. Feder, G., Just, R. E., & Zilberman, D. (1985). Adoption of agricultural innovations in developing countries: A survey. *Economic Development and Cultural Change, 33*, 255–298.

19. Hassebo, A. Global models of smart cities and potential IoT applications: A review. https://www.mdpi.com/2624-831X/4/3/17. Accessed 30 Oct 2024.
20. Lee, C.-L. (2024). Factors predicting innovation-decisions: The effects of performance expectancy, social influence, and facilitating conditions on U.S. Extension's promotion of precision agriculture technologies. https://doi.org/10.1080/27685241.2024.2420111. Accessed 30 Oct 2024.
21. Misra, R. (2024). Factors impacting behavioural intentions to adopt the electronic marketplace: Findings from small businesses in India | Electronic Markets. https://doi.org/10.1007/s12525-022-00578-4. Accessed 30 Oct 2024.
22. McCrudden, C. (2024). Human dignity and judicial interpretation of human rights | European Journal of International Law | Oxford Academic. https://academic.oup.com/ejil/article/19/4/655/349356. Accessed 30 Oct 2024.
23. Brown, D. K., Deardorff, A. V., & Stern, R. M. The determinants of child labor: Theory and evidence.
24. Bharadwaj, P., Lakdawala, L. K., & Li, N. Perverse consequences of well intentioned regulation: Evidence from India's child labor ban.
25. Smart Farming. Internet of Things (IoT)-based sustainable agriculture. https://www.mdpi.com/2077-0472/12/10/1745. Accessed 30 Oct 2024.
26. Regulatory challenges of robotics: Some guidelines for addressing legal and ethical issues. https://doi.org/10.1080/17579961.2017.1304921. Accessed 30 Oct 2024.
27. Artificial intelligence in Internet of things—Ghosh—2018—CAAI transactions on intelligence technology—Wiley Online Library. https://doi.org/10.1049/trit.2018.1008. Accessed 30 Oct 2024.
28. (2021). How to regulate tech: A technology policy framework for online services. In: Center for American Progress. https://www.americanprogress.org/article/how-to-regulate-tech-a-technology-policy-framework-for-online-services/. Accessed 30 Oct 2024.
29. Mana, A. A., Allouhi, A., Hamrani, A., Rehman, S., el Jamaoui, I., & Jayachandran, K. (2024). Sustainable AI-based production agriculture: Exploring AI applications and implications in agricultural practices. *Smart Agricultural Technology, 7*, Article 100416. https://doi.org/10.1016/j.atech.2024.100416
30. Varigonda, V., Agrawal, B., & Annamalai, V. K. (2020). IoT based automatic fault identification and alerting system for unmanned aerial vehicles.
31. AI for agriculture. How Indian farmers are harnessing emerging technologies to sustainably increase productivity | World Economic Forum. https://www.weforum.org/impact/ai-for-agriculture-in-india/. Accessed 30 Oct 2024.
32. Implementation of artificial intelligence in agriculture for optimisation of irrigation and application of pesticides and herbicides - ScienceDirect. https://www.sciencedirect.com/science/article/pii/S258972172030012X. Accessed 30 Oct 2024.
33. Applying IoT sensors and big data to improve precision crop production: A review. https://www.mdpi.com/2073-4395/13/10/2603. Accessed 30 Oct 2024.
34. Drones in agriculture: A review and bibliometric analysis - ScienceDirect. https://www.sciencedirect.com/science/article/pii/S0168169922003349. Accessed 30 Oct 2024.
35. A review on the use of unmanned aerial vehicles and imaging sensors for monitoring and assessing plant stresses. https://www.mdpi.com/2504-446X/3/2/40. Accessed 30 Oct 2024.
36. Hejazipoor, H., Massah, J., Soryani, M., Asefpour Vakilian, K., & Chegini, G. (2021). An intelligent spraying robot based on plant bulk volume. *Computers and Electronics in Agriculture, 180*, Article 105859. https://doi.org/10.1016/j.compag.2020.105859
37. Understanding the potential applications of Artificial Intelligence in Agriculture Sector - ScienceDirect. https://www.sciencedirect.com/science/article/pii/S277323712200020X. Accessed 30 Oct 2024.
38. Significance of sensors for industry 4.0: Roles, capabilities, and applications—ScienceDirect. https://www.sciencedirect.com/science/article/pii/S2666351121000310. Accessed 30 Oct 2024.

39. Anomaly Detection on Data Streams for Smart Agriculture. https://www.mdpi.com/2077-0472/11/11/1083. Accessed 30 Oct 2024.
40. Goldfarb, A., & Lindsay, J. R. (2022). Prediction and judgment: Why artificial intelligence increases the importance of humans in war. *International Security, 46*, 7–50. https://doi.org/10.1162/isec_a_00425
41. Configuring the digital farmer: A nudge world in the making? https://doi.org/10.1080/03085147.2021.1876984. Accessed 30 Oct 2024.
42. Sustainable agriculture by the Internet of Things—A practitioner's approach to monitor sustainability progress—ScienceDirect. https://www.sciencedirect.com/science/article/pii/S0168169922005403. Accessed 30 Oct 2024.
43. Farooq, M. S., Riaz, S., Helou, M. A., Khan, F. S., Abid, A., & Alvi, A. (2022). Internet of things in greenhouse agriculture: A survey on enabling technologies, applications, and protocols. *IEEE Access, 10*, 53374–53397. https://doi.org/10.1109/ACCESS.2022.3166634
44. Challenges and solutions for sustainable food systems: The potential of home hydroponics. https://www.mdpi.com/2071-1050/16/2/817. Accessed 30 Oct 2024.
45. Balkrishna, A., Pathak, R., Kumar, S., Arya, V., & Singh, S. K. (2023). A comprehensive analysis of the advances in Indian Digital Agricultural architecture. *Smart Agricultural Technology, 5*, Article 100318. https://doi.org/10.1016/j.atech.2023.100318
46. (2011). ITC's E-Choupal: Improving the lives of India's Rural Farmers—Integra LLC. https://www.integrallc.com/itcs-e-choupal-improving-the-lives-of-indias-rural-farmers-2-2/. Accessed 30 Oct 2024.
47. Frontiers | Recommendations for ethical and responsible use of artificial intelligence in digital agriculture. https://doi.org/10.3389/frai.2022.884192/full. Accessed 30 Oct 2024.
48. Role of IoT technology in agriculture: A systematic literature review. https://www.mdpi.com/2079-9292/9/2/319. Accessed 30 Oct 2024.
49. Indigenous Peoples' lands are threatened by industrial development; conversion risk assessment reveals need to support Indigenous stewardship—ScienceDirect. https://www.sciencedirect.com/science/article/pii/S2590332223003408. Accessed 30 Oct 2024.
50. Exploring the role of social protection interventions in preventing child labour: Reinforcing the case for a human rights-based model of intervention | Journal of Human Rights Practice | Oxford Academic. https://academic.oup.com/jhrp/article/15/1/262/7046235. Accessed 30 Oct 2024.
51. UNICEF-Innocenti-Child-labour-schooling-India-Report-2024.pdf.
52. Mavunga, R. (2014). A critical assessment of the minimum age convention 138 of 1973 and the worst forms of child labour convention 182 of 1999. *Potch Elekt regs, 16*, 121. https://doi.org/10.4314/pelj.v16i5.3
53. Principle 5 | UN Global Compact. https://unglobalcompact.org/what-is-gc/mission/principles/principle-5. Accessed 30 Oct 2024.
54. Srinivasan, S. P., & Shruthi, S. (2024). A comparative review of UNCRC and Indian legislation from the child mental health perspective. https://doi.org/10.1177/02537176241226714. Accessed 30 Oct 2024.
55. Goal 8 | Department of economic and social affairs. https://sdgs.un.org/goals/goal8. Accessed 30 Oct 2024.
56. Transforming our world: The 2030 agenda for sustainable development | department of economic and social affairs. https://sdgs.un.org/2030agenda. Accessed 30 Oct 2024.
57. (2016). A law that allows child labour. The Hindu.
58. Ganotra, K. Flawed child labour law amendment.
59. A battle of rights: The right to education of children versus rights of minority schools. *Oxford University Commonwealth Law Journal, 16*(2). https://doi.org/10.1080/14729342.2017.1285103. Accessed 30 Oct 2024.
60. Bonded Labour.pdf.
61. (2024). Kenya marks its commitment to ending child labour | International Labour Organization. https://www.ilo.org/resource/article/kenya-marks-its-commitment-ending-child-labour. Accessed 30 Oct 2024.

62. European farmers exempted from rules on land lying fallow. In *European commission—European commission.* https://ec.europa.eu/commission/presscorner/detail/en/ip_24_781. Accessed 30 Oct 2024.
63. (2024). Child labour in agriculture | International Labour Organization. https://www.ilo.org/international-programme-elimination-child-labour-ipec/sectors-and-topics/child-labour-agriculture. Accessed 30 Oct 2024.
64. Fallon, P., Tzannatos, Z. *Child labor: Issues and directions for the World Bank, 2. print.* World Bank, Washington, DC.
65. Kirkbride, J. B., Anglin, D. M., Colman, I., Dykxhoorn, J., Jones, P. B., Patalay, P., Pitman, A., Soneson, E., Steare, T., Wright, T., & Griffiths, S. L. (2024). The social determinants of mental health and disorder: Evidence, prevention and recommendations. *World Psychiatry, 23*, 58. https://doi.org/10.1002/wps.21160
66. Sharma, A., Sharma, A., Tselykh, A., Bozhenyuk, A., Choudhury, T., Alomar, M. A., & Sánchez-Chero, M. (2023). Artificial intelligence and internet of things oriented sustainable precision farming: Towards modern agriculture. *Open Life Sciences, 18.* https://doi.org/10.1515/biol-2022-0713
67. Dreier, L., Nabarro, D., & Nelson, J. Systems Leadership for Sustainable Development:
68. Privacy-centric AI and IoT solutions for smart rural farm monitoring and control. https://www.mdpi.com/1424-8220/24/13/4157. Accessed 30 Oct 2024.
69. (2018). "Negotiating the algorithm": Automation, artificial intelligence and labour protection | International Labour Organization. https://www.ilo.org/publications/negotiating-algorithm-automation-artificial-intelligence-and-labour. Accessed 30 Oct 2024.
70. IoT-enabled smart agriculture: architecture, applications, and challenges. https://www.mdpi.com/2076-3417/12/7/3396. Accessed 30 Oct 2024.
71. AI in anomaly detection: Use cases, methods, algorithms and solution. https://www.leewayhertz.com/ai-in-anomaly-detection/. Accessed 30 Oct 2024.
72. Dara, R., Fard, S. M. H., & Kaur, J. (2022). Recommendations for ethical and responsible use of artificial intelligence in digital agriculture. *Frontiers in Artificial Intelligence, 5*, Article 884192. https://doi.org/10.3389/frai.2022.884192
73. Frontiers | Protecting farmers' data privacy and confidentiality: Recommendations and considerations. https://doi.org/10.3389/fsufs.2022.903230/full. Accessed 30 Oct 2024.
74. (2022). Beneficial ownership information reporting requirements. In: Federal register. https://www.federalregister.gov/documents/2022/09/30/2022-21020/beneficial-ownership-information-reporting-requirements. Accessed 30 Oct 2024.
75. IoT solutions with artificial intelligence technologies for precision agriculture: Definitions, applications, challenges, and opportunities. https://www.mdpi.com/2079-9292/13/10/1894. Accessed 30 Oct 2024.
76. AI revolutionizing industries worldwide: A comprehensive overview of its diverse applications—ScienceDirect. https://www.sciencedirect.com/science/article/pii/S2773207X24001386. Accessed 30 Oct 2024.
77. IoT in manufacturing: Applications and benefits of smart factories | Digi International. https://www.digi.com/blog/post/iot-in-manufacturing. Accessed 30 Oct 2024.
78. Admin. (2023). IoT in logistics and supply chain management. In *Rishabh software.* https://www.rishabhsoft.com/blog/iot-in-scm-and-logistics. Accessed 30 Oct 2024.
79. Bhaskaran, R., Nathan, D., & Upendranadh, C. Vulnerable workers and labour standards (non-)compliance in global production networks: home-based child labour in Delhi's garment sector.
80. Joyce, A., & Javidroozi, V. (2024) Smart city development: Data sharing vs. data protection legislations. *Cities, 148*, 104859. https://doi.org/10.1016/j.cities.2024.104859
81. Ethical dilemmas and privacy issues in emerging technologies: A review. https://www.mdpi.com/1424-8220/23/3/1151. Accessed 30 Oct 2024.
82. The fine balance—surveillance, security, and the right to privacy | SCC Times. https://www.scconline.com/blog/post/2023/08/03/the-fine-balance-surveillance-security-and-the-right-to-privacy/. Accessed 30 Oct 2024.

83. Poncibó, C., Howells, G., & DiMatteo, L. A. (2024). Harm. In *The cambridge handbook of AI and consumer law: Comparative perspectives*. Cambridge University Press, pp 161–212.
84. Brous, P., Janssen, M., & Herder, P. (2020). The dual effects of the Internet of Things (IoT): A systematic review of the benefits and risks of IoT adoption by organizations. *International Journal of Information Management, 51*, Article 101952. https://doi.org/10.1016/j.ijinfomgt.2019.05.008
85. Consumer IoT and its under-regulation: Findings from an Australian study—Harkin—2022—Policy & Internet—Wiley Online Library. https://doi.org/10.1002/poi3.285. Accessed 30 Oct 2024.
86. Forney, J., & Epiney, L. (2022). Governing farmers through data? Digitization and the question of autonomy in agri-environmental governance. *Journal of Rural Studies, 95*, 173–182. https://doi.org/10.1016/j.jrurstud.2022.09.001
87. Chauhan, M., & Sahoo, D. R. (2024). Towards a greener tomorrow: Exploring the potential of AI, blockchain, and IoT in sustainable development. *Nature Environment and Pollution Technology, 23*, 1105–1113. https://doi.org/10.46488/NEPT.2024.v23i02.044
88. Jahanbin, P., Wingreen, S. C., Sharma, R., Ijadi, B., & Reis, M. M. (2023). Enabling affordances of blockchain in agri-food supply chains: A value-driver framework using Q-methodology. *International Journal of Innovation Studies, 7*, 307–325. https://doi.org/10.1016/j.ijis.2023.08.001
89. Montt-Blanchard, D., Najmi, S., & Spinillo, C. G. (2023). Considerations for community engagement in design education. *She Ji: The Journal of Design, Economics, and Innovation, 9*, 234–263. https://doi.org/10.1016/j.sheji.2023.05.004
90. Frontiers | Ethics and responsible AI deployment. https://doi.org/10.3389/frai.2024.1377011/full. Accessed 30 Oct 2024.
91. Smart drone surveillance system based on AI and on IoT communication in case of intrusion and fire accident. https://www.mdpi.com/2504-446X/7/12/694. Accessed 30 Oct 2024.
92. A comprehensive review on smart and sustainable agriculture using IoT technologies—ScienceDirect. https://www.sciencedirect.com/science/article/pii/S2772375524000923. Accessed 30 Oct 2024.
93. preparing_for_the_future_of_ai.pdf.
94. A discussion of hazardous child labor—PMC. https://pmc.ncbi.nlm.nih.gov/articles/PMC1497784/. Accessed 30 Oct 2024.
95. The path to smart farming: innovations and opportunities in precision agriculture. https://www.mdpi.com/2077-0472/13/8/1593. Accessed 30 Oct 2024.
96. Drivers, barriers and impacts of digitalisation in rural areas from the viewpoint of experts—ScienceDirect. https://www.sciencedirect.com/science/article/pii/S0950584921002469. Accessed 30 Oct 2024.
97. Frontiers | Prioritizing factors for the adoption of IoT-based smart irrigation in Saudi Arabia: a GRA/AHP approach. https://doi.org/10.3389/fagro.2024.1335443/full. Accessed 30 Oct 2024.
98. Takyar, A. (2023), AI for regulatory compliance: Use cases, technologies, benefits, solution and implementation. In *LeewayHertz—AI Development Company*. https://www.leewayhertz.com/ai-for-regulatory-compliance/. Accessed 30 Oct 2024.
99. Dunsin, D., Ghanem, M. C., Ouazzane, K., & Vassilev, V. (2024). A comprehensive analysis of the role of artificial intelligence and machine learning in modern digital forensics and incident response. *Forensic Science International: Digital Investigation, 48*, Article 301675. https://doi.org/10.1016/j.fsidi.2023.301675
100. (2022). Law in the era of artificial intelligence. eddy.se ab, Visby.
101. Generative artificial intelligence in supply chain and operations management: a capability-based framework for analysis and implementation. https://doi.org/10.1080/00207543.2024.2309309. Accessed 30 Oct 2024.
102. Public-Private-Partnership.pdf.
103. Public-private partnerships for sustainable development: exploring their design and its impact on effectiveness. https://www.mdpi.com/2071-1050/11/4/1087. Accessed 30 Oct 2024.

104. European Parliament. Directorate General for Parliamentary Research Services. (2022). *AI and digital tools in workplace management and evaluation: An assessment of the EU's legal framework*. Publications Office.
105. Al-Hasan, M., Lutfi, A., Mowafi, O., Abdeldayem, B., Eskandarany, A., & Alshira'h, A. F. (2024). The governance of non-governmental organizations: Critical lens for laws and regulations. *Journal of Open Innovation: Technology, Market, and Complexity, 10*, 100299. https://doi.org/10.1016/j.joitmc.2024.100299

Harnessing AI and IoT: Transforming Agriculture for Sustainable Solutions

C. V. Suresh Babu and M. Thrisha

Abstract The integration of artificial intelligence with the Internet of Things in agriculture is transforming agriculture because it has incorporated sustainable and efficient ways to tackle the global challenges in the food security sector. This paper discusses how AI-driven data analytics, allied with IoT-based smart sensors, and smart devices can create precision farming techniques that help optimize resource use efficiency, reduce the environmental impacts, and enhance crop productivity. The use of technologies such as real-time monitoring of soil health, weather conditions, and crop growth to predictive analytics for pest control and irrigation management-all of these advance decision-making and automation of farming practices. AI and IoT also contribute to sustainable agriculture through reduced waste, water conservation, lesser chemical inputs, and climate resilience. In fact, these technologies have a high transformation potential, meaning that they are very crucial tools in addressing the increasingly rising demand for food in an environment-friendly and economically viable way. Abstract: In recognition of AI and IoT developments and adoption, it is therefore an imperative necessity for future innovation requirements to continue driving the future of sustainable agriculture.

Keywords Artificial intelligence · Internet of things · Precision farming · Sustainable agriculture · Food security · IoT sensors · AI-driven analytics · Climate resilience · Crop productivity · Irrigation management

C. V. S. Babu (✉) · M. Thrisha
Department of Information Technology, Hindustan Institute of Technology and Science, Padur, Chennai, India
e-mail: pt.cvsuresh@hindustanuniv.ac.in; dr.c.v.suresh.babu@gmail.com

S. R. Kumar et al. (eds.), *AI-Driven IoT Solutions in Agriculture: New Insights*, Studies in Computational Intelligence 1236, https://doi.org/10.1007/978-3-032-05603-0_11

1 Introduction

1.1 Overview of AI and IoT in Agriculture

Advances in worldwide population are making it inevitable for smart agriculture techniques to be implemented. This coupled with the reduction of natural resources, scarcity of fertile land, and unpredictable climatic conditions raise concerns of food safety and security in many nations. It is raising the need for the implementation of IoT and DA solutions to enhance productivity and operational efficiency in agriculture operations [1].

The World Resource Institute expects a shortfall between food produced and the amount needed to feed an estimated 10 billion people by 2050. The available arable land has also dwindled along with the adverse limiting factors like water scarcity and increased challenges of the soil quality, infestation by pests and weeds, there is a need for innovative approaches to food production so as to optimize agricultural practices. This paper presents a systematic literature review that focuses on the use of Artificial Intelligence and the Internet of Things in agriculture [2].

1.2 Importance of Addressing Agricultural Challenges

The IoT technology emphasizes advancing the transformation of labor-intensive traditional agriculture into data-driven smart farming in providing technological and theoretical support to address the challenges of food shortages caused by rapid population growth and to meet the sustainable development challenges posed by the scarcity of natural resources due to climate warming. IoT smart farming is a multidisciplinary area that incorporates knowledge of agriculture, communication technologies, automation, artificial intelligence, and computing [3].

It is in these approaches that issues confronting the Agri-challenge must be targeted if global food security and sustainable development are to be guaranteed. These are high hurdles that have to be overcome considering factors such as climate change, soil degradation, water scarcity, and increased food demand. CSA represents an important approach integrating sound farming techniques to increase productivity, reduce greenhouse gas emission, and enhance the resilience of agriculture to climate change. Studies have shown that CSA practices promote soil health and carbon sequestration, which is a key mechanism for adapting to climatic variability [4]. Makate et al. says that CSA practices improve soil health and sequester carbon.

Environmental sustainability in agriculture is threatened by pollution, deforestation, and poor land management. The adoption of eco-friendly agricultural practices and reduced reliance on chemical inputs can mitigate these challenges, preserving both agricultural productivity and environmental health [5].

1.3 Objectives of the Chapter

- **AI and IoT Technologies**

Explain and describe the basic concepts of AI and IoT, and how they can be applied to agriculture.

- **Exploring Applications of AI in Agriculture**

Engage them on how AI is applied in agriculture, such as predictive analytics, crop health monitoring, precision farming, and automated decision-making.

- **IoT in agriculture, Applications**

The IoT devices include, for example, sensors, drones, and GPS to collect real-time farm data and ensure effective admin of the farm.

- **Resource Optimization Efficiency through AI and IoT**

Examine how AI and IoT can optimize water, fertilizers and pesticides for waste minimization and sustainable practices.

- **Green Agriculture**

What AI and IoT bring: greener, and much more environmentally-friendly agriculture with its reduced footprint to the environment?

- **Data-driven Farming**

Use data analytics to make improvements in crop yield and monitor health conditions of the soils, and predict weather conditions way before they happen.

- **Overcoming Obstacles and Hurdles**

Present the challenges and limitation of this integration, such as high cost, and not everyone has access to the opportunity given out here, and putting data at security risk.

- **Case Studies and Real-World Applications**

Use examples of successful cases about the implementation of AI and IoT in agriculture that can change the face of the sector.

- **Future Trends in Agri-Tech**

Discuss future trends in AI and IoT in agriculture and how these can potentially be used towards contributions to food security and sustainable development.

2 Research Methodology

1. **Research Design**

Goal: The primary aim here will be to examine ways on how AI and IoT technologies can be used to develop sustainable agricultural solutions. These include boosting productivity, reducing the utilization of resources, enhancing environmental sustainability, and enhancing the economic resiliency of farmers.

Approach: This research will employ the mixed-method approach in terms of collecting and analyzing qualitative data as well as quantitative data.

2. **Literature Review**

Scope: The depth of literature review in existing literature on AI, IoT, precision agriculture, and sustainability. It will be based on academic papers, government reports, and case studies of successful AI and IoT implementation in agriculture.

Purpose: To identify gaps within the current research, the potential benefits, and challenges and best practice. This also gives theoretical support to the study and helps develop hypotheses.

3. **Data Collection Methods**

Primary Data Collection

Surveys or structured interviews should be conducted with the farming stakeholders, such as farmers, agronomists, and agricultural technologists. Interviews will provide key information on how AI and IoT are used currently, what challenges are perceived, and what expectations are.

Implementation of AI and IoT solutions in a selected agricultural environment, for example, farm or greenhouse, to collect first-hand, real-time data about productivity, resource use (water, fertilizers), crop health.

Secondary Data Collection

Datasets already in existence: Utilize datasets involving crop yields, climate data, water usage, quality of the soil etc. already available from public sources like FAO, NASA.

Case Studies: Identify documented case studies concerning farms or agricultural companies that already have adopted AI and IoT solutions.

4. **Technology and Tools**

AI Models and Algorithms

Develop or leverage existing AI algorithms in predictive analytics of crop disease, yield forecasting, decision making, and automation related to irrigation and pest control.

Machine learning techniques including neural networks, regression models, and reinforcement learning applied to optimize farm management.

IoT Infrastructure

Network of sensors (soil moisture, temperature and humidity, etc.) and connected devices like drones and automated irrigation systems to monitor and control the farming environment.

Use cloud-based platforms to integrate AI and IoT systems, data storage, and real-time decision-making.

5. **Data Analysis**

Qualitative Analysis

Use thematic analysis of interview responses to find common themes or concerns about adopting AI and IoT.

Carry out SWOT (Strengths, Weaknesses, Opportunities, Threats) analysis so that these technologies can be assessed for their potential in agriculture.

Quantitative Analysis

Carry out statistical methods of evaluation for impacts of AI and IoT intervention on the farm productivity, resource efficiency, and sustainability outcomes.

Time series analysis of sensor data for patterns or anomalies in environmental variables that affect crop growth.

6. **Sustainability Analysis**

Environmental Impact: Assess how AI and IoT decrease the water usage, beneficial parameters of soil, pesticide use, and carbon emissions in crops.

Economic Impact: Analyze ROI of farmers with AI and IoT systems through savings in manpower and other inputs in comparison with higher yields or income.

Social Impact: Discuss how these technologies are impacting smallholder farmers, rural employment, and food security.

7. **Ethical and Policy Considerations**

Discuss the ethical issues of data protection of farmers, dependency on technology, and equity in access to AI and IoT technologies.

Analyze the extant government policies and frameworks that promote AI and IoT in agriculture with gaps or where they ought to be enhanced.

8. **Limitations and Future Research**

Discuss the constraints, which emerged while conducting the study, such as data unavailability, technical constraints or regional variations in farming, etc.

Discuss the possible areas of future research, such as the application of AI and other emerging technologies, such as blockchain and satellite imaging, or more detailed work on the long-term impacts of AI and IoT on ecosystems.

3 Background Information

3.1 *Historical Context of Technology in Agriculture*

These three constantly acting forces of change over the last 50 years are agriculture development, technology, and people. Globalization is the force that shifts with increasingly larger emphasis from the domestic to international opportunities due to easier access of the world markets.

Improved technologies are a force that represents improvements in the ability of production and delivery of what consumers want, and people are exerting their influence, either directly as consumers or indirectly as custodians of the environment within which food and fiber products are produced. Of course, these three forces do not act in isolation but rather influence one another.

In addition, the relative significance of the three forces differed, both, over time and in different regions and/or countries [6].

3.2 *Evolution of AI and IoT Technologies*

There is smart farming coming together in the confluence of IoT and AI, smartening agriculture by infusing it with unparalleled precision and sustainability. This is because these two technologies converge, wherein sensors of IoT methodically collect information regarding the soil, weather, and livestock, and algorithms of AI decipher this information to deliver practical insights into this technological symphony.

This article captures the essence of smart farming: a transforming method that maximizes the use of resources, improves efficiency, and secures a resilient future for agriculture. Come with us as we travel where technology meets the earth to build a more sustainable era in the age-old art of farming [7].

3.3 *Current State of Agricultural Practices*

An increased population in later decades will most probably raise much more food demand, characterized by an increase in standards of living, which constitutes a huge challenge, especially in terms of environmental impacts of agricultural systems. On the basis of agricultural data over the last 50 years, we can claim that current agriculture will hurt global food production because: total agricultural area has declined since 2000, fertilizer and pesticide consumption increased with declining efficiency, and water sources are already used for irrigation.

All the risks of a global food crisis over the next few decades will continue to gain fuel from the expansion of biofuel production, limits of cultivation of genetically modified organisms, and the trend of increasing living standards. Then, crop yield and environmental protection have to be traded off with no feasible solutions at country or regional scale; the increasing crop yield while reducing the environmental impacts of agriculture intensification are important in terms of agricultural sustainability [8].

4 Key Challenges in Agriculture

4.1 Climate Change Impacts on Agricultural

- **Agricultural Productivity and Crop Yield**

Climate changes the productivity of agriculture through changes in temperature, precipitation and frequency of extreme events. With an increased in temperatures that enhance growth, yield becomes suppressed with increased heat stress. Changes in rainfall patterns increase water deficiency and excessive flooding thereby adversely affecting crop productivity [9, 10].

- **Water resources**

Availability of water for irrigation will depend on agriculture's development and usage, particularly when the region relies on guaranteed water supply. Climate change indirectly affects water availability through modifications in precipitation, higher evaporation, and varying river flows [11]. This affects directly the viability of irrigation-based farming systems.

- **Pests and Diseases**

Plant diseases, fungi, and insects may cause havoc on crops. Developing and implementing effective pest control strategies while minimizing environmental impact is a constant challenge [12].

- **Socio-Economic Impacts**

Socio-economic impacts of these agricultural impacts have further implications in the broader socio-economic context; loss of revenue to farming communities, and uncertainties for rural economies because of food insecurity. These would be better managed by adopting agricultural adaptation strategies related to crop types and modes of irrigation [13] (Fig. 1).

4.2 Resource Scarcity and Efficiency Issues

For the research paper on "Resource Scarcity and Efficiency Issues," the content touches on the pressing need for better resource management across various sectors.

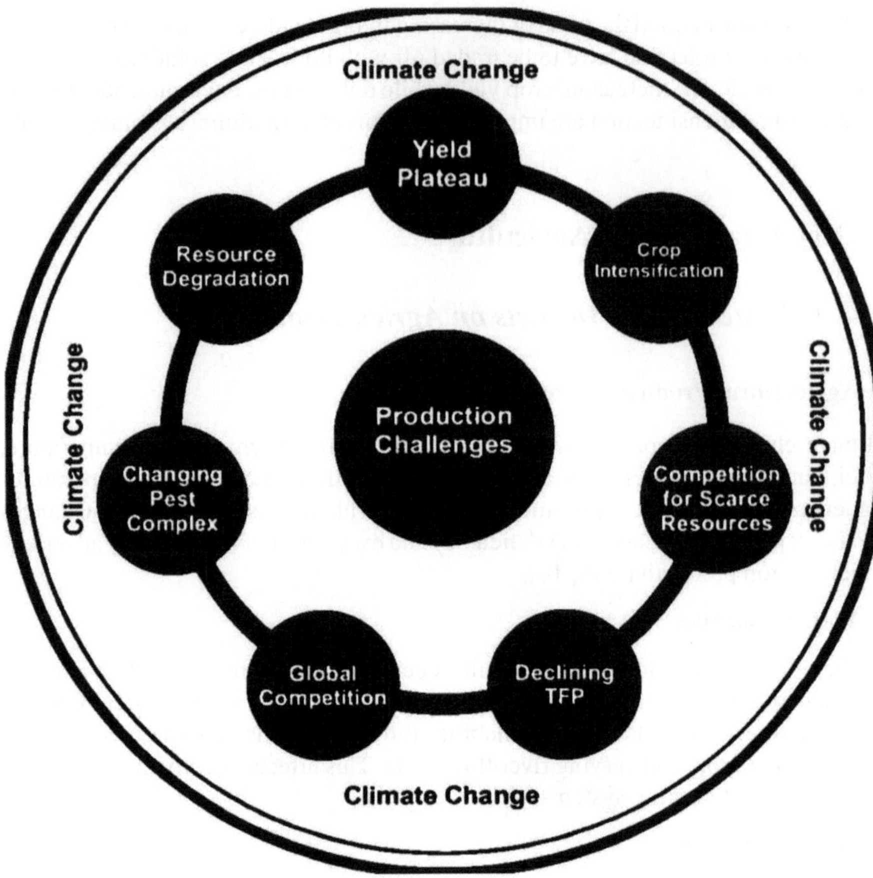

Fig. 1 Challenges of climate changes in agriculture

One significant study by Edward B. Barbier and Joanne C. Burgess explores water scarcity and efficiency. Their research discusses how increasing freshwater demand in developing regions, coupled with inadequate market incentives for technological advancements, worsens the global water crisis. They find that countries with poor water efficiency tend to experience higher productivity growth, especially when they focus on innovation and institutional development. They also recommend improvements in water markets and international trades to alleviate the problem [14].

Another relevant paper by Anne P. M. Velenturf and Phil Purnell addresses resource scarcity within the broader context of waste management. They argue that the transition to a circular economy is critical to managing resource demand sustainably. Their work focuses on the necessity of integrating technological, environmental, and social factors to promote resource recovery and reduce waste overload [15].

4.3 Challenges and Overcome in Water Management

Water management faces several critical challenges, primarily driven by growing populations, climate change, and unsustainable practices. One of the main challenges is water scarcity, which affects many regions due to increasing demand for water in agriculture, industry, and domestic use. This scarcity is exacerbated by over-extraction of groundwater and diminishing freshwater sources. Solutions to this problem include promoting efficient water use through technologies like drip irrigation, rainwater harvesting, and desalination to make seawater usable.

Another major challenge is water pollution, where industrial waste, agricultural runoff, and untreated sewage contaminate water bodies, reducing the availability of clean water. To overcome this, stricter regulations on industrial discharge, improved wastewater treatment systems, and reducing the use of harmful chemicals in agriculture are essential. Additionally, the mismanagement of water infrastructure, such as inefficient distribution systems and aging pipes, results in significant water losses. By investing in modern infrastructure and implementing smart water management systems, these challenges can be better addressed, helping to ensure a sustainable water supply for future generations.

For additional view in the challenges and overcome in water management as given in the Table 1.

5 AI and IoT Solutions

5.1 Overview of AI in Agriculture

1. **Machine Learning Applications**

With the advent of big data technologies and high-performance computing, machine learning has emerged as a new opportunity for data-intensive science in the multidisciplinary agri-technologies domain. The goal of this paper is to present an overview of state-of-art results dedicated to applications in machine learning in agricultural production systems, including those analysed in

(a) crop management including applications such as yield prediction, crop disease, weed detection crop quality, and species recognition.
(b) Livestock Management Including Applications in Animal Welfare and Livestock Production
(c) water management and
(d) soil management.

Table 1 Causes and overcome of water management

Author	Title	Issues	Challenges	Solutions
McIntosh [16]	Asian water supplies reaching the urban poor	Poor governance, low consumer awareness, intermittent water supply, low water accountability, poor construction and operation maintenance, polluted water ways, exploitation of ground water	Demand management and water conservation, managing water resources to meet megacity needs, conflict of water users	Transparent policies, paradigm shift in tariffs, organization development, waste water recycling, demand management, rainwater harvesting, urban sanitation improvements, private sector participation
Kumar [17]	Water management problems and challenges in india: an analytical review	Groundwater quality deterioration, growing demand, increasing competition and growing conflicts, surface water	Availability of scientific database, estimates of water availability and use, water quality, water demands and use rates	Centralised and segmented approach, regulatory approach to management, sectoral and segmented approaches, institutional frameworks and market instruments
Kumar [18]	Water resources issues and management in India	Water for agriculture, water quality, management of ground water resources, drought and flood, groundwater contamination		Groundwater legislation and aquifer recharge, groundwater resources: assessment and planning, participatory water management
Jaiswal [19]	Sustainable water management: issues and challenges in India	Sectoral demand of water, sectoral consumption of water, demand supply and shortage, waste water generation and its treatment	Rainwater harvesting, key obligations for various stakeholders in the water sector	

(continued)

Table 1 (continued)

Author	Title	Issues	Challenges	Solutions
Mishra [20]	Urban water resource management for sustainable development	Increasing population—migration, land-use change, climate change, urban flash flood		Low impact development, water sensitive urban design, urban water supply management, urban water demand management, rain water harvesting, water demand management
Panwar and Antil [21]	Issues, challenges and prospects of water supply in urban India	Water availability, water supply infrastructure, waste water disposal	Water quality, financing of infrastructure for water supply, lack of regulatory framework, water utilization and infrastructure management	Rain water harvesting, water reuse or water from waste water, conservation of supplied water, improving water distribution and its use, ground water recharge, public private partnership
Cronin et al. [22]	Water scarcity, need of new indices to measure water budget, climate change, population pressure, industrial growth, water quality and pollution, water, gender and equity	Role and accountability of key stakeholders, Governance deficit		All stakeholders need to work together to achieve desired goals, Government should harness community participation especially for sanitation and for a safe sustainable water supply

From the filtering and categorization of the articles herein presented, one will find an opinion about what machine learning technology may mean to agriculture [23].

2. **Computer Vision and Robotics:**

- **Computer Vision**

It's the most widely applied technology in identifying diseases and crop health monitoring among others. Computer vision, for instance, can monitor the growth rate of plants and the amount of weeds in a field, thus saving time taken for manual inspection. For instance, fruit counting and fruit grading enhance harvesting and sorting with much accuracy [24].

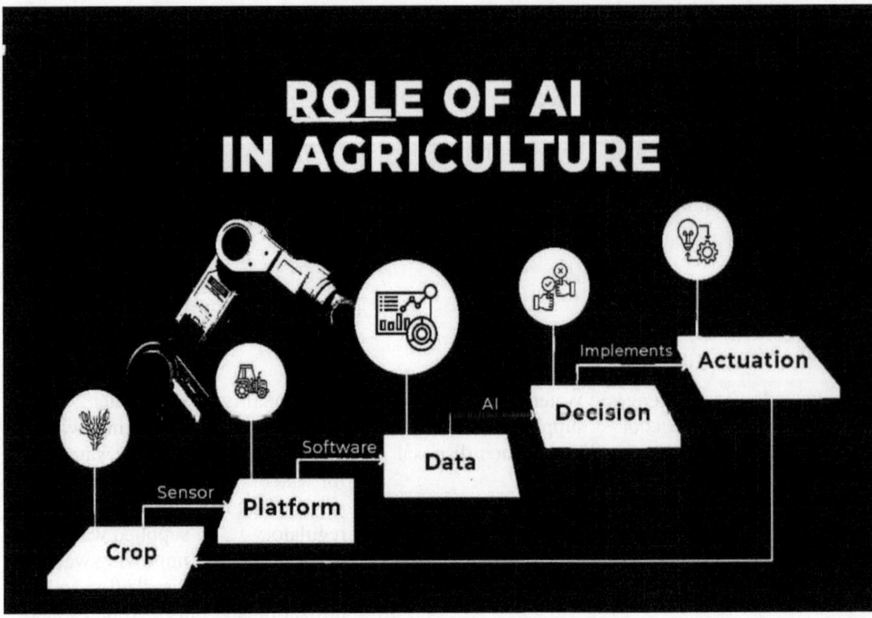

Fig. 2 Role of AI in agriculture

- **Robotics**

Agriculture is employing more agrobots which are robots used for agricultural functions such as planting, weeding, and harvesting. The machines use AI to navigate and interact with crops. There is selective harvesting and targeted pesticide applications, and minimal waste yields a maximum amount. Some of the agrobots use lasers for more precise removal of weeds without causing damage to crops [25] (Fig. 2).

5.2 Overview of IoT in Agriculture

- **Sensors and Smart Devices**

With a wide application in robotics, navigation, automation, remote sensing, underwater imaging, etc., modern sensors have played a crucial role in the field of remote sensing and smart agriculture within the last years. AI-enabled sensors work as smart sensors and besides, the advent of the Internet of Things (IoT) has resulted into very useful tools in the field of agriculture by making different types of sensor-based equipment and devices available.

Silvia Liberata Ullo, Ganesh Ram Sinha highlights the comprehensive study of progress in smart sensors and IoT, which have been integrated in remote sensing and agriculture applications, including weather conditions and soil evaluation, crop monitoring, use of robots for harvesting and weeding, and employment of drones [26].

- **Data Collection and Analysis**

The global population is demanding the practice of more smart agriculture. Coupled with this is the fact that natural resources are depleting, arable land is scarce, and unpredictable weather conditions are on the rise, which has led to making food security a basic necessity for most countries. As a result, IoT and DA are employed in the sector of agriculture to improve operational efficiency and productivity. The paradigm has shifted from the use of wireless sensor networks as a primary driver for smart agriculture to the use of IoT and DA. The IoT integrates several existing technologies that include WSN, RFID, cloud computing, middleware systems, and end-user applications [1] (Fig. 3).

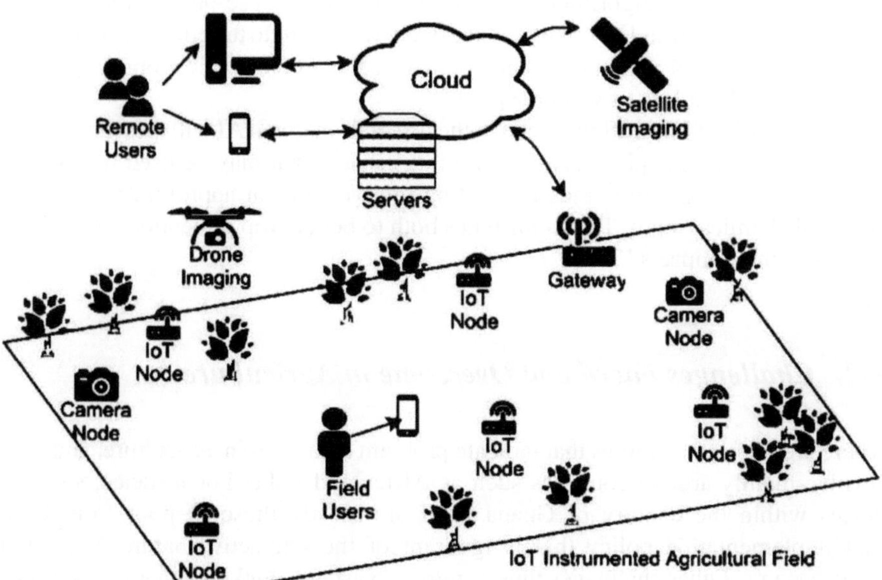

Fig. 3 IoT enabled agricultural field

6 Case Studies

6.1 *Successful Implementation of AI and IoT in Agriculture*

Many case studies have proven that AI and IoT can be effectively implemented within agriculture, thus revolutionizing farming concerning productivity, sustainability, and precision. An obvious one is the use of smart irrigation systems based on IoT. These systems use sensors to measure soil moisture, thus optimizing the use of water and preventing waste-so crucial in regions with shortages of water. Such systems have helped to provide real-time automatic control of irrigation that has led to better crop yield and saved water [27].

The Wangree Plant Factory in Thailand is yet another example that has followed the spot-on use of smart technology-IoT and AI-to optimize plant production. Sensors and machine learning algorithms deployed ensure resource use-once again, water and energy-is optimized and so is the produce quality. This led to the emergence of a more sustainable mode of agriculture that aimed at reducing the application of damaging chemicals without loss of yield [28].

An example of integration is seen in the use of drones and AI with pest and disease monitoring. For example, UAVs incorporate sensors that may be used in detecting early signs of crop stress, thus facilitating interventions at appropriate times with minimal chemical input. This contributes both to better crop protection and reduced environmental impacts [29].

6.2 *Challenges Faced and Overcome in Agriculture*

There are numerous studies that indicate problem areas within agriculture, and these revolve mainly around countries such as Africa and India. For instance, systemic issues within the country of Ghana itself are mostly those of poor formulative and implementation policy that is ignorant of the interactive nature of agricultural systems. Other challenges thus require a "systems thinking" that would reflect on holistic insight into the agricultural sector, taking into account complexities of handling the system by the policymakers. An effective approach in such a manner can leads to the discovery of deep cause, minimization of risk, and enrichment of decision-making processes towards sustainable agricultural growth [30].

One of the major issues Indian farmers are facing is financial insecurity, which is often aggravated by high interest rates from informal lenders, pushing many into debt traps. This is further worsened by unpredictable weather conditions and the lack of access to modern agricultural technologies. Government loan waivers, improvements in formal financial institutions, and crop insurance schemes are advocated as recommendations in easing these financial burdens. It would minimize labour cost and increase productivity by adopting agricultural automation and shared machinery services. For example, pilot projects have been noticed in Haryana [31].

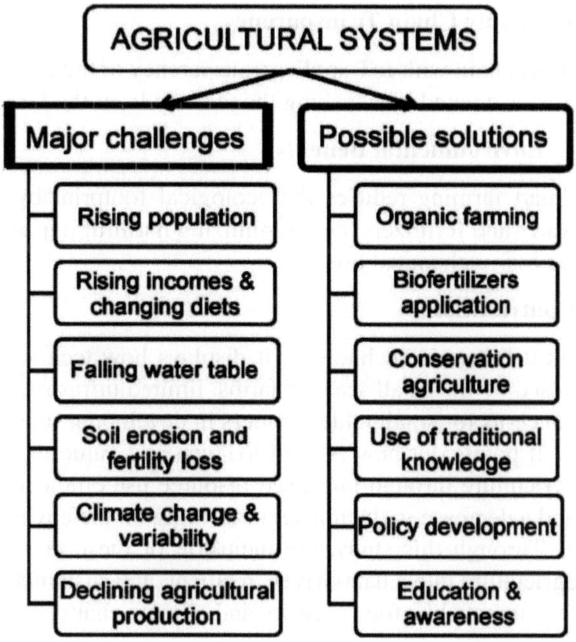

Fig. 4 Challenges and solutions of agricultural practices

These case studies illustrate that while the challenges in agriculture had given in the Fig. 4 that are multifaceted, including financial, environmental, and policy-related issues, there are strategies in place that, if implemented effectively, could significantly improve the resilience and sustainability of agricultural systems.

7 Findings and Contributions

Findings

- **Increased Efficiency and Accuracy**

AI and IoT can make sense of all kinds of conditions for the crop plants, but the present technology, in this case, makes possible "precise" farming by collecting real-time data concerning soil moisture, temperature, and pest pressure with the help of IoT sensors to assist farmers to control water consumption and minimize chemical inputs, hence bringing down waste and increase crop yields through targeted interventions, such as selective pest control or the adjustment of irrigations.

- **Predictive Analytics**

Applying AI-driven models, farmers would be able to anticipate pest and negative climate risks that can affect the farm. Such would then give way to preventive action. Predictive analytics enhance resilience and resource efficiency.

- **Supply Chain Transparency**

Blockchain with IoT applies transparency to the food supply chain, while consumer trust is created by checking the journey from the farm to the table.

- **Environmental Benefits**

Smart farming reduces the ecological footprint by optimizing resource use, like water and fertilizers, and minimizing harmful impacts such as soil degradation and excessive chemical use.

Contributions:

As it brings in technology, it displays how technology can bridge the challenges posed by the small scale of farms, limited infrastructure, and related environmental concerns for smallholder farmers in developing regions.

It points out how AI and IoT can add value to the development of sustainable agriculture through improved resource use efficiency, increased farm productivity, and a decrease in the impact of farming on the environment.

Through this study, the inclusion of these technologies transforms traditional agriculture into a data-driven, resilient, and sustainable system that can strategically face threats like food security and climate change.

8 Suggestions and Recommendations:

- **Precision Agriculture**

AI for Predictive Analytics: Apply AI to study data feeds from multitudes of sources, such as weather, soil health, and historical yields, to know when best to plant and what resources to use.

IoT Sensors for Data Collection: IoT sensors in fields monitor soil moisture and temperature and crop health in real-time to decide optimal irrigation and fertilization.

- **Automated Irrigation System**

AI-Based Water Management: Utilize algorithms to control irrigation through governing based on IoT sensor data, hence assuring adequate water supply to the crops without any wastage.

- **Pest and Disease Management**

Early Detection using AI: Use AI combined with image recognition technology to detect the symptoms of pests and diseases so that vigilance can be taken at an early stage.

IoT-based Crop Monitoring: Sensors through IoT devices monitor the crops all the time and alert the farmer instantly if those problems occur.

- **Intelligent Devices and Robotics**

Autonomous Machines: Invest AI-driven tractors and drones for equipment applied in planting, weeding, and harvesting so as to achieve highest efficiency and labour cost.

IoT Connectivity: Connect all the machines with IoT to monitor and optimize operations in real-time.

- **Resource Efficiency Efficiently sustainably**

Optimization of Energy through AI: Use AI to optimize energy use through agricultural processes, including greenhouse lighting and heating.

IoT for Carbon Footprint Monitoring: Utilize IoT technology in tracking and lowering the emissions of greenhouse gases and resource consumption.

- **Supply Chain Optimization**

AI for Demand Forecasting: Utilize AI to predict the market trends and consumer behaviour. Then, apply this knowledge in forecasting the demand for your products. This will help you, the farmer, plan production consequently.

IoT for Traceability: Utilize IoT sensors that track farm-to-mark produce for transparency and food safety.

- **Climate-Smart Agriculture**

AI for Adaptation Strategies: Utilize AI models in simulating climate scenarios and suggest optimum crop options and farming practices to be resilient under climate change.

IoT Climate Monitoring: Integrate IoT sensors to monitor local climate conditions, thus allowing farmers to adapt their practices in real time.

- **Financial Tools for Farmers**

AI Risk Management Solutions: Create AI tools that can estimate market risks and will establish insights so that farmers make the best decisions financially.

IoT for Insurance Automation: Merging IoT data with AI to automate crop insurance processes so that it pays out faster as per real-time conditions.

- **Support to Sustainability and Policy**

AI-based Data-Driven Policy Design: Artificial simulations of the agricultural policy effects can allow governments to design policies for sustainability initiatives better.

IoT Data for Regulation: Establish an IoT network of devices providing real-time data so that regulations pertaining to resource use and environmental impact can be developed hand-in-hand.

- **Education and Training for Farmers**

AI-Based Training Platforms: Availing of AI-based learning experience for the training of farmers on novel technologies and practices that will be tailored for their specific needs.

Demonstration Projects: Pilot projects on AI and IoT technologies to be set up for demonstrating these concepts to the farmers on how to exploit them.

9 Future Directions and Emerging Trends

- **Precision Agriculture**

Data-Driven Decisions: The AI algorithms are used for analysing data collected by the IoT sensors and drones toward scheduling optimum planting, watering, and harvesting activities.

Soil Health Monitoring: The soil's moisture, pH level, and nutrient content can be monitored in real-time using sensors to allow farmers to apply the right kinds of fertilizer and right amounts of water at the right time.

- **Smart Irrigation Systems**

Automated Water Management: IoT devices can sense weather predictions and soil moisture levels to automate irrigation, thus preventing wastage of water and increasing crop yields.

Drones for Irrigation Monitoring: Imaging equipped drones can monitor crop health and provide the best irrigation decisions.

- **Crop Monitoring and Disease Detection**

AI-Powered Image Recognition: applying machine learning on image crops taken by drones or smartphones to alert early cases of disease or pest attacks.

Predictive Analytics: Using environmental data, AI models can predict pest outbreaks or disease spread and help in proactive steps.

- **Supply Chain Optimization**

Smart Logistics: AI will be able to optimize logistics such that the agricultural supply chain from the farm to the table ensures freshness and less spoilage.

Blockchain for Transparency: Introduction of blockchain with IoT brings transparency to the supply chain ensuring food safety and authenticity.

- **Robotics and Automation**

Autonomous Farming Machinery: Robotic farming equipment for planting, weeding, and harvesting. Such farming machinery may well work independently with less input from the human workforce.

AI Smart Harvesting: Utilizing AI to evaluate crop ripeness and scheduling the time at which such crops should be harvested. The quality and quantity of yield are therefore on the increase.

- **Resource Management for Sustainable Agriculture**

Optimal Utilization of Inputs: AI and the IoT will ensure optimal usage of fertilizers, pesticides, and water. This means further reduction in damaging the environment and promoting efficient farming practices.

Carbon Footprint Monitoring: IoT can be deployed to monitor and regulate the farming practices that produce greenhouse gases and will go towards carbon–neutral missions.

- **Climate Resilience**

Climate-Smart Agriculture: AI will help farmers understand how weather patterns change so they could adapt accordingly, while maintaining safety in food supply.

Predictive Modelling: The future trend can be gauged with the help of climate scenarios to determine by AI models, and aid the farmers in decision making or crop management and choice.

- **Consumer Engagement and Smart Marketing**

Personalized Customer Experience: Big data analytics helping organizations understand the customer preference for targeted marketing of sustainable products.

Direct-to-Consumer Sales: With an IoT-enabled device, farms get connected to consumers, and freshness and quality can be tracked.

- **Interconnection and Integration**

IoT Ecosystems: Such an ecosystem is made of interconnecting devices in order to share relevant data that has a better possibility to provide insights and automate agricultural systems in appropriate manners.

Open Data Platforms: Staking holder collaboration toward the creation of an open platform for agriculture data sharing, furthering research and innovation.

- **Education and Skill Development**

Training Programs: Training programs on newly adopted technologies for farmers, so the techniques get effectively implemented.

Community Initiatives: Promotion of community projects through AI and IoT for the local development of agriculture.

10 Conclusion

This is an important leap toward the development of sustainable agricultural practices, which incorporates AI and IoT in agricultural processes. Advanced technologies will enable farmers to work more effectively with greater productivity, optimal resource use, and lesser environmental impact. Precision farming techniques help in the betterment of decision-making through the analysis of real-time data, increasing crop yields and reducing waste by making use of the given resources effectively. The connectivity and automation features offered through IoT devices result in the effective monitoring and management of agricultural processes.

With an ever-rising demand for food globally, AI and IoT are sure to unlock new issues associated with climate change, resource constraints, and environmental friendliness. Embracing such technologies encourages sustainable solutions in addition to empowering the farmer with the right tools to thrive amidst these increasing complexities in agriculture. The Future of Agriculture: Innovation and Acceptance of Techniques Embracing AI and IoT for a Resilient, Efficient, and Sustainable Food System.

References

1. Elijah, O., Rahman, T. A., Orikumhi, I., Leow, C. Y., Nour Hindia, M.H.D. (2018). *IEEE Internet of things Journal, 5*(5), 3758–3773.
2. De Abreu, C. L., & van Deventer, J. P. (2022). Southern African conference for artificial intelligence research (pp. 32–46).
3. Huo, D., Malik, A. W., Ravana, S. D., Ur Rahman, A., & Ahmedy, I. (2024). *Renewable and Sustainable Energy Reviews, 189*, 113858.
4. Makate, C., Wang, R., Makate, M., & Mango, N. (2019). CSA adoption, impacts, and implications for sustainable development. *Mitigation and Adaptation Strategies for Global Change.*
5. Afolabi, O. O., Wali, E., Ihunda, E. C., Orji, M. C., Emelu, V. O., Bosco-Abiahu, L. C., & Ogbuehi, N. C. (2022). Environmental sustainability in Nigeria: Challenges and remedies. *Environment, Development and Sustainability.*
6. Van Keulen, H. (2007). *Science for agriculture and rural development in low-income countries* (pp. 7–26).
7. Mandapuram, M., Mahadasa, R., Surarapu, P. (2019). *Global Disclosure of Economics and Business, 8*(2), 165–178.
8. Liu, Y., Pan, X., & Li, J. (2015). *Journal of Agricultural and Environmental Ethics, 28*, 203–216.
9. Kim, C.-G., Na, Y.-E., et al. (2009). Impacts of climate change on agricultural productivity. *Environmental Science & Pollution Research.*
10. Parry, M. L., Rosenzweig, C., Iglesias, A., Livermore, M., & Fischer, G. (2004). Effects of climate change on global food production. *Global Environmental Change, 14*, 53–67.

11. Falloon, P. D., & Betts, R. A. (2006). The impact of climate change on global river flow in HadGEM1 simulations. *Atmospheric Science Letters, 7*, 62–68.
12. Suresh Babu, C. V., Ganesh, B. S., Kishoor, T., & Khang, A. (2023). Automatic irrigation system using solar tracking device. In A. Khang (Ed.), *Handbook of research on AI-equipped IoT applications in high-tech agriculture* (pp. 239–256). IGI Global. https://doi.org/10.4018/978-1-6684-9231-4.ch013
13. Iglesias, A., & Rosenzweig, C. (2009). Effects of climate change on global food production from SRES emissions and socioeconomic scenarios. *NASA Socioeconomic Data and Applications Center (SEDAC).*
14. Barbier, E. B., & Burgess, J. C. (2024). Economics of water scarcity and efficiency. *Sustainability, 16*(19), 8550. https://doi.org/10.3390/su16198550
15. Velenturf, A. P. M., & Purnell, P. (2017). Resource recovery from waste: Restoring the balance between resource scarcity and waste overload. *Sustainability, 9*(9), 1603. https://doi.org/10.3390/su9091603
16. McIntosh, A. C. (2003). *Asian water supplies reaching the urban poor*. Asian Development Bank.
17. Kumar, M. D., & Ballabh, V. (2000). Water management problems and challenges in India: An analytical review. Working Paper. Retrieved from https://www.researchgate.net/publication/312525643_Water_Management_Problems_nd_Challenges_in_India_An_Analytical_Review
18. Kumar, C. P. (2018). Water resources issues and management in India. *Journal of Scientific and Engineering Research, 12.*
19. Jaiswal, M. (2017). Sustainable water management: Issues and challenges in India. *JETIR, 5.*
20. Mishra, P. K., Khare, D., Singh, L. L., & Chandrakar, A. (2016). Urban water resources management for sustainable development. *Convergence, 13*, 35–51.
21. Panwar, M., & Antil, S. (2015). Issues, Challenges and prospects of water supply in urban India. *IOSR Journal of Humanities and Social Science IOSR JHSS, 205*, 68–73. https://doi.org/10.9790/0837-20526873
22. Cronin, A. A. (2014). *Water in India: Situation and prospects*. IWA Publishing.
23. Liakos, K. G., Busato, P., Moshou, D., Pearson, S., & Bochtis, D. (2018). *Sensors, 18*(8), 2674.
24. Xu, B., & Jia, W. (2024). Computer vision and artificial intelligence in agriculture. *Special Issue in Agriculture Journal, MDPI*. Available at: https://www.mdpi.com/journal/agriculture/special_issues/Computer_Vision_AI_Agriculture
25. Jha, K., Doshi, A., Patel, P., & Shah, M. (2019). A comprehensive review on automation in agriculture using artificial intelligence. *Artificial Intelligence in Agriculture, 2*, 1–12. https://doi.org/10.1016/j.aiia.2019.06.001
26. Ullo, S. L., & Sinha, G. R. (2021). *Remote Sensing, 13*(13), 2585.
27. Farooq, M. S., Riaz, S., Abid, A., Abid, K., & Naeem, M. A. (2019). A survey on the role of IoT in agriculture for the implementation of smart farming. *IEEE Access, 7*, 156237–156271. https://doi.org/10.1109/ACCESS.2019.2949703
28. Tippayawong, K. Y., & Tamvimol, K. (2020). The role of smart technology in sustainable agriculture: A case study of Wangree plant factory. *Sustainability, 12*(11), 4640. https://doi.org/10.3390/su12114640
29. Goap, A., Sharma, D., Shukla, A. K., & Krishna, C. R. (2018). An IoT-based smart irrigation management system using machine learning and open-source technologies. *Computers and Electronics in Agriculture, 155*, 41–49. https://doi.org/10.1016/j.compag.2018.09.041
30. Banson, K. E., Asare, E. A., Bansah, K. A., & Fennell, J. (2014). Systems thinking in agricultural sustainability. *Agricultural Systems, 126*, 1–7. https://doi.org/10.1016/j.agsy.2014.01.005
31. Rakhra, M., Brar, S. K., & Kumar, S. (2022). Agricultural automation and challenges in India. *International Journal of Finance and Management Research (IJFMR), 7*(3), 1–10.

Computer Vision and AI-Based Apple Plant Disease Detection Framework in Smart Farming

R. Karthick Manoj and S. Aasha Nandhini

Abstract In this chapter, a novel method of apple plant disease detection based on computer vision and the ensemble machine learning methods is presented. Apple diseases like apple scab, fire blight and cedar rust are serious problems for agriculture that early detection can help reduce hits to the bottom line. This work presents a multistage approach combining state-of-the-art image preprocessing, segmentation and feature extraction to detect disease areas on apple plants from images. Preprocessing methods that smooth input resolution and equalize the appearance across its histogram, such as Gaussian smoothing or adaptive histogram equalization are applied first. K-means clustering has been used to group pixels with similar features, and Watershed algorithm is utilized for refining boundaries of disease-affected regions. From the segmented regions, informative features such as texture, color and shape are extracted which serve as input to a machine learning classifier that consists of an ensemble (SVM-KNN-RFC) set. An ensemble model uses voting techniques to leverage these classifiers' individual strengths, which gives more accuracy, precision and recall. Experiments show that the ensemble obtains it 97.2% accuracy rating which clearly leading among other traditional methods and also deep learning models currently available Its robustness and scalability makes the system suitable for real-time deployment in agricultural scenarios, thereby providing a dependable early disease detection and crop management strategy. Future efforts will focus on extending the dataset, and using techniques such as hyperspectral imaging or stacked ensemble methods to enhance detection capabilities.

Keywords Apple plant disease detection · Image segmentation · Multi-feature extraction · AI in agriculture

R. K. Manoj (✉)
Department of EEE, AMET University, Chennai, Tamil Nadu, India
e-mail: karthickmanoj.r@gmail.com

S. A. Nandhini
Department of Electronics and Communication Engineering, Sri Sivasubramaniya Nadar College of Engineering, Chennai, India

S. R. Kumar et al. (eds.), *AI-Driven IoT Solutions in Agriculture: New Insights*, Studies in Computational Intelligence 1236, https://doi.org/10.1007/978-3-032-05603-0_12

1 Introduction

Agriculture is as old as human civilization, which provided the basic needs of life and increased population. With world population increasing, there is more and more need for agricultural production. In response, agriculture has seen some major shifts as well—from traditional farming to modernized agricultural practices heavily reliant on technology for operational efficiency and productivity. As a result, smart agriculture as we know it today (the practice of integrating various technological innovations to optimize crop management and resource use) took form. The integration of artificial intelligence (AI) and machine learning (ML) technologies has been one of the driving forces in this evolution, which is proving to be an invaluable tool for disease detection, pest management and optimising yields among several challenges being addressed across the agronomic scenario [1]. One of most useful field where AI and ML has highly affected is in the early plant disease detection. Over time, farmers used automation techniques to monitor and determine their crops from disease outbreaks manually which most of the times was a delay in diagnosis leading to substantial crop loss. Plant diseases are recognized to cause about 20–40% of global crop loss annually. Part of the challenge is that many conditions have only minor, non-specific symptoms when a disease starts out. To solve this problem, AI systems have been proposed in which the visual symptoms of a diseased plant can be detected using an automatic identification approach associated with image recognition and intervention made in its early stage [2]. These could be game-changers for both the environmental and economic implications of crop diseases.

To monitor the health of plants, computer vision system implementation partnered with AI is a vital milestone—computer vision and ML based plant health monitoring the systems rely on images taken by high resolution sensors and cameras installed in the field or mounted to drones, which are analyzed for abnormalities like spots, discolouring, or wilting. Recent advances in deep learning algorithms based on convolutional neural networks (CNNs) have permitted the accurate identification of plant diseases by utilizing vast annotated image datasets [3]. Various models have been developed with better accuracy than traditional methods and some of them achieved more than 90% detection rate. Not only does automatic detect diseases more accurately, but it saves time and labor also to be used as a must-have tool in agriculture nowadays.

In addition, the use of ensemble machine learning methods bolstered both accuracy and robustness in plant disease detection systems. Relying on the benefits of different classifiers like SVM, KNN and RFC respectively ensemble model improve handling outcomes about the heterogeneous nature and entropy of real-world agricultural settings. Ensemble methods have been demonstrated to outperform single models by alleviating overfitting and enhance generalization in a wide variety of data sets. Given the increasing demands on agriculture due to climate change, pest infestations and limited resources for required crop management activities; AI innovation in disease identification is likely to play a major role toward ensuring sustainable food production.

2 Literature Survey

A comparison of simple and few shot analysis for leafy crops using convolutional neural networks (CNNs) against disease detection rates achieved with traditional image processing techniques, where the study CNN provided an accuracy rate between 92% in order to detect pages affected by these diseases. According to the study authors, deep learning models are a powerful tool for automatic disease detection in large apple orchards [4].

It is observed that [5], used ensemble learning for apple disease detection to improve the classification accuracy while study done by was focused on developing fuzzy prioritization method. In order to combine the results of multiple classifiers (Random Forest, Decision Trees and SVM) they used a majority voting mechanism. Our results show an F1 score increase over most (fire blight and black rot average accuracy improved by 5–7%) of the diseases compared with individual classifiers using ensemble method.

In the method proposed by [5] a full apple diseases detection system was developed using image processing techniques with deep learning. Among other things, their work emphasized the necessity for preprocessing steps like noise reduction and image enhancement before deep learning models (known in this particular case as CNNs) can be applied to an imaging task. This platform could identify multiple diseases including rust and leaf spot simultaneously by 89%. This approach allows the use of largely traditional image processing methods and may make up for what you can lose by not being able to add more training data (or already using every last one), so increasing that area under ROC curves.

A transfer learning technique [6] to apply pre-trained deep learning models in solving apple disease detection. For instance, the models they used included ResNet and VGG16. However, these models were images apples, more specifically. The authors fine-tuned them using a small data set image of apple diseases. The findings of indicated that transfer learning improved the classification accuracy metric in comparison with existing methodologies. The identified architecture reached a 93.7% accuracy level, especially when identifying diseases like apple scab, rust, and black rot. Based on the test, the model can be used even in small orchards where there is no access to relatively large apple disease images.

An IoT embedded with AI for apple disease in their study [7]. In their study, the AI integrated system should be automatic and real-time when monitoring and detecting apple diseases. Accordingly, the two high technologies were integrated into an embedded device. The team applied sensors and cameras while installed after some distances in the apple fields. After data capturing, machine learning algorithms were used to classify and detect the data. The application reduced the duration of time required to detect the disease by 50%, which ultimately reduced the time for disease spread.

The experiment showed that the classification accuracy rose with pre-train models as VGG16 and fine-tuning on apple leaf data. Transformers, as they are very easy to re-use and thus benefit from transfer learning abstraction while keeping the model generalized across multiple tasks that improves their overall performance. This may in particular be beneficial for agriculture as it can facilitate the scarcity of rich, labeled datasets [8].

The CNN was also used in the automated classification of apple leaf diseases [9]. So, for this project they create a data set called apple leaf images and apply CNN model to detect diseases like Apple blotoch, rust, and frogeye spot. Other image classification techniques resulted in significantly lower accuracies but even the system most successful at this task came up short of our 95% accuracy rate. They also reinforced the use of data augmentation techniques, such as flipping and rotation, to improve model results regardless of different environmental conditions. Deep learning has become a popular tool for automatic feature extraction and classification with plant disease detection. A study [10] reported 97.4% accuracy in detecting different diseases on tomato leaves using a deep learning model. However compared to other traditional machine learning models such as naïve Bayes and logistic regression this CNN-based approach significantly outperformed as the no handcrafted features are required yet again meanwhile speed of classification was better with improved accuracy. Transfer learning has also been used with the pre-trained models particularly to fit into specific plant diseases datasets, since it solves problem of scarce training data supports quite limited for certain disease classes.

In a study [11] the Soybean diseases detection which was based model EFficientNet Model. The use of this showed transfer learning, leading to more superior robustness and generalisation over smaller datasets getting a precision upto 98.1%. This is particularly beneficial for tasks of disease detection, since the model can leverage rich and robust features learned from a large-scale dataset to identify pathology in cohorts which have very few annotated images.

In plant disease detection, a fusion of convolutional neural networks (CNNs) and recurrent neural networks (RNNs) has been evaluated on sequential image analysis. The proposed a hybrid CNN-RNN [12] model to learn spatio-temporal features in the time-series images of diseased plants. The model was tested on a dataset for the detection of grapevine diseases and it achieved an accuracy of 95.6%. The RNN component helps in capturing the temporal evolution of disease, leading to improved classification accuracy as compared with only using CNN based independent models.

In the same concept, recent studies have studied about attention mechanisms in deep learning models which helps to better attend at certain parts of an image such as diseases area in plant images. The proposed integrated attention modules in CNN framework allowing the model to highlight only important areas while suppressing unnecessary background information [13]. The results presented an accuracy of 98.2% for prediction plant wheat rust by our approach. Thus the attention mechanism not only helps in improving the performance with better results but it also provides some visual indication of what our model is actually focusing on and thus enhances its interpretability.

Recently, ensemble learning strategies were followed being a tendency to the classification and robustness in plant disease diagnosis as it combines plethora classifiers. Apple scab was detected using a voting ensemble of CNN, SVM and Random Forest classifier with weighted votes obtaining 99.1% accuracy in detection [14]. In addition, ensemble methods can improve performance on a wide range of types by decreasing the impact of individual model errors.

The aerial monitoring of crop health requires the useof unmanned aerial vehicles (UAVs) that are equipped with high-resolution cameras. A study [15] signals the addition of yet another click in this story. Discussed a RL-CNN model based on UAV-captured images for real-time detection of rice diseases. The system had an accuracy of 96.8% showing the capability of UAVs in disease detection on large field scales at early stages. It is especially useful to monitor more remote areas in large agricultural fields this way.

Previous research has employed multispectral and hyperspectral imaging for the early detection of plant diseases, gathering more information over various wavelengths. A developed system based on hyperspectral imaging and deep learning framework to detect early symptom of citrus greening [15]. It made fewer than 4% errors on average and easily out performed conventional RGB-based methods, which often make more than twice as many (note: conventional outlier detection in weakly supervised learning). By measuring the reflected and transmitted radiation of biological material in plant tissues, hyperspectral imaging can identify previously unobserved disease-related perturbations affecting changes in detectable tomic scale physiological processes.

Real-time monitoring and detection of plant diseases through sensor networks driven by Internet of Things (IoT) technologies. A study [6] introduced an IoT-based system combined with deep learning algorithms in wheat crop under fungal disease detection. To achieve this, edge computing was utilized within the system on site and used to process sensor data locally so that diseases could be identified almost instantaneously. While in the lab IoT with deep learning was more accurate at 97.8% and in field trial it reached an accuracy of about only 94.3%.

As plant disease detection leans more and more towards deep learning, explainable AI (XAI) has been identified as a key mechanism to make decisions truly transparent. Patel et al. that utilized an XAI based explanation framework to explain CNN predictions for disease detection in maize. It applied saliency maps to localize the importance of regions that drive model decisions, which can lead agronomists trust and use AI-based solutions. It featured the model that was 96.5% accurate and showed visual explanations of how decisions are driven.

In a world where you might not have access to really small datasets during your deep learning project and also for the purpose of better generalization—data augmentation comes handy. A study [18] leveraged additional augmentation such as rotation, flip and GAN-based synthetic image generation to improve performance of plant disease detection model in advance. The paper cited aimed to enrich tomato leaf image but detection accuracy was used as 97.8% by improving the variation of training samples.

Since mobile application is available in every smartphone, it has become a simple way to detect plant diseases. The author [19] introduced a CNN based mobile app for detecting diseases real-time in crops like potato and tomato. The app worked without any internet connection, had 94.2% classification accuracy and resulted in a high adoption rate among farmers with small land holding areas in India. All this shows the value of mobile apps in translating sophisticated AI solutions to on-the-ground usage by agri end-users.

To deal with the limited number of labeled datasets for plant disease detection, GANs was used to generate artificial training data [20]. It increased the accuracy with which deep learning models could be trained on a dataset of infected wheat images by augmenting that data using a GAN. Using synthetic data then further allowed the model to reach an accuracy of 98.6% on wheat rust and leaf blight detection based on a CNN-based architecture Generative Adversarial Networks (GANs) have been widely used to generate high-fidelity natural, physically realistic images.

Thermal images could be used to detect early stages of disease due to temperature differences associated with the infection and has been explored as a non-invasive detection method [21]. Detection of Fusarium wilt by using thermal images captured in the day time as documented a deep learning model was proposed with early stage detection satisfying results adopted to resist stressful environment and efficacy at detecting disease onset, meanwhile effective null experiment show trememdous diversity. The model achieved 94.7% accuracy (which turned out to be better than the existing visual inspection method It can provide disease-related changes in plant physiology which is shown before appearance of visible symptoms [22].

Recently, self-supervised learning achieved some promising results in making big labeled datasets less imperative by allowing a model to learn meaningful feature representations without supervision on unlabeled data. It used self-supervised learning to pre-train a model on unlabeled plant images and then fine-tuned the model for disease classification [24]. It achieves state of the art accuracy compared to other methods and obtained a 95.2% classification score on a challenging grapevine disease dataset Self-supervised learning comes with the promise of making headway in domains which do not have access to enough labeled data and that is expensive to get (fine-tuning on smaller datasets; finetuning low resource languages [25]).

Federated learning: Leveraged in case of multiple devices that want to train a common model without bringing the raw data at one point, hence keeps privacy and security intact. In. federated learning for disease detection on farmland without sharing raw images of the crop; By applying this federated model to the potato disease detection, they managed to achieve an accuracy of 97.1% and success was secured towards privacy-preserving AI in agriculture [26].

Although the usage of quantum computing for plant disease detection is in an exploratory phase, some impressive results have been published. It utilized a quantum-inspired machine learning model for maize disease classification, which yielded an accuracy of 98.9%, thereby decreasing the computation time approximately by half in comparison to classical models [27]. QC may be able to accelerate disease classification for large datasets and complex data, but much of this needs research.

Incredible researches in taking advantage of the graph neural networks (GNNs) computation, to model relations between diverse plant diseases and environmental factors as shown by other two works [28]. For example GNN was applied to identify disease propagation for rice plants from spatio-temporal data. The model demonstrated an accuracy of 95.4%, providing a new method to unveil the dynamics of disease and for early interventions plan.

Edge AI makes it possible to detect plant disease directly on low-power devices like smartphones and IoT sensors, saving cloud-based computations [29]. It used an edge AI system for disease detection in soybean with a 93.8% accuracy rate. It uses a sparse-optimized, lightweight CNN for edge deployment on resource-constrained devices and is suitable to be deployed in remote agricultural fields.

3D imaging is to realize the complete structure of leaves and fruits, 3D imaging techniques have been investigated which might enhance accuracy for plant disease detection. CNNs to identify apple scab with an accuracy of Predicting Apple Scabs. Compared to traditional 2D images, the model provides more accurate disease detection by capturing additional spatial information i.e., comparing spatial structures of ROI's [30].

Multi-modal systems perform plant disease detection by encoding data from multiple sources (for example, RGB and thermal images together with sensor input), which enables us to increase the system accuracy. A study [31] utilised data from both visual and non-visual sources in order to classify late blight for the potato, with a final classification accuracy of 98.3%. Multi-modal approaches implicitly combine information from different sensor types, leading to improved detection performance particularly in earlier stages of the disease process.

3 Proposed Approach

In Fig. 1, the proposed novel methodology to detect apple plant diseases in multi-stage process using computer vision and AI are integrated. The architecture of this system starts when image acquisition and preprocessing, images are noisy reduced, the contrast is intensified and or they were resized o normalized in useful to improve clarity. Then, segmentation separates the diseased area, and further we extract features to represent texture color or shape of diseased region. We finally feed the extracted features to ensemble machine learning model which classify using SVM, KNN and RFC where we choose majority prediction. This approach provides high accuracy in identifying multiple diseases, as well as computational efficiency and scalability for large-scale deployment under real-world agricultural conditions. This method is a promising way to detect and diagnose apple plant diseases at an early stage by using machine learning along with computer vision together.

In this chapter, we present an ensemble machine learning framework which utilizes computer vision and multiple image processing stages to detect multiple diseases in apple plants. The methodology was stepwise in nature, first acquiring images and then heavily preprocessing to remove noise as well as register for digitalization artifacts aimed at image restoration. Then, segmentation and feature extraction methods are used to find significant characteristics of diseased areas. The ensemble classification process in this work employs hard voting and soft voting, as

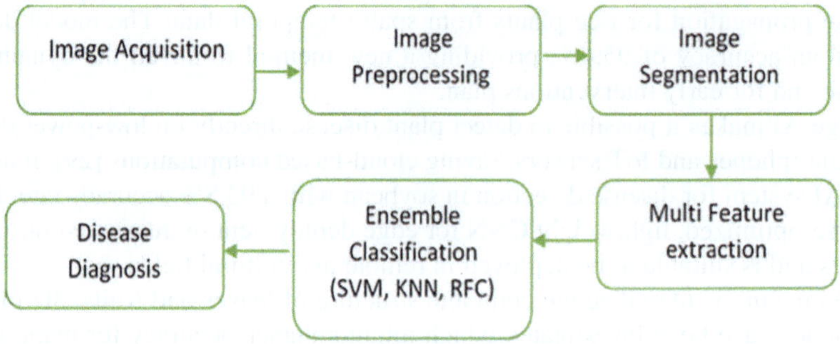

Fig. 1 Apple plant disease detection ensembling machine learning model

well as performs weighted averaging (as described on the proposed model section) based on a combination of classifiers using SVM, KNN and Random Forest Classifier to achieve high accuracy results among identity of different disease models across apple plant. Furthermore, used approach is computationally efficient and scalable in a practical agricultural environment.

3.1 Image Acquisition

The first key stage in methodological implementation is consisting the collection of photographs with a high resolution images of apple plants compatible to some technologies mainly RGB cameras. Images capture based on the RGB channels red, green and blue to effectively identify symptoms of diseases (spots, unusual coloration or wilting) related generally with apple plant disease. The natural color swings in plants are understood with an RGB imaging that makes it easy to draw the distinction between healthy and diseased areas by detecting symptoms like discoloration, leaf spots which indicate diseases such as Apple Scab, Fire Blight and Cedar Apple Rust. RGB cameras in varied environmental conditions (e.g. lighting and shadow exposure), and a wide variety of image types form an invaluable resource for training/testing models with the ideal attributes to meet goal requirements.

3.2 Image Preprocessing

Apple leaves are captured and then passed through a preprocessing pipeline, consisting of several stages processing to enhance its quality and correct the distortions affecting future detection accuracy. While shadows, clutter in the background and widely differing lighting conditions are problems that can be avoided or easily retried if you have a lot of time, it is very difficult to solve all three at once when dealing with images collected from apple orchards. In apple plant disease detection, it is very important to expose the critical characteristics of diseases such as spots and lesions introduced by multi-rescaling while removing irrelevant information. This begins, of course with the application of Noise Reduction smoothing image using a Gaussian or Median filters to reduce high frequency noise originating from environment variables—while preserving connectivity hairs related disease features. The contrast between healthy and diseased regions is improved by applying Contrast Enhancement such as histogram equalization to highlight the disease symptoms which are difficult to visualize, or not at all visible through normal imaging processes i.e., making spots/lesions more prominent. Color Normalization is crucial with apple orchards, where the lighting conditions vary greatly from one capture to another: this further mitigates color balance changes and does not let disease detection rely on shadows or different degrees of light in successive images for make consistent predictions. The last step is resizing and cropping, since this helps to have a fixed image size with targeted areas (i.e. apple leaves and fruits in the case of apple scab/fire blight or powdery mildew). This multi-step preprocessing step gets the images ready to be used for later stages in a machine learning pipeline, which helps make apple plant disease detection more accurate.

3.3 Image Segmentation

The preprocessing is followed by image segmentation. The most crucial preprocessing stage is the segmentation which separates Image into Regions of Interest (ROI) where ROI contain disease-affected areas and rest of them are background. Segmentation is performed on the plant and hence it reduces too much complexity in terms of analysis which results leads to increase detection accuracy. K-means clustering: K-means is an unsupervised machine learning algorithm that segments pixels into a certain number of clusters with similar pixel intensity values. This technique distinguishes the healthy organs (from apple plant) from the diseased regions, associating to each cluster those pixels sharing particular characteristics of colour and textures. If your feature space primarily consists of lesions, spots or discoloration such as the symptoms above due to disease—K-means clustering can be particularly effective.

During the k-means process we are minimizing an objective function, given by:

$$\arg\min_{s} \sum_{i=1}^{k} \sum_{x \in s_i}^{n} (||x - \mu_i||^2) \tag{1}$$

For each cluster S_i are the clusters x is a pixel and μ_i represents the centroid of cluster S_i. It minimizes the distance of pixels and their respective cluster center, which means regions with similar color patterns will be put into one group (for example, diseased areas).

Watershed Algorithm—K-means clustering is good to identify the general regions of interest but it may over-segment or under segment. To improve the segmentation, we use watershed algorithm. The watershed algorithm sees the image as a topographical map, with pixel intensities representing elevation. By flooding the low-intensity areas (valleys) and thus assigning labels to different regions, it enables a highly accurate discrimination of relevant boundary between healthy and unhealthy clear tissues. The input to the watershed algorithm is prepared by calculating the gradient magnitude of image using Sobel filters,

$$G = \sqrt{G^2{}_x + G^2{}_y} \tag{3}$$

where G_x and G_y are the gradients in x, y respectively. With the Watershed algorithm, we can build lines between sick and non-sick areas to provide a more accurate segmentation. By using these two segmentation techniques together, the regions of interest are efficiently separated which provide a clean input for feature selection phase.

3.4 Multi Feature Extraction

Feature extraction is an important step in machine learning where the raw image data are transformed into measurable features that may help classifiers to segment diseased regions from apple plants. This method uses the three most important group attributes extracted from each segment region: texture, color and shape features. Texture features detect surface-parameter, e.g., contrast, energy and homogeneity for certain apple diseases (apple scab identified by fruit blotch out), too. Color features are able to capture the color appearance of changes within an image (such as in yellowing, browning) by relying on properties such as histograms and higher-lower descriptors—e.g. 1st-order statistics or moments, which further help differentiate disease patters). Shape features provide shape and morphology information of the lesions, which are useful for distinguishing diseases with different shapes such as oblong and irregular (e.g. fire blight) from circular spots on the leaves (such as apple scab). Taken all these properties about texture, color and shape into account makes us have a great understanding of the apple plant disease symptoms which reflect in a more accurate model that detects different diseases shown on an apple.

3.5 *Classification Using Ensemble Machine Learning*

In this chapter, we have proposed an ensemble machine learning based method for apple plant disease detection to improve the accuracy of classification and robustness. It is an ensemble model that combines several classifiers and utilizes strengths from different models to improve overall performance. In short, the ensemble consist of SVM (Support Vector Machine), KNN (k Nearest Neighbor), and RFC (Random Forest Classifier). These classifiers when combined together using methods like hard voting, soft voting and weighted averaging gives the model more accurate predictions. Ensemble classification makes use of SVM, KNN and RFC machine learning classifiers that work collectively on the features extracted from apples to accurately identify apple plant diseases. In the case of high-dimensional data, SVM performs particularly well because it finds the best hyperplane to differentiate healthy and sick regions such that any boundary—whether linear or not from non-linear kernel functions. K-Nearest Neighbor, on the other hand, classifies new instances based off how close they are to its nearest training examples which is highly beneficial in agricultural imbalanced data domains. On the other hand, RFC is an ensemble of classifiers that are gather together to make final outputs produced by decision trees that overfit less and can increase generalization across many types of data sets (in particular high-dimensional feature space). The ensemble helps improve both the accuracy and robustness of disease detection system by combining these classifiers.

Ensemble voting model used to aggregate the predictions from SVM, KNN and RFC for it's final classification. It is implemented using hard voting that each classifier votes for a class, outputting the majority vote. Soft voting, on the other hand, averages predicted probabilities from all classifiers and then selects class with highest average. Weighted averaging is used, which differs in the way that it weighs higher accuracy or reliability classifiers allowing them to have a bigger say on the final prediction. This way, the ensemble model can actually benefit from best of both classifiers in order to make disease detection more precise and reliable. Once all the ensemble classifiers are in agreement about a given disease diagnosis for apple leaves (i.e., none or both Apple scab and Fire blight, who knows powdery mildew too). This diagnosis is then used to trigger other behaviors (e.g. a recommendation for treatment, or an estimate of disease spread). The system is computationally efficient and scalable, making it ideal for rapid disease diagnosis in a large-scale agricultural setting.

4 Result and Discussion

The proposed computer vision and ensemble machine learning model for apple plant disease detection. Start with the discussion on dataset and how individual model performing, next by seeing performance of Ensemble using various Voting techniques. The rest of the section is structured as follows: Sect. 4 compares with existing models and reports results quantitatively using performance measures such as accuracy, precision, recall F-1 score etc.

4.1 Dataset Description

A few high-resolution images of apples plants with apple scab, fire blight, and cedar after rust diseases as well healthy samples that we used for this chapter are presented in Table 1. It consists of plant images taken in the field: PlantVillage dataset (publicly available) and more than five thousand pictures for each data category using RGB cameras. These environmental conditions provided a wide range of images in order to obtain a complete and non-biased dataset for training the machine learning models. The preprocessing included Gaussian smoothing, contrast enhancement and color normalization to improve the quality of images. Lastly, all images were re-sized to 256×256 pixels for even input dimension against the mode used.

4.2 Ensemble Machine Learning Classification

Raw input images with large differences between lighting conditions, background elements and camera perspectives. It is crucial to catch these variations before preprocessing the data because this could result in inaccuracies during model training. For example, in diagnosing a disease like Apple Scab the spots on apple leaf discoloration changes and texture will be crucial so other background information should not bring disturbance. Figure 2 depicts this visually with one of the said raw images in dataset, featuring disease symptoms as dark spots on apple leaves. Pre-processing is mandatory for improving the input images and enabling to focus on disease-specific features in a model. In order to remove randomly generated high-frequency noise and preserve higher frequencies representing relevant patterns of global feature such as leaf spots, the natural gaussian smoothing is employed. This makes lost noise without changing the important details. Equalization helps in improving the color and lesion visibility through contrast enhancement making them more distinct, especially when the images are taken under different lighting conditions.

Table 1 Dataset summary of apple leaves

Disease type	Number of images
Apple scab	1200
Fire blight	1100
Cedar apple rust	900
Healthy	1000

Fig. 2 Sample raw apple leaf images

Figure 3 sample preprocessed image to bring out disease features and reduce background noise due to minor environmental change. This preprocessing prepares the image for other components such as making information stand out in following steps of detection pipeline. An important further to preprocessing is the process of segmentation, where a region-of-interest (ROI) that contains lesion symptoms appears visible. Segmentation separates these parts from the rest of the image and leaves them in place so that training can take into account only diseased areas while removing irrelevant information. It can be broken down into two basic concepts: K-means clustering which is an algorithm where a predetermined number of cluster centers are prepared based on texture or colour similarity to differentiate between healthy and diseased regions either as spots, discoloration; Watershed supervision mean smoothens the boundary lines separating disease afflicted parts updating lesion edges with 100% certainty. Secondary cropping algorithm specifically region-wise crops infected regions and supports feature extraction that can further improve the model accuracy with fine-grained segmentation. The sample diseased areas are isolated in Fig. 4, which present segmented images.

Once we have identified the disease regions, informative attributes are derived to describe visible symptoms that an input data will feed into machine learning classifiers. In this regard, three key features have been defined: leaf texture features that capture side leaf surface change as a result of Apple Scab symptoms where lesions produce rough spots; Color Features that account for color change due to yellowing variation or brown shadow in some instance and Spotting also while the third feature is Shape Feature relaying on lesion geometry whereby different leaf disease leaves round spot etc. By using all above-mentioned features we make an clear-cut description of disease and then forwarded it to classifiers in machine learning for final confirmation.

The features are aggregated to a general presentation of the disease which is used as an input for deep learning model. This model essentially aggregates predictions of a bunch of classifiers (SVM, KNN & RFC). This way the ensemble model gets a benefit of every classifier, which in turn compensates each other's weaknesses and yield to high precision while keeping robust performance. It can be assured that the

Fig. 3 Sample preprocessed image

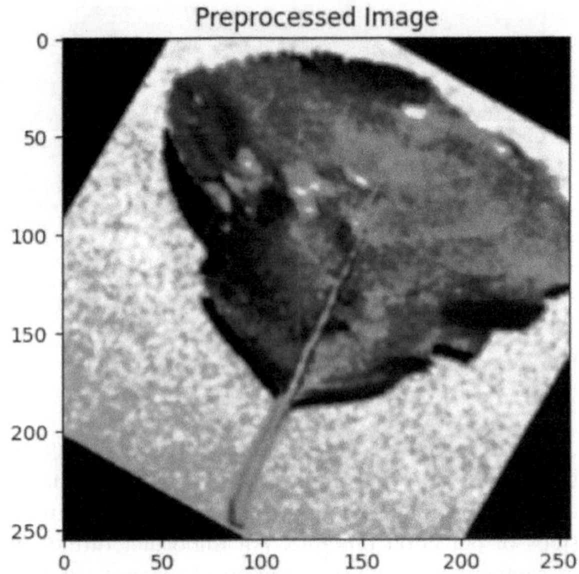

Fig. 4 Sample segmented Image

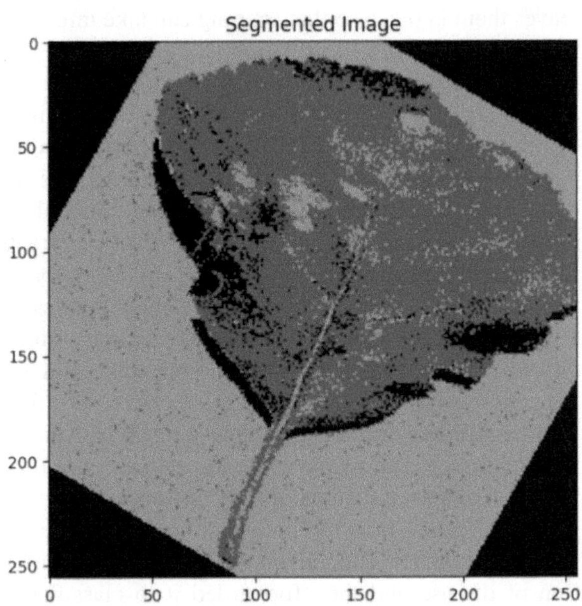

ensemble way performs better than individual models, since voting techniques are utilized to further polish this model's predictions.

The results of individual models in classifying apple plant diseases are shown on Table 2. The SVM achieves a high accuracy of 92.5% and is particularly suitable for

Table 2 Performance of individual models

Model	Accuracy (%)	Precision (%)	Recall (%)	F1-score (%)
Support vector machine (SVM)	92.5	93.2	91.8	92.5
K-nearest neighbor (KNN)	89.7	90.5	88.9	89.6
Random forest classifier (RFC)	94.1	94.5	93.7	94.1

Table 3 Performance of ensemble models

Ensemble voting technique	Accuracy (%)	Precision (%)	Recall (%)	F1-score (%)
Hard voting	95.6	95.8	95.3	95.5
Soft voting	96.3	96.7	96.1	96.4
Weighted averaging	97.2	97.5	97.1	97.3

complex non-linear decision boundaries, which can be used to differentiate between different diseases but are computationally expensive. The accuracy of KNN is the lowest 89.7% as it depends on surrounding data patterns, hence when your reference points are noisy then you get worst values like in this competition and with the same classifier uses that to classify those testing images. In contrast, RFC had the highest classification accuracy (94.1%), in part due to its capability of aggregating individual predictions from multiple decision trees and thus generalizing well across a variety of disease types. Although in Table 3 show the ensemble model outperforming all other models individually as it achieves an accuracy of 97.2% using weighted averaging, which consolidates classifier ensembles are good for better final performance. Soft voting (96.3%): is another technique applied with the support of confidence levels, thus making better predictions in comparing to hard-voting. This is effective, but less accurate than soft voting and weighted averaging.

Table 4 reavels the proposed ensemble model to detect apple plant disease outperforms existing models. It gives 97.2 percent accuracy on the test data, outperforming Convolutional Neural Network (CNN) it gave as wrap up of 94.8%, Deep Neural network (DNN) ending at 93.5% generated this way. A more enhanced ensemble model also eclipses both precision and recall rates outperforming the base differentiable version in correctly identifying diseased plants at a lesser rate of false positives. The theme of improved performance was expressed during multiple runs with the ensembles.

Table 4 Comparison with existing models

Model	Accuracy (%)	Precision (%)	Recall (%)	F1-score (%)	References
Convolutional neural network (CNN)	94.8	95.2	94.6	94.9	[17]
Deep neural network (DNN)	93.5	93.8	93.2	93.5	Patel et al. (2024)
Proposed ensemble model	97.2	97.5	97.1	97.3	This work

5 Conclusion

The proposed model achieved excellent results on the apple plant disease detection task with respect to individual classifiers as well as state-of-the-art models like Capture Convolutional Neural Networks (CNN) and DNN. The ensemble model combining Support Vector Machine (SVM), K-Nearest Neighbor (KNN) and Random Forest Classifier with advanced voting techniques had more accurate, precision, recall, F1-score for various apple plant diseases. The model leverages the capability of preprocessing images, segmenting diseased areas and extracting relevant features in a quite efficient way which led it to demonstrate a good performance achieving in accuracy rate of 97.2% and F1-scoreequalto 97.3%. By utilizing the ensemble technique, they capitalized on votes of all classifiers which results in better generalization and fewer errors. Moreover, the computational time was within reasonable bounds such that this model can be deployed in real-time to agricultural fields where early disease detection is essential for timely action and management of crops.

References

1. Yang, Y., Guan, S., Ou, Z., Li, W., Yan, L., & Situ, B. (2023). Advances in AI-based cancer cytopathology. *Interdisciplinary Medicine, 1*(3), Article e20230013.
2. Ahmad, A., Saraswat, D., & El Gamal, A. (2023). A survey on using deep learning techniques for plant disease diagnosis and recommendations for development of appropriate tools. *Smart Agricultural Technology, 3*, Article 100083. https://doi.org/10.1016/j.atech.2022.100083
3. Lu, J., Tan, L., & Jiang, H. (2021). Review on convolutional neural network (CNN) applied to plant leaf disease classification. *Agriculture, 11*(8), 707. https://doi.org/10.3390/agriculture11080707
4. Liu, J., & Wang, X. (2021). Plant diseases and pests detection based on deep learning: A review. *Plant Methods, 17*(1), 22.
5. Mehmood, A., Ahmad, M., & Ilyas, Q. M. (2023). On precision agriculture: Enhanced automated fruit disease identification and classification using a new ensemble classification method. *Agriculture, 13*(2), 500.

6. Yan, Q., Yang, B., Wang, W., Wang, B., Chen, P., & Zhang, J. (2020). Apple leaf diseases recognition based on an improved convolutional neural network. *Sensors, 20*(12), 3535.
7. Assad, A., Bhat, M. R., Bhat, Z. A., Ahanger, A. N., Kundroo, M., Dar, R. A., & Dar, B. N. (2023). Apple diseases: Detection and classification using transfer learning. *Quality Assurance and Safety of Crops & Foods, 15*(SP1), 27–37.
8. Polder, G., Blok, P. M., van Daalen, T., Peller, J., & Mylonas, N. (2025). A Smart Camera With Integrated Deep Learning Processing for Disease Detection in Open Field Crops of Grape, Apple, and Carrot. Journal of Field Robotics.
9. Pacal, I., Kunduracioglu, I., Alma, M. H., Deveci, M., Kadry, S., Nedoma, J., & Martinek, R. (2024). A systematic review of deep learning techniques for plant diseases. *Artificial Intelligence Review, 57*(11), 304.
10. Farmanesh, A., Sanchis, R. G., & Ordieres-Meré, J. (2025). Comparison of Deep Transfer Learning Against Contrastive Learning in Industrial Quality Applications for Heavily Unbalanced Data Scenarios When Data Augmentation Is Limited. *Sensors, 25*(10), 3048.
11. Hassan, S. M., Maji, A. K., Jasiński, M., Leonowicz, Z., & Jasińska, E. (2021). Identification of plant-leaf diseases using CNN and transfer-learning approach. *Electronics, 10*(12), 1388.
12. Galabuzi, C., Abdullah, H., Ahmad, N., & Kaidi, H. M. (2024, September). EfficientNet-Based Deep Learning Neural Network for Accurate Plant Disease Detection. In 2024 5th International Conference on Smart Sensors and Application (ICSSA) (pp. 1–6). IEEE.
13. Lee, S. H., Goëau, H., Bonnet, P., & Joly, A. (2020). Attention-based recurrent neural network for plant disease classification. *Frontiers in Plant Science, 11*, Article 601250.
14. Mi, Z., Zhang, X., Su, J., Han, D., & Su, B. (2020). Wheat stripe rust grading by deep learning with attention mechanism and images from mobile devices. *Frontiers in Plant Science, 11*, Article 558126.
15. Ashmafee, M. H., Ahmed, T., Ahmed, S., Hasan, M. B., Jahan, M. N., & Rahman, A. A. (2023, February). An efficient transfer learning-based approach for apple leaf disease classification. In 2023 International Conference on Electrical, Computer and Communication Engineering (ECCE) (pp. 1–6). IEEE.
16. Gülmez, B. (2024). Advancements in rice disease detection through convolutional neural networks: A comprehensive review. Heliyon, 10(12).
17. Deng, X., Zhu, Z., Yang, J., Zheng, Z., Huang, Z., Yin, X., & Lan, Y. (2020). Detection of citrus huanglongbing based on multi-input neural network model of UAV hyperspectral remote sensing. *Remote Sensing, 12*(17), 2678.
18. Khan, A. T., Jensen, S. M., Khan, A. R., & Li, S. (2023). Plant disease detection model for edge computing devices. *Frontiers in Plant Science, 14*, 1308528.
19. Kaler, B., & Kaur, A. (2025). A systematic survey on explainable artificial intelligence (XAI) for plant health monitoring: Challenges and opportunities. *Applied Intelligence, 55*(12), 889.
20. Deng, H., Luo, D., Chang, Z., Li, H., & Yang, X. (2021). Rahc_gan: A data augmentation method for tomato leaf disease recognition. *Symmetry, 13*(9), 1597.
21. Yulita, I. N., Amri, N. A., & Hidayat, A. (2023). Mobile application for tomato plant leaf disease detection using a dense convolutional network architecture. *Computation, 11*(2), 20.
22. Ramadan, S. T. Y., Sakib, T., Haque, M. M. U., Sharmin, N., & Rahman, M. M. (2023, July). Wheat leaf disease synthetic image generation from limited dataset using gan. In International Conference on Human-Centric Smart Computing (pp. 501–511). Singapore: Springer Nature Singapore.
23. Da Silva, P. E. C., & Almeida, J. (2024). An Edge Computing-Based Solution for Real-Time Leaf Disease Classification Using Thermal Imaging. IEEE Geoscience and Remote Sensing Letters.
24. Elon Correa da Silva, P., & Almeida, J. (2024). An Edge Computing-Based Solution for Real-Time Leaf Disease Classification using Thermal Imaging. arXiv e-prints, arXiv-2411.
25. Yilma, G., Dagne, M., Ahmed, M. K., & Bellam, R. B. (2025). Attentive Self-supervised Contrastive Learning (ASCL) for plant disease classification. *Results in Engineering, 25*, Article 103922.

26. Wang, Y., Yin, Y., Li, Y., Qu, T., Guo, Z., Peng, M., & Li, F. (2024). Classification of plant leaf disease recognition based on self-supervised learning. *Agronomy, 14*(3), 500.
27. Mamba Kabala, D., Hafiane, A., Bobelin, L., & Canals, R. (2023). Image-based crop disease detection with federated learning. *Scientific Reports, 13*(1), 19220.
28. Attri, I., Awasthi, L. K., & Sharma, T. P. (2025). EQID: Entangled quantum image descriptor an approach for early plant disease detection. *Crop Protection, 188*, Article 107005.
29. Ahammad, M., Babu, M. A., & Bhatnagar, V. (2024, April). Classification of Rice Leaf Diseases Using Deep Residual Graph Neural Networks. In International Conference on Innovations in Computational Intelligence and Computer Vision (pp. 55–67). Singapore: Springer Nature Singapore.
30. Joshi, H. (2024). Edge-AI for agriculture: lightweight vision models for disease detection in resource-limited settings. arXiv preprint arXiv:2412.18635.
31. Chadoulis, R. T., Livieratos, I., Manakos, I., Spanos, T., Marouni, Z., Kalogeropoulos, C., & Kotropoulos, C. (2025). 3D-CNN detection of systemic symptoms induced by different Potexvirus infections in four Nicotiana benthamiana genotypes using leaf hyperspectral imaging. *Plant Methods, 21*(1), 15.
32. Zhu, H., Shi, W., Guo, X., Lyu, S., Yang, R., & Han, Z. (2025). Potato disease detection and prevention using multimodal AI and large language model. *Computers and Electronics in Agriculture, 229*, Article 109824.

IOT and UAV Integrated System for Proactive Crop Disease Prediction

K. Krishna Sowjanya and K. P. Bindu Madavi

Abstract The productivity of agriculture is often affected by the crop diseases incurring economic loss and reduced food production. Detecting these crop diseases in the early stages is the need of the hour to mitigate the spread of the disease and manage the crop yield effectively. This paper presents a proactive crop disease prediction system by employing the technologies such as Unmanned Aerial Vehicles (UAVs) and Internet of Things (IoT) to detect and mitigate the crop disease at early stages. The employed UAVs have a high resolution camera to capture the aerial images of crops and leaves. The images captured via UAVs are transferred to a computing environment dynamically for analysis purposes. The analysis is performed on the images to predict the disease of the leaf along with its intensity. After analyzing and predicting the leaf disease and its intensity, the pesticide information is passed to the IoT sensors which are fixed in the fields to spray the recommended pesticide. The system analyzes the captured images by leveraging the machine learning algorithms in the realtime to identify different diseases and predict the spread of the disease. The proposed proactive approach helps the farmers to take preventive actions to identify the crop disease at the early stage and mitigate them in a timely manner to produce a better crop yield. Moreover, the proposed method provides a cost-effective solution for disease prediction and is easily accessible to farmers to enhance crop productivity.

Keywords Disease prediction · Internet of things · Machine learning · Unmanned aerial vehicles

---

K. K. Sowjanya (✉)
CMR Institute of Technology, Bengaluru, India
e-mail: sowjanya.k@cmrit.ac.in

K. P. B. Madavi
Christ University, Bengaluru, India
e-mail: bindu.madavi@christuniversity.in

S. R. Kumar et al. (eds.), *AI-Driven IoT Solutions in Agriculture: New Insights*, Studies in Computational Intelligence 1236, https://doi.org/10.1007/978-3-032-05603-0_13

1 Introduction

In today's modern agricultural world, farming has become increasingly difficult since farmers struggle more and more to maintain the health of their produce and yield. However, discovering the potential of the crops early on can help them overcome these issues rather than being slow on the response like most farming issues. When combined, IoT (Internet of Things) and UAVs (Unmanned Aerial Vehicles) allow farmers to transform their cropping patterns completely, as they can have instant access to all areas and an up-to-date view of their fields.

IoT-driven crop disease prevention and management solutions leverage advanced technologies to enhance agricultural productivity and sustainability. These systems combine several components, including sensor technologies that allow for the monitoring of environmental conditions, cloud computing services that allow for data analysis, and predictive modeling using machine learning. At the same time, deep learning models analyze images of crops to detect diseases early, as demonstrated in studies focusing on fruits and greenhouse environments [1–3]. Also, the cloud-connected architectures enhance the decision-making process among farmers because of the insights provided as a result of past registration and current processes, which lead to diminished damages due to soil-borne pathogens [4] Overall, these IoT solutions improve disease management and promote sustainable agricultural practices by minimizing chemical pesticide use and optimizing resource allocation [5].

The system includes UAV components and other agricultural and IT systems that work as subsystems and allow the improvement of farming practices. The core of this system features drones equipped with 'ultra-high definition' cameras. Using several camera systems, images are automatically captured for remote review, allowing advanced imaging processing of the crop fields [6]. To interpret these images, machine learning techniques, especially convolutional neural networks, are used to learn features such as color histograms and textures to assist in the identification of diseases [7, 8]. Furthermore, a multi-vision monitoring framework can monitor the health status of the targeted crop; concurrently, it can monitor farmers' activities using multiple sensor systems such as hyperspectral and thermal cameras [9]. This is complemented by the predictive modeling aspects of this integration, such as crop recommendations, disease control methods, yield prediction, etc. This ensures farmers can access the right information at the right time and make the right decisions [10]. All these parts make a robust system that aids in minimizing response time to disease outbreaks and offers improved management of crops, consequently enhancing food production systems and sustainable agricultural development.

With IoT in agricultural and UAV systems, predictive crop disease can now occur. It leads to prompt intervention, thus reducing loss to crops due to better utilization of means as shown in (Fig. 1).

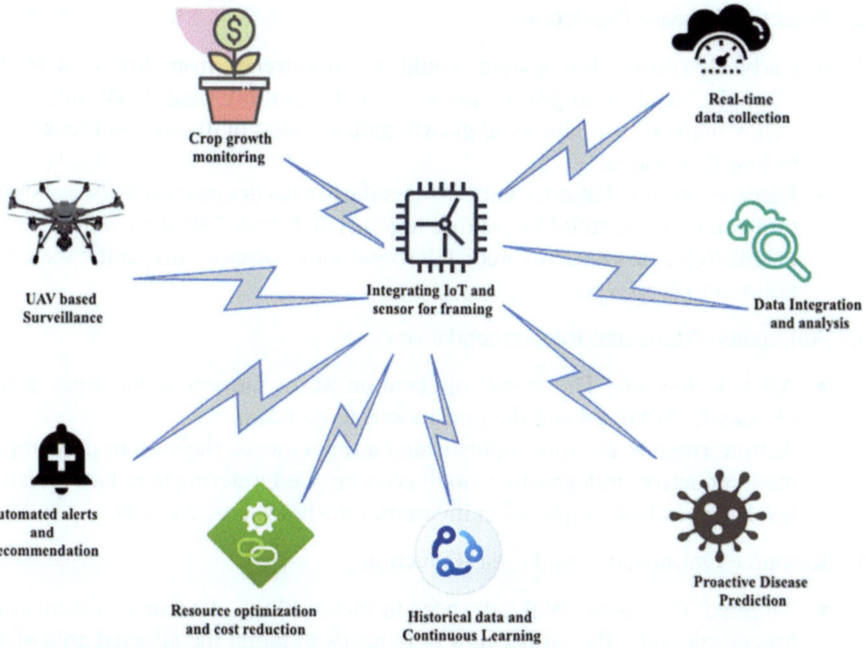

Fig. 1 Application of integrated IoT and sensor for framing

1. Real-time data acquisition
   - IoT Sensors involve soil sensors: moisture, pH, nutrients, temperature, humidity, wind conditions, weather stations along with air quality and level of sunlight from environmental sensors, that collect data unceasingly on-ground.
   - UAV-based Imaging Sensors: Drones equipped with high-resolution cameras, multispectral or thermal cameras, can capture detailed images of the crop area, providing details about crop stress, health, and beginning signs of diseases.
2. Data Integration and Analysis
   - Common Data Platform: It includes a cloud or on-site data platform that consolidates data from various IoT sensor types and images from a UAV, giving an integral view of field conditions.
   - AI & ML Models: Predictive Algorithms Analyze Integrated Data For disease patterns from leaf color changes, temperature, moisture levels, and pest activities. Machine learning models for the early signs of various crop diseases based on historical trends, weather trends, and real-time field conditions on the farm.

3. Proactive Disease Prediction

    - Early Detection: The system would monitor trends from IoT sensors, for example, sudden surges in moisture and humidity, and UAV imagery - yellowing leaves or abnormal growth patterns alert of disease outbreaks even before they spread.
    - Disease-Specific Patterns: Patterns specific to a particular crop's disease signature can be understood by putting together IoT and UAV data, which would enable the same system to predict the presence of diseases in specific scenarios tailored to crop type.

4. Automated Alerts and Recommendations

    - Alerts to farmers: The mobile application warns farmers about zones at risk of disease outbreaks and the precautions to be taken.
    - Action triggers: In more sophisticated arrangements, the system may implement proactive strategies like modified controlled watering to reduce moisture levels or pesticide application if certain thresholds are achieved.

5. Resource Optimization and Cost Reduction

    - Targeted Treatment: With advances in meteorology, disease treatment is no longer guessing; the farmer now only needs to target the affected area of the field instead of spraying the entire field with chemicals, thus minimizing the chemicals, labor, and equipment in widening the scope of the work.
    - Less Crop Loss: With the combination of early detection of diseases and targeted interventions, there is less contagiousness of the diseases and, therefore, minimal suffering in crop yields and quality.

6. Historical Data and Continuous Learning

    - Data Logging for Future Insight: Because the IoT and UAV systems operate on a seasonal basis, the information they accumulate year after year establishes a strong historical perspective that is useful in enhancing the accuracy of such predictions with time.
    - Constant Model Updates: Machine Learning models rely on new data to continuously learn and improve on targets that enable the models to make effective forecasting around disease conditions.

7. UAV Surveillance with Advanced Imaging

    - Drones equipped with high-resolution cameras, multispectral, hyperspectral, and thermal imaging capture real-time data on crop health and growth progress. The aerial view allows for large-scale monitoring, helping to spot early signs of disease or stress across extensive fields.

8. Real-time Crop Growth Prediction

- Data Fusion and Analysis: The data from UAVs (captured imagery) is combined with IoT data on soil and environmental conditions to assess plant health and growth patterns. Machine learning models analyze this data to predict growth rates and identify areas showing signs of growth issues.
- Growth Modeling: By analyzing multiple data points, including temperature, light exposure, soil conditions, and weather, growth models can forecast crop development stages. Predictive algorithms based on these factors help forecast crop yield, providing early insights into potential productivity.

Smart Agriculture Information Exchange Model

Figure 2 illustrates a data exchange and analysis workflow in a smart farming or precision agriculture system. Here's an explanation of each component and the overall process:

1. Data Acquisition by Sensors

The icon with a thermometer represents data acquisition through sensors. In a smart farming system, various types of sensors collect critical information about the environment and crops, such as temperature, humidity, soil moisture, and nutrient levels. These sensors are positioned across the farm to continuously monitor conditions, helping detect changes or issues like plant stress, disease, or pest presence.

2. Communication Gateway

The central icon with arrows pointing in all directions and a globe represents the communication gateway. This gateway is responsible for collecting data from sensors

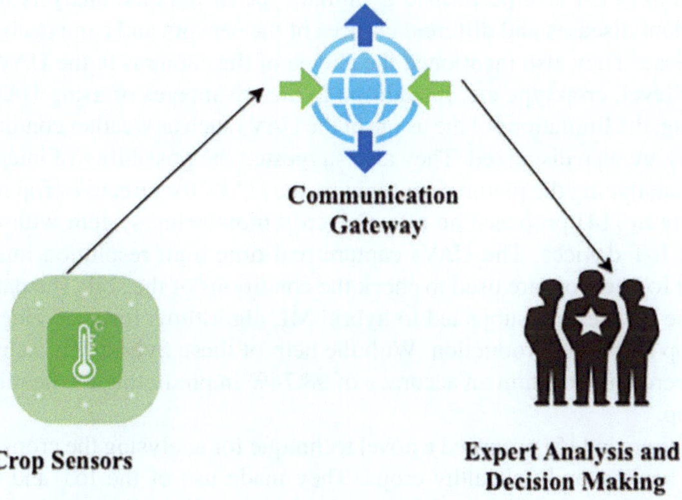

Fig. 2 Smart farming data flow and decision support system

and transmitting it to a centralized system, often through satellite, Wi-Fi, or cellular networks. The gateway ensures that data from the field reaches the system for further analysis, and it enables the exchange of information between different components of the system.

3. Expert Analysis and Decision-Making

The group of people represents expert analysis and human decision-making. This stage involves agronomists, data scientists, or other experts analyzing the collected data to derive insights and recommendations. Experts might look for patterns in the data to detect diseases early or to identify areas that need more water, nutrients, or other interventions.

2 Literature Survey

With the help of UAVs and IoT, authors in [11] came up with a novel framework that is suitable for predicting the occurrence of disease based on the climatic conditions. The authors made use of solar panels to provide energy required for the UAVs. These UAVs are employed to capture the real time images of the crops and the captured data is sent to the cloud repository to analyze the damage occurred to the crop.

Authors in [12] have done an elaborative comprehensive analysis of the usage of drones in detecting the disease in plants in early stages to reduce the expenditure incurred for pesticides. The authors have given more emphasis on UAVs for capturing the images and monitoring the health of a plant regularly. The detection of the disease and its spread is done via image processing techniques with the help of Machine Learning (ML) and Artificial Intelligence (AI).

Authors in [13] also performed a similar type of detailed analysis focusing on various plant diseases and different choices of the sensors and cameras based on the plant disease. They also mentioned the choice of the cameras in the UAVs based on the detail level, crop type etc. In addition to the advantages of using UAVs for crop monitoring, the limitations of the usage of the UAVs such as weather conditions, flight conditions are also discussed. They also suggested the possibility of integrating ML and AI in analyzing the pictures captured by the UAVs for effective crop monitoring.

Authors in [14] proposed an effective crop monitoring system with the help of UAV and IoT devices. The UAVs capture real-time high resolution images of the crops and IoT sensors are used to check the conditions of the soil. The data captured from these devices are subjected to hybrid ML algorithms for analyzing the health of the crop and yield production. With the help of these hybrid ML techniques, the authors were able to attain an accuracy of 98.74% in predicting the health condition of the crop.

The authors in [15] proposed a novel technique for analysing the crops and differentiating healthy and unhealthy crops. They made use of the IoT and UAV technologies to monitor the health of the crop by dynamically collecting the high-quality pictures of the crops via UAVs. The data generated is used to train various ML models for better precision and accuracy. The main objective of their study is to provide a smart solution in monitoring agriculture.

In [16], authors proposed a novel crop management and monitoring system via UAV and ML algorithms. The monitoring system of the proposed method mainly focuses on the crop and also the climatic conditions. UAVs are used to extract the images of the crop and are pre-trained using Convolutional Neural Networks (CNN) to identify and extract the features. By pre-training the model, the authors were able to detect abnormalities in the crop growth. They also categorized the abnormality stages based on the damage occurred to the crops. The authors also tried to differentiate the crops and the weeds in the images.

Authors in [17] have emphasized on the importance of using the advanced technologies such as UAVs, IoT, AR in the field of agriculture. A detailed explanation was given about the usage of IoT in providing precise farming by leveraging the benefits of IoT. Cloud-based solutions/options are provided for storing the data collected by the sensors. The farming methods can be enhanced by using some more technologies such as UAV and Augmented Reality (AR). The usage of UAVs in agriculture involving crop monitoring, spraying the pesticides, observing the soil conditions are also discussed. The adaptation of ML algorithms is also emphasized for the smart prediction of the crop for the better yield.

Authors in [18], proposed a low-cost model for predicting the disease in pearl millet crop. The method detects and classifies the disease in crops automatically with the help of IoT and Deep Learning (DL) framework. The automation of the disease prediction system is done by collecting the data via IoT sensors and sending them to the centralized cloud server. Authors have conducted experiments on the data with the help of several DL algorithms and identified that Grad-CAM algorithm is identifying the crop disease by an accuracy of 98.78 by extracting the relevant features.

Authors in [19] have identified the importance of integrating AI and IoT in the field of agriculture. Using these two technologies there is potential to manage the agriculture in a smoother way to increase the crop field. The key role of AI in agriculture is to collect the data and perform analysis on this vast data to provide any insights on the crop conditions or anomaly detection to maximize the productivity and minimize the losses. Data that is provided to the AI algorithms are collected by various IoT devices. AI-driven solutions provided by these authors also focus on the early identification of any kind of plant disease and treating them in a timely manner.

Authors in [20] have proposed a hybrid approach for monitoring crop health using IoT, ML and UAVs. They also pointed out the challenges and benefits of integrating these technologies as they produce huge volumes of data. The proposed method collects the environmental factors such as weather, humidity of soil via IoT devices and the images of the crop via UAVs. Based on the collected data, authors implemented an ML model to generate indices for analyzing the crop health to identify healthy and unhealthy crops.

3 Proposed Methodology

The proposed method of crop disease prediction employs the usage of UAVs that captures the pictures of the crop and its leaves. The capturing of the leaf picture is done on a regular basis. Soon after capturing the pictures, the UAV sends the captured data to a computing environment for the storage purpose. All the pictures collected and transmitted by the UAV are stored in the database and are used for analysis purposes.

All the pictures that are collected and stored in the database are subjected to image analysis to predict the infection and its intensity on the leaves. To perform the image analysis the proposed model makes use of opencv tools to preprocess the image and identify the Region of Interest (ROI) to find any patches, lesions or infected area on the leaf. In addition to that, the portion or the intensity of the infection of the leaf is also identified by considering the total area of the leaf versus infected area of the leaf.

After identifying the intensity of the infected area on the leaves, based on the type of the crop, a suitable combination of the medicine is chosen automatically by the system based on the level of intensity. And the same pesticide information is passed to the IoT Sensors that are available in the crop field. These IoT sensors will take care of spraying the pesticides on the plants thereby reducing the spread of the infection in the crop. The proposed method is shown in (Fig. 3).

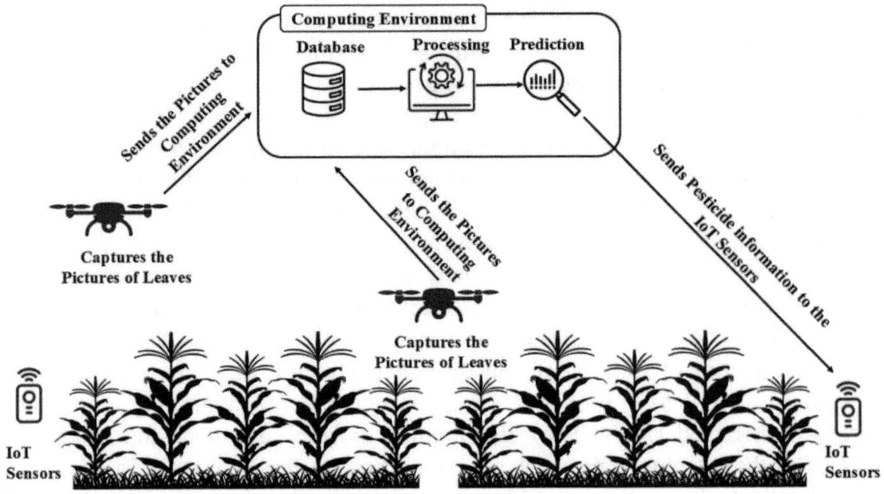

Fig. 3 Proposed method for crop disease prediction healthy apple leaf

4 Results and Discussions

The proposed method involves UAVs to capture the data and transmit the data to a central repository dynamically. The collected data is analyzed and the level of infection in the leaves is detected via ML algorithms and based on the intensity of the leaf disease, corresponding pesticides are also suggested. To perform the analysis of the leaves and predict the disease, a dataset containing healthy and unhealthy apple leaves are considered. The unhealthy apple leaf dataset is considered belong to a specific disease, scab. The images of healthy and unhealthy leaves of apple are shown in Figs. 4 and 5.

Fig. 4 Healthy apple leaf

Fig. 5 Infected apple leaf (scab disease)

Fig. 6 Apple leaf containing scab disease

The process of identifying the disease and its intensity in the leaf is done in the following steps/phases. A sample image shown in (Fig. 6) the output obtained of the same image in all the phases are shown in the following sections.

The process of disease and its intensity detection in a leaf is done in the following steps:

Step 1: Image preprocessing The process of identification starts with scanning the image and preprocessing it by converting it to grayscale by counting the number of white pixels. If the count of the white pixels are more than 10%, then it is converted to grayscale. This preprocessing step helps to reduce the interference with white colored pixels. It also allows one to focus only on the leaf portion to identify the features and characteristics of the leaf for analyzing the leaf. A sample leaf having the scab disease is taken (Fig. 6) and is converted into image grayscale and shown in (Fig. 7).

Once the image is converted into gray scales, the noise is reduced by using Gaussian blur with a kernel of the size (3, 3). The Gaussian blur used averages the pixel

Fig. 7 White pixels converted to grayscale

values around each pixel by considering the neighborhood. This step is also useful in finding the edges to enhance the edge detection algorithms used in the later steps. This step is important for identifying the shape of the leaf accurately to identify any disease on the leaf.

Step 2: Filtering by Mean-Shift: This step involves segmenting the image into regions based on the color. The regions are segregated based on comparing the color of a certain pixel to its neighboring pixels. The comparison is done by considering the average of the pixels with radius set to spatial and color. The spatial radius is a count of neighboring pixels and the color radius is the distance between the colors to consider as a group. The image after performing mean-shift averaging is shown in (Fig. 8).

Step 3: Leaf Border Detection: Once the image is segmented with colors, it is necessary to identify the shape and borders of the leaf. Canny edge detection method is employed to do the same. By identifying the borders, the leaf can be isolated from the background. This step also provides a Region of Interest (ROI) on the leaf area to enhance the analysis of the probable disease. The detected borders of the leaf are represented in (Fig. 9).

Step 4: Conversion of RGB to HLS: After identifying the border of the leaf, the RGB (Red, Green, Blue) image is converted into HLS (Hue, Lightness, Saturation) to effectively identify the health of a leaf. The type of the color is represented with Hue, brightness is indicated with Lightness and the intensity of various colors is represented with saturation. Using this step, it can easily isolate the portions of the leaf that are healthy or unhealthy. In Particular, the parameter hue is used to extract the color information to differentiate healthy green tissue and unhealthy leaf by identifying lesions or patches on the leaf. The conversion of RGB to HLS is shown in (Fig. 10).

Step 5: Setting a threshold on Hue Value: To distinguish the healthy and unhealthy/infected leaf, a threshold of 28 for the hue value is considered. Any values below this threshold are converted to white and above are converted to black. This step will

Fig. 8 Filtered image by mean-shifting

Fig. 9 Leaf border detection

Fig. 10 Conversion from RGB to HLS

highlight the portions of the infected areas based on the color characteristics. The converted binary image is used to extract the infected areas from the ROI. In addition to that, this step also eliminates any unwanted/irrelevant details from the image. The image conversion after setting a threshold value is shown in (Fig. 11).

Step 6: Contour Detection for Leaf: The process of contour detection for leaf helps in identifying the shape and borders of the leaf in the preprocessed image. The contours detected help to isolate the leaf border from the background. This is very crucial to maintain the shape of the leaf. This also provides a ROI for identifying the health of the leaf. The leaf image after identifying the contours is represented in (Fig. 12).

Fig. 11 Setting a threshold based on hue value

Fig. 12 Contour detection in leaf

By implementing the above steps, the designed model is able to identify the spots, patches or diseased areas on the leaf by analyzing the color changes during the process. The exact identification is done as follows:

- **Analyzing the color**: The image is processed and a HLS color space version of the images is created. The hue values are considered to distinguish the differences in the color. The differences in the color is generated based on a threshold value that is setup during the processing phase. This will be able to differentiate the healthy and unhealthy portion of the leaves.
- **Leaf Border Detection**: After converting the image based on the threshold, leaf border and edge detection helps to find the areas that have the infection. These regions are highlighted separately by allowing to view the infection shape and intensity.

- **Infection Intensity**: The intensity of the infection on the leaf is identified by calculating the area of the infected portions. The intensity is calculated based on the ratio of the unhealthy portion versus the healthy portion of the leaf.

After applying all these steps, the results of identifying the infection and its intensity of the leaf is represented in Table 1.

The parameters considered to represent the results of the identification of the leaves are:

- **Perimeter**: The perimeter of the detected contour of the leaf is calculated with the help of Eq. 1 The contour represents the boundary of the image having the same color or intensity.

$$P = \sum_{I=1}^{n} \sqrt{(x_{i+1} - x_i)^2 + (y_{i+1} - y_i)^2} \tag{1}$$

- **Total and Infected area**: The total area of the leaf is calculated by considering the area covered by the contour. Given a set of pixels in the image, (×1,y1), (× 2,y2),..., (xn,yn) the total area is calculated with the help of Eq. 2. The infected region of the leaf is also calculated using the same formula represented in Eq. 2. by considering the contour region based on the selected hue value.
- **Percentage of Infection**: The percentage of infection is the ratio of the total infected area and total leaf area. This is mainly used to assess the intensity of infection on the leaf. The formula to calculate the percentage of infection is given in Eq. 3. This approach provides the total area of infection on the leaf, which can then be used to calculate the **percentage of infection** relative to the total leaf area: This percentage gives a useful metric to assess the severity of the infection.

$$A = \frac{1}{2}\left|\sum_{i=1}^{n}(x_i.y_{i+1} - y_i.x_{i+1})\right| \tag{2}$$

$$Percentage\ of\ Infection = \frac{Total\ Infected\ Area}{Total\ Leaf\ Area} * 100 \tag{3}$$

The data represented in the table is generated for a huge collection of images containing healthy and unhealthy leafs and the features are generated in a.csv file. Based on the infected area, the leaf is automatically labeled as 1 if the leaf is infected,

Table 1 Parameters considered for identifying the infection on leaf

S. no.	Parameter	Value (mm)
1	Perimeter	217.38
2	Total area	2170.00
3	Infected area	653.00
4	Percentage of infection	30.99

Table 2 Fungicides for apple scab control

Fungicide	Description	Application timing
Captan	Protective fungicide, applied preventatively to control scab before symptoms become severe	Early stages of leaf development; repeated during the season, especially after rain
Mancozeb	Protective fungicide, applied preventatively to control scab before symptoms become severe	Early spring during bud development; continue through the season as needed
Myclobutanil	Systemic fungicide are absorbed by leaves, providing residual control with fewer applications	First sign of disease; can also be used early in season preventatively
Copper-based fungicides	Organic option to prevent initial infection by reducing fungal spores	Dormant season or very early spring

Table 3 Recommended concentration and application guidelines for fungicides to control apple scab

S. no.	Fungicide	Recommended concentration (%)
1	Captan	0.25–0.5
2	Mancozeb	0.2–0.3
3	Myclobutanil	0.04–0.1
4	Copper-based fungicides	0.04–0.1

else labeled as 0 if the leaf is healthy. The generated csv file can be subjected to ML algorithms for an accurate classification.

Apple scab, a widespread apple tree fungi disease, can be effectively treated with appropriate pesticides, such as fungicides targeting *Venturia inaequalis* specific to the disease. Here are some of the most common fungicides recommended and their methods of application as shown in Table 2.

The concentration of pesticides applied to control apple scab on leaves is usually expressed as a dilution ratio, a multiplied figure in percentage representing the amount of the active ingredient in a volume amount of water as shown in Table 3.

5 Conclusions and Future Works

The crop disease prediction system predicts the presence of infection in the crops by analyzing the pictures of the leaves. The pictures of the leaves are taken by the UAVs periodically which are stored in a central computing repository. To identify the infection in the leaf, the proposed method employs image processing techniques such as HLS to have the hue, light and saturation parameters, canny edge detection for identifying the border and establishing a ROI to perform the analysis. With help of these steps, the infected region can be easily differentiated with the healthy portion of the leaf. The intensity of the infection is also identified in the leaf. A sample of apple leaves are taken to identify the infection and the proposed model is able to identify the disease scab along with the intensity of the infection. The area of the infection is represented in millimeters (mm). Based on the type of infection and the intensity of the infection, a suitable pesticide is suggested to maintain the health of the crop. The proposed method helps to identify the infection in the early stages and proactively curbs the growth of infection by suggesting the pesticides in a timely manner.

References

1. Zhibin, W., Xiaojun, Q., Ying, W., Hao, Y., Cuixia, M. (2024). IoT-based system of prevention and control for crop diseases and insect pests. *Frontiers in Plant Science, 15*. https://doi.org/10.3389/fpls.2024.1323074
2. Kavitha, R., Shanthi, T., Maithili, P., & Kumar, N. (2023). *Deep learning and internet of things based detection of diseases and prediction of pesticides in fruits* (pp 1133–1140). https://doi.org/10.1109/ICOEI56765.2023.10125946
3. Komal, K., Aliza, A., Noorain, M., Radu, T., Noje, D. (2023). *Sustainable agriculture: An IoT-based solution for early disease detection in greenhouses* (pp. 1–4). https://doi.org/10.1109/EMES58375.2023.10171676
4. Shireshi, S. S., & Raman, R. (2024). *Soil-borne pathogens management in cloud-connected agriculture: An IoT and decision tree algorithmic approach* (pp 1–5). https://doi.org/10.1109/esci59607.2024.10497395
5. (2022). *Smart farming solution for crop disease prediction and protection* (pp. 282–298). https://doi.org/10.2174/9789815049251122010019
6. Sagar, R., Arthi, B., Prasad, P., Prajwal, R. S., & Sanjana, V. (2024). Drone based crop disease detection using ML. https://doi.org/10.1109/icdcece60827.2024.10548111
7. Mishra, V., Naik, N. S., Kumar, S., Alsamhi, S. H., Saif, A., & Edward, C. (2023). *Maize plant disease prediction of UAV images for precision agriculture using fusion of multimodal* (pp. 353–358). https://doi.org/10.1109/iccit58132.2023.10273966
8. Slimani, H., El Mhamdi, J., & Jilbab, A. (2023). Drone-assisted plant disease identification using artificial intelligence: A critical review. international journal of computing and digital systems. https://doi.org/10.12785/ijcds/1401112
9. De Ocampo, A. L. P., & Montalbo, F. J. P. (2024). A multi-vision monitoring framework for simultaneous real-time unmanned aerial monitoring of farmer activity and crop health. *Smart agricultural technology*, 100466–100466. https://doi.org/10.1016/j.atech.2024.100466
10. Shetty, S. N., Karanth, N. N., & Hegde, V. (2023). *Enhancing crop productivity: Integrated solutions for crop yield, crop recommendation, and crop disease management* (pp. 1–6). https://doi.org/10.1109/csitss60515.2023.10334167
11. Gao, D., Sun, Q., Hu, B., & Zhang, S. (2020). A framework for agricultural pest and disease monitoring based on internet-of-things and unmanned aerial vehicles. *Sensors, 20*(5), 1487.
12. Abbas, A., Zhang, Z., Zheng, H., Alami, M. M., Alrefaei, A. F., Abbas, Q., Naqvi, S. A. H., et al. (2023). Drones in plant disease assessment, efficient monitoring, and detection: a way forward to smart agriculture. *Agronomy, 13*(6), 1524.
13. Kouadio, L., El Jarroudi, M., Belabess, Z., Laasli, S.-E., Roni, M. Z. K., Amine, I. D. I., Mokhtari, N., Mokrini, F., Junk, J., & Lahlali, R. (2023). A review on UAV-based applications for plant disease detection and monitoring. *Remote Sensing, 15*(17), 4273.

14. Narayanappa, G. B. C., Abbas, S. H., Annamalai, L., Meenakshi, R., Singh, M., Yadav, T. N., & Kumar, A. R. (2024). Revolutionizing UAV: Experimental evaluation of IoT-enabled unmanned aerial vehicle-based agricultural field monitoring using remote sensing strategy. *Remote Sensing in Earth Systems Sciences* 1–15.
15. Shukla, R., Dubey, G., Malik, P., Sindhwani, N., Anand, R., Dahiya, A., & Yadav, V. (2021). Detecting crop health using machine learning techniques in smart agriculture system. *Journal of Scientific & Industrial Research, 80*(08), 699–706.
16. Vijayalakshmi, K., Al-Otaibi, S., Arya, L., Almaiah, M. A., Anithaashri, T. P., Sam Karthik, S., & Shishakly, R. (2023). Smart agricultural–industrial crop-monitoring system using unmanned aerial vehicle–internet of things classification techniques. *Sustainability, 15*(14), 11242.
17. Ponnusamy, V., & Natarajan, S. (2021). Precision agriculture using advanced technology of IoT, unmanned aerial vehicle, augmented reality, and machine learning. *Smart Sensors for Industrial Internet of Things: Challenges, Solutions and Applications* 207–229.
18. Kundu, N., Rani, G., Dhaka, V. S., Gupta, K., Nayak, S. C., Verma, S., Ijaz, M. F., & Woźniak, M. (2021). IoT and interpretable machine learning based framework for disease prediction in pearl millet. *Sensors, 21*(16), 5386.
19. Hassan, M., Malhotra, K., & Firdaus, M. (2022). Application of artificial intelligence in IoT security for crop yield prediction. *ResearchBerg Review of Science and Technology, 2*(1), 136–157.
20. Shafi, U., Mumtaz, R., Iqbal, N., Zaidi, S. M. H., Zaidi, S. A. R., Hussain, I., & Mahmood, Z. (2020). A multi-modal approach for crop health mapping using low altitude remote sensing, internet of things (IoT) and machine learning. *IEEE Access, 8*, 112708–112724.

MIX
Papier aus verantwortungsvollen Quellen
Paper from responsible sources
FSC® C105338

If you have any concerns about our products,
you can contact us on
ProductSafety@springernature.com

In case Publisher is established outside the EU,
the EU authorized representative is:
Springer Nature Customer Service Center GmbH
Europaplatz 3, 69115 Heidelberg, Germany

Printed by Libri Plureos GmbH
in Hamburg, Germany